Biology of
Subterranean Fi

Biology of Subterranean Fishes

Editors

Eleonora Trajano
Departamento de Zoologia
Instituto de Biociências da USP
São Paulo – SP
Brazil

Maria Elina Bichuette
Departamento de Ecologia e Biologia Evolutiva
Universidade Federal de São Carlos
São Carlos – SP
Brazil

B.G. Kapoor
Formerly Professor of Zoology
Jodhpur University
India

CRC Press
Taylor & Francis Group
Boca Raton London New York

CRC Press is an imprint of the
Taylor & Francis Group, an **informa** business

Science Publishers
Enfield, New Hampshire

CRC Press
Taylor & Francis Group
6000 Broken Sound Parkway NW, Suite 300
Boca Raton, FL 33487-2742

First issued in paperback 2017

ISBN-13: 978-1-57808-670-2 (hbk)
ISBN-13: 978-1-138-11547-7 (pbk)

Cover Illustrations

A small troglobitic catfish, *Rhamdiopsis* sp., from caves in Chapada Diamantina, Northeastern Brazil" and "Poço Encantado Cave, habitat of *Rhamdiopsis* sp., in Bahia State, Northeastern Brazil" Reproduced by permission of Adriano Gambarini.

Library of Congress Cataloging-in-Publication Data
Biology of subterranean fishes / editors, Eleonora Trajano, Maria Elina Bichuette, B.G. Kapoor. -- 1st ed.
       p. cm.
    ISBN 978-1-57808-670-2 (hardcover)
  1. Amblyopsidae. 2. Fishes--Ecophysiology. 3. Fishes--Adaptation. 4. Fishes--Habitat. 5. Deep sea fishes. I. Trajano, Eleonora. II. Bichuette, Maria Elina. III. Kapoor, B.G.
       QL638.A366B56 2010
       597.17584--dc22
                                                                    2009051455

Visit the Taylor & Francis Web site at
http://www.taylorandfrancis.com

and the CRC Press Web site at
http://www.crcpress.com

This book is dedicated to Dr. Thomas L. Poulson, in appreciation for his pioneering and inspiring work on cave fishes and other subterranean creatures

Preface

Subterranean fishes are among the most instigating, scientifically relevant and yet fragile organisms. Their specialization to the selective regime typical of subterranean habitats, highly contrasting to epigean (surface) ones, results in unique combinations of features, of which the most conspicuous are the reduction of eyes and dark pigmentation. Envisaged as bizarre creatures, cave fishes have been known and are the object of interest for centuries, but studies intensified after the 1920's, with the advent of modern genetics and, more recently, molecular biology, phylogeny and conservation.

Due to that selective regime, with permanent darkness as a major directive force throughout the hypogean realm, diverging rates of populations genetically isolated from their epigean colonizers are generally high. At the same time, homoplastic evolution of cave-related characters is quite common, producing sets of convergent character states in more or less distantly related taxa. High divergence rates in relation to close epigean relatives, homoplastic evolution of several characters, and the simplified nature of subterranean ecosystems, with fewer trophic levels, fewer species interacting and a tendency toward environmental stability (i.e. fewer varying abiotic factors), render subterranean organisms excellent subjects for evolutionary and ecological studies. Models of regressive evolution of characters, a rather frequent phenomenon among epigean animals, as Darwin himself was aware of, may be tested using troglobitic (exclusively subterranean) species. Likewise, physiological and behavioral models, in areas such as chronobiology and stoichiometry, benefit from the study of hypogean animals.

However, these characteristics, which we find so interesting, may impose serious problems for survival under conditions of unnatural environmental stress. Small population sizes due to geographic restriction and adaptations to cope with food scarcity, including K-selected life cycle traits and low population densities, result in low population turnovers and, consequently, a reduced capacity to recover from population losses.

Moreover, many troglobites, having evolved under relatively constant abiotic conditions, lost their ability to cope with environmental fluctuations, such as those provoked by human disturbance. A robust knowledge of the ecology and biology of these animals is a necessary and most urgent step for efficient actions that target their protection.

Knowledge about subterranean fishes progresses at a steady pace, with novelties arising every day. This book begins with an overview of the main evolutionary and ecological concepts applied to the study of subterranean fishes, in the context of a retrospective and prospective of the field (Chapter 1). In the following chapters, we present updated approaches, integrating new and previously known data, on general topics such as evolutionary genetics (Chapter 5) and development (Chapter 6), behavior (Chapter 7), biodiversity (Chapter 2), and conservation (Chapter 3). A worldwide checklist with 164 troglomorphic species represents a significant increase in relation to the most recent one, also by Proudlove (104 species described up to 2003) -- an increase that is mainly due to additions to the subterranean ichthyofauna of China and Brazil. Regional faunas, *in totum* or focusing on selected taxa, of countries such as China (Chapter 11), India (Chapter 12), Brazil (Chapter 9), Mexico (*Poecilia mexicana*; Chapter 8), and the African continent (Chapter 10), are discussed in terms of biodiversity and its causes, natural history and distribution, and ecology. A comprehensive chapter on the Amblyopsidae, a small North American family with one epigean, one troglophilic and four troglobitic species, summarizes information on almost every aspect of the systematics and biology of this family, which are currently the best known cave fishes (Chapter 7).

The contributors to this book are from leading speleobiologists dedicated to the study of subterranean fishes around the world. The editors of this project acknowledge the prompt response and enthusiasm of these scientists, who were essential to its success as a new and vibrant publication showing the utmost progress in the field. We also thank the specialists who reviewed the texts, in particular Drs. David C. Culver and John Holsinger.

This book is dedicated to Dr. Thomas L. Poulson, who earned our deepest admiration and respect for his pioneering and inspiring work on amblyopsid fishes and other subterranean animals, always keeping a fresh enthusiasm for these wonderful creatures. Thank you, Tom, for being such an example of a scientist and a person.

Eleonora Trajano and Maria Elina Bichuette

Foreword

Tom Poulson and Speleobiology—A Reflection

David C. Culver

Department of Environmental Science
American University
4400 Massachusetts Avenue NW
Washington, DC 20016, USA
E-mail: dculver@american.edu

I have been asked by the editors to summarize Tom Poulson's contributions to speleobiology and I am honored to do so. I mentioned this honor to a student who replied, "Oh yes, he is the one who works with Kathy Lavoie on the beetle-cricket interaction."[1] I was reminded of the often quoted comment about the 1960's iconic rock and roll band, The Beatles. Oh yes, someone from the next generation said, "Isn't that the band Paul McCartney was in before Wings?" I realize that the anecdote dates me as well, but the point is that I want to place Poulson's contributions, which have spanned generations, in a larger context. Tom himself, both in this volume and elsewhere (Poulson 2001a) has already commented on his contributions and their relevance to the research questions facing hypogean fish researchers. I certainly don't intend to comment on his comments, but

---

[1] For the record, Kathy Lavoie is an accomplished cave microbiologist and a former graduate student of Poulson's.

Photograph of Tom Poulson reproduced by Courtesy of Dante Fenolio

rather to focus on what I think is the real reason we celebrate Poulson's contributions. Simply put, he put not only the study of subterranean fishes, but speleobiology in general, in a neo-Darwinian context by his seminal work on adaptation in the amblyopsid cavefishes.

In most habitats, adaptations is the single most obvious aspects of an organism's phenotype. Cheetahs are adapted for running and capturing prey; monarch butterflies are brightly coloured to warn potential predators of their toxicity; etc. The theme of most nature films is the adaptation of organisms to their environment, and by implication the triumph of evolution by natural selection. Subterranean animals are different in this respect. The most obvious feature of many subterranean animals are losses, not adaptations. Even Darwin saw subterranean animals as degenerates: examples of eyelessness and loss of structure in general. For him, the explanation was a straightforward Lamarckian one, and one that did not involve adaptation and the struggle of existence.

"It is well known that several animals which inhabit the caves of Carniola [Slovenia] and Kentucky, are blind....As it is difficult to imagine that eyes, though useless, could be in any way injurious to animals living in darkness, their loss may be attributed to disuse" (Darwin 1859)[2].

Small wonder then that for decades following Darwin, adaptation was not associated with subterranean organisms. Much confusion followed, which I will only briefly consider.

At the end of the 19th century, one of the leaders of the neo-Lamarckian school of evolution was Packard, who was also the leading American speleobiologist of his time. He was convinced that use and disuse governed the evolution of subterranean animals and gave virtually no role to natural selection and adaptation. He also held that evolution of what we would now call troglomorphy was rapid. Packard knew cave animals well. He had visited several dozen caves in North America and described many species (Packard 1888).

In 1907 Banta supported what seems to a modern biologist a bizarre theory—that of orthogenesis:

"Animals do not possess degenerate eyes and lack pigment because they are cave animals....They are cave animals because their eyes are degenerate and because they lack pigment....They are isolated in caves and other subterranean abodes because they are unfit for terranean life..." (Banta 1907).

In other words, animals are not blind because they are in caves, they are in caves because they are blind! The major proponent of orthogenesis in

---

[2] This quotation remained the same through all six editions of the book *On the Origin of Species.*

subterranean species was Vandel (1964). Vandel also minimized the role of natural selection and adaptation with the following analogy, which links the idea of aging individuals to senescent phyletic lines:

"The idea of adaptation has grown to the point where it has been written that depigmentation and anophthalmy represent 'adaptations to subterranean life'. This is like saying that cararrh [common colds], rheumatism, and presbyopia [far-sightedness] are adaptations to old age".

The final approach to the evolution of the morphology of cave animals that is not selectionist came from the Kosswigs. Working in the 1930's they were very interested in genetic polymorphism and believed it held the key to understanding regressive evolution. Based on their studies of the highly polymorphic isopod *Asellus aquaticus* in the Postojna Planina Cave System in Slovenia (Kosswig and Kosswig 1940), they believed that mutation was the key to understanding this variability, and they held that the presence of highly polymorphic populations of stygobionts was the result of accumulating mutations that were not subject to selection.

In the late 1950's, when Poulson began his graduate studies on amblyopsid fish, it was hard to find anyone who studied cave animals that looked to natural selection as an explanation of morphology, life history, or behavior. In fact, it was hard to find anyone who looked at any aspect of cavefish biology other than eye and pigment degeneration. The integration of Darwinian ideas of evolution by natural selection with population genetics (neo-Darwinism) was very much at the forefront in biology at that time (see Gould 2002 for an extended discussion of the neo-Darwinian synthesis and what he calls its hardening). Cave life was being left behind.

What Poulson did, and it truly transformed the study of subterranean animals, was to focus on adaptations, not on losses of eyes and pigment. Using the comparative method, he analyzed all the known species of Amblyopsidae (Poulson, 1963). The five[3] species ranged in habitat from freshwater swamps to streams deep in caves. This work resulted in what is probably the most quoted single research paper on cave biology ever[4]. In the paper, Poulson did a thorough comparative study of the morphology, physiology, and behavior of the five species of fish and demonstrated that the differences could only be explained by differences in the selective environment. In particular, Poulson showed that the food-poor, aphotic environment of caves imposed a selective regime that

---

[3] Since his initial work, a sixth species has been discovered—*Spleoplatyrhinus poulsoni.*

[4] According the Web of Science©, his paper has been cited 106 times since 1983 (as far back as digital records go). I can find only one research paper (some reviews are cited more) that is cited even 50 times—Wilkens (1971), which was cited 71 times.

resulted in an increase in extra-optic sensory structures, reduced metabolic rate, increased longevity, larger eggs, and a host of other features. The key result of this work (also summarized in Poulson and White 1969) is that cavefish are adapted to their environment, just as the cheetah is to its environment. I believe this is why his work is so often cited. Now, it is a matter of course to read in taxonomic descriptions of cave animals from all over the world and in a variety of languages that the species under consideration is adapted to the subterranean environment because of a set of elaborated or elongated morphological features. Poulson made the idea of adaptation to the cave environment respectable.

Of course, he said more than this in his 1963 paper, and has substantially expanded this work since that time (Poulson 1985, 2001b). I will consider some of these expansions and elaborations below, but first I want to complete the context of his work in the 1960's. Two other North American speleobiologists working in the 1960's added to and complemented the neo-Darwinian view of cave animals. Kenneth Christiansen (1961, 1965) also used a comparative approach to study the morphology and behavior of cave Collembola, wingless hexapods. He also gave a convincing demonstration of adaptation in caves. The primary difference (except, of course, for the choice of organisms) was that Christiansen emphasized the problem Collembola faced in moving across wet mud in water in caves rather than on food scarcity or darkness. Thomas Barr (1967, 1968), coming from a more taxonomic tradition, developed the dominant paradigm for speciation in caves.

There was more to Poulson's seminal paper. In particular, he argued that the degree of eye and pigment loss was a measure of the length of time the different amblyopsid species have been isolated in caves. It may seem quaint to try to measure time of isolation using morphological rather than molecular measures, but of course no molecular techniques for sequencing or measuring genetic distance were available at the time. It is an indication of how much he changed the agenda of studies of cave organisms that his use of regressive features was only in this context. He then argued, with considerable data to back it up, that the degree of morphological, physiological, and behavioral change exhibited by the cave amblyopsid fish was correlated with the amount of eye and pigment reduction, and that the greater the change, the greater the adaptation to cave life. The interesting question of how long cave animals have been isolated in caves continues to resonate and continues to be controversial (e.g. Trontelj 2007). The argument that more troglomorphic species are more highly adapted is likewise controversial because different species may be adapted to different subterranean environments (e.g., Noltie and Wicks 2001). Whatever the final answers to these questions, it was Poulson who set the agenda and first asked them.

I would be remiss if I left the impression that all or even most of Poulson's contributions to speleobiology have involved fish. For more than 30 years, he has championed the idea that the amount, nature, and distribution of food in caves were key elements in organizing terrestrial cave communities and in determining the selective environment. This is not the appropriate place to review this work, but I want to note that he trained and was assisted by a remarkably talented group of Ph.D. students, including the late Thomas C. Kane (Kane and Poulson 1976), Kathleen Lavoie (Poulson and Lavoie 2000), David Griffith (Griffith and Poulson 1993), and Kurt Helf (Poulson, Lavoie, and Helf 1995). I was also honored to be his graduate student, and we did joint work on the determinants of species of richness in caves (Poulson and Culver 1969) and metabolic rates in cave amphipods (Culver and Poulson 1971).

My own experiences with him in the field remain vivid and important, now nearly forty years since most of them transpired. I have never encountered a more careful observer of cave life—he was and is a great naturalist. I have also never encountered a better critic of bad ideas and a more enthusiastic supporter of new ideas and hypotheses. But more than anything it is his enthusiasm for studying cave life that dominates time in the field with Tom.

His influence on graduate students extended far beyond his own and far beyond his published papers. Several generations of American graduate students had the experience of being in the field with Poulson, or hearing one of his inimitable seminars. And their work was better as a result. Mine certainly was.

Literature Cited

Banta, A.M. 1907. The fauna of Mayfield's Cave. *Carnegie Institution of Washington Publications* 67: 1-114.

Barr, T.C. 1967. Ecological studies in the Mammoth Cave system of Kentucky. I. The biota. *International Journal of Speleology* 3: 147-204.

Barr, T.C. 1968. Cave ecology and the evolution of troglobites. *Evolutionary Biology* 2: 35-102.

Christiansen, K.A. 1961. Convergence and parallelism in cave Entomobryinae. *Evolution* 15: 288-301.

Christiansen, K.A. 1965. Behavior and form in the evolution of cave Collembola. *Evolution* 19: 529-537.

Culver, D.C. and T.L. Poulson. 1971. Oxygen consumption and activity in closely related amphipod populations from cave and surface habitats. *American Midland Naturalist* 85: 74-84.

Darwin, C. 1859. *On the Origin of Species by Means of Natural Selection, or the Preservation of Favoured Races in the Struggle for Life,* John Murray, London, first edition.

Gould, S.J. 2002. *The Structure of Evolutionary Theory.* Harvard University Press, Cambridge.

Griffith, D.M. and T.L. Poulson. 1993. Mechanisms and consequences of intraspecific competition in a carabid cave beetle. *Ecology* 74: 1373-1383.

Kane, T.C. and T.L. Poulson. 1976. Foraging by cave beetles: spatial and temporal heterogeneity of prey. *Ecology* 57: 793-800.

Kosswig, C. and L. Kosswig. 1940. Die Variabilität bei *Asellus aquaticus* unter besonderer Berucksichtigung der Variabilität in isolierten unter- und oberirdischen Populationen. *Revue de Facultie des Sciences (Istanbul), ser.* B, 5: 1-55.

Noltie, D.B. and C.M. Wicks. 2001. How hydrogeology has shaped the ecology of Missouri's Ozark cavefish, *Amblyopsis rosae,* and southern cavefish, *Typhlichthys subterraneus:* insights of the sightless from understanding the underground. *Environmental Biology of Fishes* 62: 171-194.

Packard, A.S. 1888. The cave fauna of North America, with remarks on the anatomy of brain and the origin of the blind species. *Memoirs of the National Academy of Sciences (USA)* 4: 1-156.

Poulson, T.L. 1963. Cave adaptation in amblyopsid fishes. *American Midland Naturalist* 70: 257-290.

Poulson, T.L. 1985. Evolutionary reduction by neutral mutations: plausibility arguments and data from amblyopsid fishes and linyphiid spiders. *Bulletin of the National Speleological Society* 47: 109-117.

Poulson, T.L. 2001a. Adaptations of cave fishes with some comparisons to deep-sea fishes. *Environmental Biology of Fishes* 62: 345-364.

Poulson, T.L. 2001b. Morphological and physiological correlates of evolutionary reduction of metabolic rate among amblyopsid fishes. *Environmental Biology of Fishes* 62: 239-249.

Poulson, T.L. and D.C. Culver. 1969. Diversity in terrestrial cave communities. *Ecology* 50: 153-158.

Poulson, T.L. and W.B. White. 1969. The cave environment. *Science* 165: 971-981.

Poulson, T.L. and K.H. Lavoie. 2000. The trophic basis of subsurface ecosystems. In: *Subterranean Ecosystems.* H. Wilkens, D.C. Culver and W.F. Humphreys (eds). Elsevier Press, Amsterdam, The Netherlands. pp. 231-250.

Poulson, T.L., K.H. Lavoie and K. Helf. 1995. Long term effects of weather on the cricket guano community in Mammoth Cave National Park. *American Midland Naturalist* 134: 126-136.

Trontelj, P. 2007. The age of subterranean crayfish species. A comment on Buhay and Crandall (2005): subterranean phylogeography of freshwater crayfishes shows extensive gene flow and surprisingly large population sizes. *Molecular Ecology* 16: 2841-2843.

Vandel, A. 1964. *Biospéologie: la Biologie des Animaux Cavernicoles.* Gauthier-Villars, Paris.

Wilkens, H. 1971. Genetic interpretation of regressive evolutionary processes: studies on hybrid eyes of two *Astyanax* cave populations (Characidae, Pisces). *Evolution* 25: 530-544.

Contents

CHAPTER 1

Cavefish: Retrospective and Prospective

Thomas L. Poulson

Professor Emeritus, Department of Biological Sciences, University of
Illinois-Chicago. E-mail: tomandliz@bellsouth.net

I. INTRODUCTION

A. My Aims

One of my aims is to take a retrospective and prospective view that will help colleagues identify areas of cavefish research that need increased attention. In this context I urge everyone to think about how to solve 'mysteries' in their research area and to explain any new insights (eurekas) they have had. My second aim is to urge everyone to place their studies in the context of ecological and evolutionary concepts so that others will use cavefish as examples. In this chapter I identify what I consider to be exemplary examples from both my own work and that of others. I pay attention to how new techniques are giving better answers to old questions. And I consider what questions can and cannot be answered with different study systems, especially *Astyanax* species vs the Amblyopsidae. I urge you to do the same in your chapters and future work. In all cases try to update what you have published in readily accessible books. And of course build on past work of others especially in The Biology of Hypogean Fishes (Romero 2001) and Subterranean Fishes of the World (Proudlove 2006) but also build on a number of fine chapters in Encyclopedia of Caves (Culver and White 2005), Subterranean Ecosystems (Wilkens, Culver, and Humphreys 2000), and the Special Issue of The National Speleological Society Bulletin on Regressive Evolution (Culver 1985).

B. Ecological Concepts

I like the following definition of ecology from Charles Krebs ecology text, with key words in capitals. ECOLOGY is the SCIENTIFIC STUDY of INTERACTIONS that determine the DISTRIBUTION and ABUNDANCE of organisms. By the process of 'SCIENTIFIC STUDY' I mean starting with a question or patterns, like distribution and abundance, and then using prior knowledge and observations to formulate a hypothesis that has simple enough predictions to be potentially falsifiable. The next step is using natural or manipulative experiments to test the predictions from each hypothesis. I strongly believe in formulating multiple hypotheses to avoid the trap of a favorite hypothesis that can at best lead to inadvertent bias and controversy and at worst lead to advertent fraud. For natural experiments I follow Jared Diamond in using both 'snapshot' and 'trajectory' experiments. 'INTERACTIONS' are potential agents of natural selection. They include all combinations of abiotic-abiotic interactions (e.g. temperature and oxygen concentration in water), abiotic-biotic interactions (e.g. flooding and injury to fish), and biotic-biotic interactions (e.g. competition, predation, parasitism, mutualism, and commensalism). 'DISTRIBUTION' includes 'address', i.e. habitat, and geographic location at all spatial scales. This includes the spatial scale of biogeographic patterns that are generated at evolutionary time scales. And 'ABUNDANCE' includes frequency across potentially suitable habitats and both density and population size in each habitat occupied. It also includes niche. Niche is a way of life or 'profession' (e.g. top predator in a food web vs omnivore).

C. Tradeoffs

At the interface of ecology and evolution I have always found it useful to consider tradeoffs and how they relate to rigor, variability, and predictability of agents of natural selection. A common misconception is that plus, i.e. advantageous, tradeoffs occur in different species or circumstances than minus, i.e. disadvantageous, tradeoffs. In fact each trait or circumstance has both plus ands minus tradeoffs. For example floods often are associated with transport of allochthonous organic matter into caves (+) but the rigor of turbulence and velocity of flood-water may injure cavefish (–). And a generalist, in habitat or feeding niche, is a 'jack-of-all-trades' (+) but a 'master' of none (–). The opposite tradeoffs occur for a specialist. Low variability in abiotic conditions or highly predictable timing of cycles of variability may favor specialization. But high variability with low predictability may favor generalization.

D. Temporal and Spatial Scale

Whether experiments, differences, or events are biologically significant depends on relevant spatial and temporal scales. Unusual ecological events may be unimportant on an evolutionary time scale. And experiments that have statistically significant results may be unrealistic at the large spatial scales of cave streams or the long generation times of troglobitic cavefish. On the other hand predictability of rare bad or good events on a generation time scale may not be statistically significant but could have huge and important effects as agents of selection. Think of catastrophic capture of a surface stream by a subterranean cave or a 50-year flood that renews allochthonous organic matter to an otherwise low nutrient cave stream.

An important issue for temporal variability and predictability is the scale of cavefish longevity, or better generation time, in relation to return time for floods. With evolution of long cavefish lives, associated with slow growth and great metabolic economies, the unpredictability of food renewal with 10-year or 25-year floods can change from unpredictable to predictable. Thus the chance of successful reproduction is increased especially since long life is usually associated with spread of risk of reproductive failure by repeated reproduction – iteroparity – with few large eggs hatching to large young (quality rather than quantity). In short lived species with high fecundity and small eggs there may be only one reproduction – semelparity – and here spread of risk is among many small young (quantity rather than quality).

An important spatial scale issue is the relation of manipulative experiments in aquaria or artificial streams to natural experiments in orders of magnitude larger cave streams and the suite of caves a species occupies in a watershed. Interpretations of the importance of chemical cues and territoriality are especially subject to using unrealistic spatial scales.

E. Evolutionary Concepts

Evolution is the central paradigm of biology. As Theodosius Dobzhansky wrote, "nothing makes sense except in the light of evolution." Evolution is change in inherited traits in a population over generations. The appropriate time unit is the generation (e.g. bacteria evolve faster than cavefish in the same environment). The unit of evolutionary change is the population and the unit of natural selection is the individual. To my knowledge nobody has considered whether or how there might be group selection in any cavefish.

Implicitly Darwin used the scientific method to develop his hypothesis of evolution by natural selection. He observed that organisms have many more offspring than needed to replace themselves and so hypothesized that there should be unlimited population growth. But tests of the prediction falsified the hypothesis because he did not observe continued exponential population growth of any species in nature. Instead he noted that population numbers of a species vary over time but tend to remain stable. So Darwin further hypothesized that there must be "a struggle for existence". And he suggested that those individuals that survive competition, predation, and other exigencies are naturally selected. These individuals will leave the most offspring. Based on the 'strong principle of inheritance', that Darwin observed in domestic animals – offspring resemble their parents – he further suggested that there will be evolution. And so species may evolve new or improved adaptations as environments change.

Over time Darwin's hypothesis of evolution by natural selection has changed from a hypothesis to a continually supported theory. First his hypothesis was tested in myriad species in many ways and could not be falsified. So his hypothesis became a theory (note the misunderstanding by creationists that 'evolution is only a theory').

It does not matter, to the validity of the theory, that Darwin did not know about genes or that mutation and recombination are sources of genetic variation. He in fact thought that degeneration or rudimentation of cavefish eyes was due to disuse. We now know that the real explanation of cavefish eye and pigments rudimentation involves both neutral mutations of loss and concurrent positive and negative pleiotropic effects of a few genes with large developmental effects.

II. GENETIC AND DEVELOPMENTAL BASES FOR RUDIMENTATION

A. Background

Our earliest model for a gene was simplistic and survives only as the reality that there are some 'Mendelian' genes with recessive and/or dominant alleles. We now know that most traits are coded by additive effects of quantitative trait loci (QTLs) in a polygenic manner. We also know that some QTL genes have very large effects by controlling development often pleiotropically by affecting several different traits. And, as Stephen J. Gould argued for years, changes in developmental control genes may have and have had huge evolutionary effects. In fact, for a complex eye in an amphipod a catalog of mutants has only one with slight increases, for amount of inter-ommatidial pigment, but dozens that are moderately to strongly deleterious (Sexton and Clark 1936). To me that means not only that accumulation of point mutations can help explain rudimentation but also that mutations that can act to increase elaboration of a structure are rare. Hence developmental control mutations that are polygenic and recombination of such genes would seem to be the main way to elaborate structures in general and in cavefish in particular. We will see that ongoing studies in the *Astyanax fasciatus* system support this scenario.

B. Alternative Hypotheses for Rudimentation

What is/are the mechanism(s) for the evolutionary mechanism of eye and pigment rudimentation seen convergently in all cavefish species? Is rudimentation due to accumulation of neutral loss mutations in the absence of stabilizing selection? And/or is it due to indirect selection against eyes and pigment by antagonistic pleiotropy, a kind of genetic hitchhiking? Rudimentation of eye and pigment systems is the pattern, evolution is the process, and either drift or selection is the mechanism.

Why do many invertebrate troglobites completely lack eyes or melanin pigment whereas this is never the case for vertebrates? I agree with many workers who suggest that any mutations that would eliminate eyes and pigment are selected against in vertebrates because of the close relation among developmental pathways. Put another way stabilizing selection maintains genes important for eye, pigment, and head morphogenesis in vertebrates. I suggest that this is not the case for invertebrates because eyes and pigment developmental pathways occur relatively late and so mutations of loss are not selected against.

I prefer the words elaboration for evolution of non-visual sense organs in cavefish and the terms reduction in size and rudimentation for changes in eyes and pigment. These terms have less implied etymological baggage than the terms constructive and degenerative or regressive. The terms – elaboration and rudimentation – do not imply mechanism and so they avoid the issues of whether there has been selection against or for traits or accumulation of loss mutations with loss of stabilizing selection. In fact evidence is strong that there are pleiotropic tradeoffs for elaboration and reduction in an amphipod, *Gammarus minus*, and in a cavefish, *Astyanax fasciatus*. However there is some disagreement about how pleiotropy is working and its relative importance for pigment vs eye rudimentation at early vs late stages of the evolution of troglomorphy.

Two recent reviews critically evaluate all mechanisms of rudimentation. Both reinforce my plausibility arguments, based on relative sizes and savings of parts and metabolic rates (Poulson 1985), that metabolic savings is not the basis for antagonistic pleiotropy (Culver and Wilkens 2000, Jeffery 2005 and this volume). I find Protas *et al.* (2008) suggestion that high cost of retinal metabolism may help account for eye reduction unconvincing and unlikely based on the kinds of analyses I have made earlier (Poulson 1985).

C. Ongoing Studies of Rudimentation and Elaboration

Jeffery (2005), Wilkens (2007), and Protas *et al.* (2008) propose three variants of the pleiotropy mechanism at the developmental control gene level for eye rudimentation in *Astyanax fasciatus*. In all cases eyes are indirectly selected against by selection for elaboration of non-visual sense organs, a kind of genetic hitchhiking.

All genetic/developmental mechanisms for rudimentation in *Astyanax fasciatus* have to account for several patterns that have been repeatedly documented. One pattern is that, compared to the surface ancestor, cavefish eye primordia are small, that eyes tissues proliferate and differentiate normally to a point, and that eyes then stop differentiation and become highly variable in size and structure with continued body growth (e.g. Figures 1 and 2 in Wilkens 2007). As Culver and Wilkens (2000) point out this high variability is common to all vestigial structures in many organisms. Another common pattern is that hybridization studies show that polygenic inheritance is predominant especially with eye size and pigment cell numbers. And hybridization between different blind cave populations shows that there can be partial recovery of eye size.

Even the best worker's research groups, each doing exciting and important work, often fail to deal with all of the general patterns. In Jeffery's (2005) review I do not believe he addresses the ontogenetic increase in variability of eye size and structure so emphasized by Wilkens. And in Wilkens' (2007) review I do not believe he addresses Jeffery's results that show that even late acting genes are still 'functioning' for eye and pigment systems. Nor does Jeffery's review address Wilkens' points about mutations that change functionality of pigment genes. And Protas *et al.'s* new hypothesis does not address either Wilkens' or Jeffery's suggestions about pleiotropy. Instead they present a new twist of 'multi-trait evolution' where both 'genes' that are associated with elaboration and rudimentation are clustered but not so closely linked to preclude recombination. The elaborated traits include metabolic economy, taste buds, and other traits important for surviving in a food-poor cave environment. I suggest that these gene clusters may be the animal analog of regulatory plant genes with wide-ranging effects on growth and metabolic economies that code for what has been called the 'stress-resistance syndrome' genes in plants (Chapin *et al.* 1993, West-Eberhard 2003).

Hybridization studies show that eye genes associated with rudimentation are different in different blind cave populations of *Astyanax fasciatus*. Wilkens (2007) showed early on that laboratory hybrids of different blind *Astyanax* populations have larger eyes than either parent. And Borowsky's group (Borowsky 2008, Protas *et al.* 2008, and Borowsky, this volume) have just shown that different eye rudimentation genes, in populations independently isolated in caves, can complement each other to restore vision. The degree of complementation, from 1-40% in different hybrids, increases with geographic distance between parental cave populations and thus with the increasing likelihood that related but different quantitative genes have been involved. These results are as we would predict for parallel evolution leading to convergent phenotypes via different genes affecting the same developmental processes.

In *Astyanax* some of these control genes, especially 'sonic hedgehog', have pleiotropic effects by both inhibiting eye development and promoting development of taste buds (Jeffery 2005). Borowsky's group (Protas *et al.* 2008) have shown that the traits that are reduced and enhanced by a control gene are clustered on *Asytanax* chromosomes but are not so closely linked that recombination is inhibited. Jeffery (2005) reviews how these control genes operate by using elegant transplant experiments of eye anlage to one side of an early embryo, using the other side of the head of the donor recipient as a control. Remarkably, transplants of early eye lens anlage, or even lens messenger RNAs, can both restore eyes and vision of blind forms, if the donor is eyed, and result in eye degeneration in eyed forms, if the donor is blind.

Microarray and differential display measures of gene expression show that as many genes are up-regulated as down-regulated (Jeffery 2005). Generally, elaborated traits are affected positively and rudimenting traits are affected negatively. Jeffery concludes that this is the mechanism for antagonistic pleiotropy. Surprisingly all genes continue to be expressed, as measured by gene arrays and messenger RNAs, even though their down-regulation results in abnormal and small eyes. As with hens teeth ("as rare as hens teeth") I expect someone to eventually find an eyed fish among a wild blind cavefish population when its suppressing control genes fail to function. In addition using such atavistic mutants, researchers are currently using genetic and developmental manipulation to resurrect teeth in birds and show that genetic programs for other reptilian traits are still present. Thus the studies resurrecting eyes in blind cavefish by Jeffery's group are just an example of an emerging generality i.e. that evolutionary losses are often a matter of modulating and suppressing genetic circuitry which is not lost. But so far in these studies workers have not been able to resurrect teeth, tails, or scales that are completely normal. To me this suggests that there has also been some loss of function in structural genes. So too, I suspect, in cavefish.

I suggest that, if we could do developmental studies, in more troglomorphic cavefish than *Astyanax* we would find that many of the latest acting eye and pigment genes would not even be expressed. I suggest this is the case for species that may even completely lose eyes and pigmented melanophores as adults (*Phreatichthys andruzzii* and probably *Speoplatyrhinus poulsoni*).

The early developmental constraints seem to me to be least for pigment system genes and so I expected and documented that they apparently show faster evolutionary rudimentation than for visual systems in the Amblyopsidae (Poulson 1985). Jeffery agrees that pigment genes at the end of a developmental cascade should be least subject to stabilizing selection and so was surprised to find that they too are expressed early on in *Astyanax*. But Wilkens (2007) documents three different kinds of Mendelian pigment genes that have apparently ceased to function in some *Astyanax* cavefish populations. Wilkens suggests that Jeffery's results may have been complicated because the Pachon population of *Astyanax* studied has had both ancient and recent introgression with surface fish. Thus Pachon fish may be less troglomorphic than other *Astyanax* cave populations and certainly much less troglomorphic than many cavefish species such as the cave Amblyopsids which have much more reduced pigment systems.

A detailed summary of data on Amblyopsids (Poulson 1985, Table 2, Niemiller and Poulson, this volume) show that visual systems and pigment systems continue to become more rudimented even though advantageous troglomorphic traits show much less further elaboration in the grades of increasing troglomorphy seen among species.

My hypothesized increasing grades of troglomorphy in Amblyopsid fishes are from *Typhlichthys subterraneus* to *Amblyopsis spelaea* to *Amblyopsis rosae* to *Speoplatyrhinus poulsoni*. For this series there is continued and clear increased rudimentation with smaller size of eyes and optic lobes, number of pigmented melanophores and their sizes and melanin densities with an increasing degree of negative allometry for each. But Figure 2c in Poulson (1985) shows that, aside from neoteny (see the section below **Is Neoteny a Route to Troglomorphy?**), there are relatively small and inconsistent increases of adaptive elaborating traits, both for size and positive allometry, among the three well-studied Amblyopsid troglobites. These traits include head size, pectoral fin length, neuromast numbers, cerebellum size, and forebrain size. Another adaptive trait that does not change is that egg volume is close to the same in all troglobites. After 23 years my interpretation is still that there has been continued accumulation of loss/ simplifying mutations for visual and pigment systems in Amblyopsid troglobites (Poulson 1985). My logic is that the early developmental constraints and stabilizing selection are not so important when eyes and pigment systems have already become so reduced evolutionarily.

A corollary is that continued rudimentation of eye and pigment systems support my hypothesis that the four troglobitic Amblyopsids indeed represent increasing evolutionary time in caves. My interpretation for elaborated traits is that with strong selection at the time of isolation in caves at least the sensory systems quickly got about as good as possible. At present, I see no independent way of testing this hypothesis because, even if molecular clocks can be shown to be accurate, we have no way of dating the time when current troglobite species were isolated in caves.

D. Sources of Genetic Variation

The relative contributions of mutation, recombination, and hybridization to genetic variation, the fuel for evolutionary change, have not been adequately considered in studies of cavefish. Recent work, especially on Galapagos finches by Peter and Rosemary Grant, has shown the importance of rare hybridization as a source of enhanced variation that increases rates of adaptation to environmental change. Thomas Barr has suggested to me that some species of his trechine cave beetles do occasionally hybridize. And Wilkens' colleagues have shown, using mitochondrial DNA, that at least one *Astyanax* cave population, Pachon, probably had both ancient and recent introgression of genes from surface fish. However neither Barr nor Wilkens speculate as to whether rare hybridization might have increased the rate or degree of adaptation to life in the cave or slowed it.

E. Population and Quantitative Genetic Approaches

Astyanax researchers and others, would benefit by taking a population genetic approach. Culver effectively uses theory to explore the actual and potential effects of neutral and pleiotropic genes in *Gammarus minus* (Amphipoda). See Tables 20.2 – 20.4 in Culver and Wilkens (2000). It would also be of interest to calculate the plausibility of and potential rates of evolutionary elaboration with Mendelian genes vs structural polygenes, vs developmental control genes with large pleiotropic effects.

III. A HIERARCHY OF ADJUSTMENT TO STRESS

I find it useful to consider a temporal hierarchy of possible adjustment modes to physical-chemical stress; first by behavior, then physiology, then development, and finally genetic. In the hierarchy for adjustment the four modes take increasing time and are decreasingly reversible from behavior to physiology to development to genetic. There is some overlap in time taken but in principle they are 'employed' in sequence if stress is not alleviated. If behavioral change does not alleviate the stress over a period of minutes to hours then physiological adjustments are made over periods of days to seasons. And if physiological adjustment does not alleviate stress then developmental change over a lifetime may work. If the stress continues or is predictable and rigorous then genetic change may alleviate the stress over generations i.e. evolutionary adaptation. Of course even the three modes that operate within a lifetime have a genetic basis.

In most cases mobile animals like cavefish can alleviate stress using just behavior and development. Sedentary plants and sedentary colonial animals often have striking developmental adjustments that are adaptive. A common example is sun leaves vs shade leaves in trees. These are of course changes in morphology within the developmental phase of a lifetime.

Looking at behavioral and physiological adjustments is especially relevant to determining whether a species is preadapted to living in the complete darkness of caves with attendant food restriction. It is useful to study these adjustments even if we do not have the real epigean ancestor of a cavefish species to study. For example *Chologaster cornuta* is the only possible living model for a preadapted Amblyopsid ancestor. And the right model for troglobitic *Astyanax fasciatus* is the surface stream *Astyanax fasciatus*.

In the context of low food supply as a stressor, assessment of behavioral and physiological adjustment are the first lines of defense. So I suggest the following research agenda: (1) Start by putting the eyed form in the dark and quantifying its food searching and acquisition behavior. (2) At the physiological time scale, assess the resistance to starvation, by rate of weight loss and ability to recover from different degrees of weight loss (Huppop 2000). This may relate to whether the intestinal epithelium and intestine length remain intact during starvation. And determine whether resistance to starvation can be prolonged by providing detritus or even cave mud. Finally, quantify the rate of and amount of fat deposition when preferred food is available ad libitum. (3) At the developmental time scale, determine whether allometric increase in sense organs with low food supply and low growth rate lead to enhanced food finding abilities.

IV. EVOLUTION OF TROGLOMORPHY IN CAVEFISH

A. Preadaptation: The First Step in Cavefish Evolution

Virtually all cavefish researchers agree that species that successfully made the transition from the surface (epigean) to caves (hypogean) were in some way preadapted. Preadaptations are not necessarily exaptations with new functions in a new environment. A problem, as Culver and Poulson (1971) first pointed out, is that we often do not know the relevant surface species to use as a comparison.

If we do know a close surface relative or, in a few cases know the actual surface ancestor, then too often we have not done the experiments to see how much of the cave species' behavior and food finding ability is just due to the surface species compensating, within a few hours. weeks, or months to being in darkness. We all know that, as humans that rely heavily on vision and are not preadapted to life in darkness, we can get better at navigating in the dark over time. See the cartoon of a cavefish using a cane (Romero 2001 page 6). And we never have trouble making love in complete darkness. Some humans that become blind at an early age develop extraordinary abilities to navigate without sight by kinesthetic sense, touch, and a crude echolocation. And some rare individuals can even do extraordinary feats of balance on a tightrope. The Great Wallenda, a world renown trapeze performer, could even walk blindfolded on a swaying rope. So it should not be surprising that epigean fish species that do not rely on sight and are nocturnal and have well developed non-visual sensory systems, might do well in caves if there is sufficient food. This is the case for most ancestors of troglobitic cavefish but not all.

B. *Astyanax fasciatus* vs the Amblyopsidae as Models for Cavefish Evolution

I suggest that each model system has advantages and disadvantages. I focus on these two systems because they are by far the most thoroughly studied of all cavefish.

Astyanax fasciatus troglobites would seem to be of relatively recent origin because there are no reproductive barriers (but see Porter *et al.* (2007) suggesting as long as 1-2 million years of isolation for some cave populations). And genetics and developmental analyses can be done because they are easy to breed in the laboratory. But *Astyanax fasciatus* is unusual for cavefish in several ways: the mode of cave isolation, heavy reliance on vision in surface streams, and relatively modest troglomorphy. Either they have not been isolated in caves long enough or have not had enough food limitation to evolve the kinds of sensory elaboration and metabolic efficiencies possible for cavefish.

As treated in detail by Niemiller and Poulson (this volume), the Amblyopsid species represent the entire spectrum of epigean to hypogean existence and extremes of troglomorphy. And the one troglophile species, *Forbesichthys agassizi*, is a possible model for the what most researchers agree is the more usual transition from epigean to hypogean life. Even the strictly surface species, *Chologaster cornuta*, has obvious preadaptations for life without light and *Forbesichthys* is somewhat more preadapted (Poulson 1961, 1963). All the Amblyopsid troglobites are very troglomorphic and their surface ancestors are no longer extant and they are apparently ancient based on molecular data (Niemiller, pers. comm.). So we do not know the extent to which any of their surface ancestors were preadapted. The lack of a surface ancestor also means that genetic studies of the evolution of troglomorphy are impossible. Even genetic studies of any Amblyopsid troglobite species are currently impossible because we do know how to get any of the species to reproduce in the laboratory.

Astyanax fasciatus seems to be an exception to my hypothesis that the habitat shift model does not apply to troglobitic fish. There are two reasons. First, its isolation in caves is likely associated with catastrophic loss of surface streams as they were 'captured' by subterranean drainage. This is apparently an unusual mode of isolation and, because it is sudden, it certainly resulted in extreme selection for traits that would allow survival and reproduction in caves. Wilkens and Huppop (personal communication) seem to agree by using the phrase "disruptive selection at an abrupt epigean to hypogean border". The second reason *Astyanax* is unusual is that initial selection in caves was especially strong because it is not 'pre-adapted' by having small eyes and being nocturnal.

If surface *Astyanax fasciatus* did not also have preadaptations of a lateral line system, that acted in concert with vision to allow striking schooling behavior and agonistic interactions, and taste buds, for discriminating edible and inedible food, then I suggest that it would not have survived in caves. When placed in the dark an eyed fish changes its body orientation and searching behavior. This immediate behavioral change allows it to find food but not quickly or efficiently compared to the cave form (Huppop 1985). Parzefall (2000) describes how the cave form, compared to the surface form, searches with a more oblique body angle that brings its chin, that has the densest taste buds, in closer contact with the substrate. In the dark the eyed form can use its lateral line system to avoid obstacles and perhaps sense living prey (Montgomery *et al.* 2001 and Coombs, pers. comm.).

Physiological preadaptations of *Astyanax* are that it can put on some fat and resist starvation for a while, though not as well as the blind cave form (Huppop 2000). I suggest that the selection for increased short-term resistance to starvation and fast fat deposition during good times were selected by periodic but not completely predictable flood inputs of allochthonous organic matter. The surface streams that were captured by underground drainage still accumulate organic litter in the dry stream bed and this litter will flush into the cave during especially heavy and prolonged rain. In between times food supply would decline quickly. If there had been slow and steady decline in food input over time after stream capture then I expect that the fish would have evolved lower metabolic rates and much more efficient food finding capacities. Even the egg size is only marginally greater than in the epigean ancestor suggesting that even post-yolk absorption hatchlings are not severely food limited.

An alternative hypothesis for the lack of lowered metabolic rate and for small eggs is that cave *Astyanax fasciatus* populations have been isolated too few generations to evolve the extreme food-finding and metabolic economies seen in Amblyopsid troglobite species (but see Porter *et al.* 2007 suggesting > 1 million year lineage age for some cave populations). Still another alternative is that the *Astyanax* caves, even those without bat colonies, are not very food-limited when averaged over several years. And still another hypothesis is that rare introgression with the surface form slows the evolution of troglomorphy. But introgression is only apparent for some cave populations: Pachon, Chica, and Micos.

An updated overview of the Amblyopsidae as a model system (from Poulson 1963, 1985) including preadaptations in behavior, physiology, and development is the chapter by Niemiller and Poulson, in this volume. This chapter covers the efficiencies of search, detection, capture, and utilization using some heretofore unpublished data. In summary, the troglophile cannot survive long if in the food-poor cave environment. The springs and spring runs where it goes from springs or caves to feed at night have especially abundant food supplies. When there is enough food they have no trouble making chance contact with live prey and grabbing them no matter what part of their body is contacted.

C. Phenotypic vs Genotypic Troglomorphy

One persisting challenge is to separate phenotypic change during development, for example due to allometric growth with different food supplies, from genotypic differences selected over generations. And we need to see if allometric increases in sense organs and brain parts during the development of the troglophile, in a low food cave, translate into better food-finding abilities.

As in *Astyanax fasciatus* cave forms, each Amblyopsid cavefish species has opposite allometry of eye and pigment systems (–) vs sensory systems (+). Eye and optic lobe size and pigment cell number show negative allometric growth and head size and non-visual sensory systems, especially neuromast number, show positive allometric growth. I spent considerable effort measuring allometric growth in my dissertation because I had been impressed by Heuts' suggestion of its importance in understanding cave adaptation in *Caecobarbus geertsi* (Heuts 1951). Subsequently I decided that my effort had been largely a waste of time. But I am now newly excited about my allometry results since I see how to relate them to the work of Jeffery's group on developmental control genes in *Astyanax fasciatus*. Remember that Jeffery's work was partly influenced by Stephen J. Gould's emphasis on the importance of developmental genes in major evolutionary change. And the interface of developmental plasticity and evolution is of great interest currently (West-Eberhard 2003).

To me, my dissertation data suggest that changes in positive and negative allometry in Amblyopsids represents the result of antagonistic pleiotropic effects. It could mean that the same processes that are being discovered by genetic and developmental studies in *Astyanax fasciatus* are reflected in allometric differences within and among species of Amblyopsids. Data on allometry are important because in Amblyopsids, and many other troglobitic fishes, we cannot do breeding experiments. This means we cannot thoroughly determine the relative contributions of developmental phenotypic flexibility and genetic determinism when we see differences between populations of a species. I now see a possible way around this conundrum.

I suggest that we could do 'common garden' experiments and measure the degree of positive and negative allometry for traits. We would start with the smallest individuals available and give them low and high food rations; unfortunately this may be practicable only with the short-lived troglophiles, like *Forbesichthys agassizi* among the Amblyopsids.

However I already have several natural experiments that show that especially low or high food supply and growth rate have opposite effects on rudimenting structures and elaborating structures. Rudimenting structures are eyes, brain optic lobes, and pigment cells. And elaborating structures are lateral line neuromasts and cerebellum. In my dissertation (Poulson 1961, Figure 6) I compared epigean and cave populations of the troglophile *Forbesichthys agassizi* in the Mammoth Cave area. A briefly surviving population in the lowest food supply area of Mammoth Cave had by far the lowest growth rate and smallest known eyes and number of pigmented melanophores each with the lowest known negative allometric coefficients for the species. And it had the most neuromasts and largest cerbellum with the highest known positive allometric coefficients of all populations.

Also in my dissertation I compared nine populations of *Typhlichthys subterraneus* (Poulson 1961, Figure 7) including three with opposite extremes of food supply. *Typhlichthys* in Blowing Cave (west central TN), with an unusual amount of allochthonous organic matter, had the highest growth rate with the largest optic lobes and most pigmented melanophores, with correspondingly highest positive allometry. They also had small heads and low numbers of neuromasts, with correspondingly low positive allometry. In contrast Missouri populations of *Typhlichthys* (River and Welch's Caves) had among the lowest growth rates, had the smallest optic lobes with the lowest positive allometry, and had either few pigmented melanophores (Welch's Cave) or the only population with no pigmented melanophores even at the smallest body size (River Cave). These two cave populations had the largest heads, longest pectoral fins, and most neuromasts of any *Typhlichthys* populations.

How much of the Missouri *Typhlichthys* morphology is due to phenotypic flexibility and how much of it is due to especially great evolutionary troglomorphy in populations that have been independently isolated in caves? Based on the analyses of cave and karst of the Missouri Salem Plateau (Noltie and Wicks 2001), I predict that Missouri *Typhlichthys* caves have a low food supply compared to Kentucky or Tennessee caves. This could have selected for faster evolutionary increases in troglomorphy and it could also have phenotypic effects on allometry during each generation via low growth rates. As a test there is a natural experiment at The Gulf, a deep karst window in Missouri with a very high food supply and a large *Typhlichthys* population (Poulson 2001). It could show us whether all Missouri *Typhlichthys* are especially troglomorphic and/r whether the low food supply presumed from the geology is partly responsible. I predict that the fast growth rates and small body size of The Gulf *Typhlichthys* (Poulson, unpublished) will be associated with relatively smaller optic lobes and eyes, and more neuromasts than for the Tennessee and Kentucky *Typhlichthys*. But the Gulf population will also have fewer neuromasts and larger eyes and optic lobe and more pigmented melanophores than any other Missouri *Typhlichthys*. New data on melanophores (Niemiller and Poulson, this volume) are consistent with this hypothesis.

D. Is Neoteny a Route to Troglomorphy?

Neoteny (aka paedomorphosis or paedogenesis) is the retention in adults of traits seen in juveniles. Based on the grades in evolution of troglomorphy I identified in my PhD dissertation (Poulson 1961) I predicted the next grade in troglomorphy for the family if evolution were to continue. In

1967 I received a photograph in a plain brown envelope of a white cavefish with an extremely large head relative to its body size (Figure 1A in Cooper and Kuehne (1974), Color plate in Niemiller and Poulson, this volume) with the cryptic question 'Is this what you predicted'? Thinking that my colleagues were putting me on, I went so far as to say it was a small ~ 10 mm *Typhlichthys* because of its relatively large head and un-branched fin rays. If I had noted the oxygen bubbles as a clue to scale I would have realized that it was a much larger fish. In fact it was the 58.3 mm holotype of *Speoplatyrhinus poulsoni* named in my honor! In discussions with John Cooper while examining the type series I suggested that it represented an even more extreme neoteny than in any other the other Amblyopsid troglobites and he and his coauthor agreed in their description of this new species (Cooper and Kuehne 1974). In fact Table 2 in my 1985 paper, where I discussed the basis for increasing troglomorphy, does not adequately show how large the head is relative to the body or the strikingly low weight to length ratio both of which, along with absence of branched fin rays in the adult, are the strongest neotenic trends of any of the troglobitic Amblyopsids (see drawings by J. E. Cooper in Niemiller and Poulson, this volume). In addition there is preliminary evidence that both visual and pigment systems are the most rudimentary of any of the troglobites. Pigmented melanophores are very small, very sparse, and disappear by adult size. And the sunken, spathulate snout and narrow head possibly reflect absence of circumorbital bones and change in other related parts of the skull. On this basis I predict that histological studies will show that there is not even an eye vestige in the adult.

The drawings in Weber (2000) of convergently large spathlulate heads of *Speoplatyrhinus poulsoni*, other highly troglomorphic cavefish, and even the most troglomorphic salamander known (*Eurycea rathbuni*) all suggest extreme neoteny to me. In addition to reflecting extreme eye rudimentation, I have suggested (as early as in Mohr and Poulson 1967) that the large head of *E. rathbuni* allows hypertrophied sensory systems, especially lateral line, that are associated with enhanced detection of moving prey. When and if we are ever able to determine the developmental – genetic basis for neoteny in extreme troglobites we can test my prediction (Poulson 2001) that the basis will be developmental genes with pleiotropic effects that work in an analogous way to the developmental genetic changes responsible for the 'stress-resistance syndrome' in higher plants (Chapin 1991). And I now suggest that the basis will be similar to the processes that Jeffery's group is so elegantly showing at the early stages of troglomorphic evolution in *Astyanax fasciatus* (see earlier section **Ongoing Studies of Rudimentation and Elaboration**).

V. ECOLOGY AND LIFE HISTORY

Generally slow growth, long life, late age at first reproduction, and iteroparity with few large young at each reproduction are life history traits associated with neoteny and troglomorphy, at least in the Amblyopsidae. Ecology of other cavefishes is especially well-treated by Trajano (2001).

A. Life History Patterns

The + and – tradeoffs of both low r and high r life history extremes have been reviewed earlier (**Spatial and Temporal Scale**) in the context of the **Tradeoffs** section. I have recently (Poulson 2001) made a detailed critique of the nature and quality of evidence, especially from analysis of scale marks and recapturing marked fish, for differences in life history patterns of deep sea fishes and epigean, troglophilic, and troglobitic Amblyopsids. In brief the extreme patterns are as follows.

Fish with a high r life history (r in the logistic equation for population growth) are characterized by fast growth, early reproduction, and short life. The main mode of risk spreading of risk is among many 'low quality' young at a single reproduction (semelparity). This pattern occurs when and where there are definite and predictable seasons. In Amblyopsids this is the case for both the epigean species, *Cholgaster cornuta*, and the troglophile *Forbesichthys agassizi* (see Niemiller and Poulson, this volume). The + tradeoff of this life history is that a species can quickly increase in numbers during good times. The – tradeoff is that populations will decline rapidly in poor times. In words this has been called boom (+) and bust (–).

Fish with a low r life history are characterized by slow growth, late age at first reproduction, long life, and repeated reproduction (iteroparity). The main mode of risk spreading, to ensure some successful reproduction, is by repeated reproduction attempts each with few high quality young. This pattern occurs when times of favorable food supply are rare or unpredictable. The – tradeoff is that populations cannot respond quickly to good times. The + tradeoff is that populations decline slowly during bad times. For students, I have called this pattern 'fizz (–) and fizzle (+)'. It is seen in all troglobitic Amblyopsids but it is accentuated in *Amblyopsis spelaea* (see Pearson, this volume) and probably in *Speoplatyrhinus poulsoni*.

B.   Abiotic Agents of Selection, Including Food Supply

I have rejected the so-called 'constant cave' paradigm that I supported early in my studies of cave organisms (e.g. Poulson 1964, Poulson and White 1969). Caves are clearly less variable physically and chemically than overlying surface environments but I suggest that rare or unusual events can have great effects, good and bad. Because water flows and mixes, aquatic habitats are much more constant spatially than fully terrestrial habitats in the same cave. But they are more variable temporally, also because of flow. Thus allochthonous input of organic matter to aquatic habitats is much more variable temporally than input to terrestrial habitats. During spring floods, water temperature and chemistry can change rapidly. For both terrestrial and aquatic cave habitats the rigor of low food supply with darkness and lack of plant primary productivity seems to be almost universal even though there is certainly temporal variability at a scale of cavefish life-spans. Thus I was not surprised, in retrospect, to find that all Amblyopsid species tested had the same very low change of metabolic rates and spontaneous activity when acclimated to temperatures from 5 to 25°C (Poulson 1985 and Niemiller and Poulson, this volume).

It will be useful to infer how much the rigor of temperature change, flooding, and food supply has varied over glaciation cycles, that encompass evolutionary time for cavefish. And whether it has changed in the past several centuries with forest clearing and agriculture and associated increased of fine sediments in cave streams. In addition reservoirs have flooded or partly flooded caves along some river valleys. Have these anthropogenic impacts had negative effects on cavefish populations?

Though it is clear that caves are food-limited compared to surface habitats, it is not clear to me that the degree of troglomorphy seen among cavefishes is related to differences in food supply. For example, Trajano (2001) suggests that troglobites from phreatic habitats are especially troglomorphic and I agree that we would expect deep phreatic habitats to have less food input than shallow caves with open channel streams. But the deepest phreatic habitat known is the Edwards Aquifer in Texas and it has perhaps the highest troglobite diversity including apparently among the highest densities of both a predatory troglobitic salamander and fish (reviewed by Trajano 2001). The Edwards Aquifer ecosystem may be unusual in being based on bacteria and fungi that somehow derive their energy in situ, maybe from hydrocarbons (Poulson and Lavoie 2000). And among and within the putative *Typhlichthys* species and *Astyanax* cave species, there is no clear relation of the degree of troglomorphy and the presence or absence of allochthonous particulate plant or bat fecal organic matter. Of course we do not know what the food supplies were at the time of isolation in caves when selection for troglomorphy would have been the most intense (Poulson 1985). However some have argued, myself included (e.g. Poulson and Lavoie 2000), that caves in the wet tropics are and were always less food limited than temperate caves.

Are caves in the wet tropics really less food limited than temperate caves and if so are troglobitic tropical fish in these areas less troglomorphic than temperate species? Almost all wet tropic caves have many bats of several species year around. And with high rainfall and warm temperatures cave formation and fragmentation are fast; this results in many entrances and so many chances for abiotic input of allochthonous organic matter both by gravity and flooding. At least in Brazil caves Trajano's (2001) review of cavefish ecology documents much organic matter, some bat populations, regular sinking streams, and Siluriform ('catfish') troglobites of several lineages that are not very troglomorphic. In addition surface sister species exist for many of the troglobites. This last observation could be interpreted to mean, as an alternative to low selection intensity with high food supplies, that the troglobites are recent and have not had long enough to evolve extreme troglomorphy. In the case of Amblyopsids in North America the lineages of the troglobitic species are ancient and the surface ancestors are no longer extant. And, in addition to apparently long evolutionary time in caves the caves occupied by Amblyopsids overall have much lower food supplies than in the Brazilian tropics.

C. Biotic Interactions as Agents of Selection

Background

Though I regard decreased interspecific competition in caves as an implausible explanation for fish actively invading caves (section below **Mechanisms of Isolation in Caves**), I do believe that competition, predation, and parasitism are likely much less intense for cavefish than for their surface ancestors. If true then there are predictions to be made about the kinds and intensities of −/− interactions (competition), +/− interactions (predation and parasitism), +/0 interactions (commensalism), and +/+ interactions (mutualisms). Culver's elegant experimental studies and theoretical analyses of aquatic stream communities in the Appalachians of USA are testimony to the surprising diversity of both direct and indirect biotic interactions including mutualism (summarized in Culver 1982).

Predation

Cavefish are either the top predator in caves or they have no predators even if they are mainly detritivores. This has three consequences. First they may evolve rudimentation of 'fright' or startle reactions as in the Amblyopsidae (Poulson 1963). For the more troglobitic Amblyopsids I can often pick them up by hand by moving slowly and carefully! It would be interesting to see if they have lost chemical skin exudates with injury i.e. 'shreckstoffs" known from Ostariophycine schooling epigean fish (*Astyanax* is an Ostriophyscine but its cave populations do not have a functioning shreckstoff). But even if cavefish have water borne chemical ways of distinguishing themselves from other species it does not mean they are biologically useful at the scale of cave streams with huge dilution and mixing. Chemical contact recognition is a more realistic possibility (e.g. Parzefall 2000 for *Poecilia*).

A second consequence of being the top predator is that, because of trophic transfer inefficiencies, food becomes increasingly limiting as one moves up a food chain. It is known that cannibalism can sometimes be important in *Amblyopsis spelaea* in population regulation (Poulson 1969) and in *Forbesichthys agassizi* with virtual absence of alternative food in caves (Hill 1966). Though many cavefish are predators it does not mean that they might not get some nutrition by eating detritus or carrion when starving. Whoever wrote 'predator by choice and saprovore by necessity' had a point.

A final consequence of being a top predator, especially with a long life, is damage by toxins including heavy metals and pesticides. If even parts per billion or trillion of mercury or DDT are present, they will biomagnify up a food chain over a few days and very slowly bioaccumulate in individuals over many years of life. It seems possible that 'broken back syndrome' discussed by Keith and Poulson (1981) for *Ambylopsis spelaea* was due to pesticide runoff from agriculture in the cave watershed.

Parasitism and Disease

Possible reduction in parasitism and disease has not been adequately studied and it has not been suggested, heretofore, as a possible advantage to life in caves. What little is known for Amblyopsids is reviewed in Niemiller and Poulson (this volume). For the troglophile *Forbesichthys agassizi*, Whittaker and Hill (1968) have documented the high frequency and high density of a cestode but there are not as extensive data on parasites for troglobitic Amblyopsids (but see Whittaker and Zobel 1978). I predict that parasitism will be found to be less in troglobites than in troglophiles and less in troglophiles than in epigean species. Another side of this coin is that immunological resistance to disease will have become lower in troglobites.

I urge all cavefish researchers to start taking data on diversity, frequency (parasitologists use the term prevalence) and abundance (parasitologists use the term intensity) of parasitism and disease. Pearson (this volume) has data showing a high prevalence of fin erosion in one *Amblyopsis spelaea* population. The fin erosion got better in marked fish over a year or two. It would be useful to know if captive cavefish are susceptible to and able to immunologically respond to some of the common aquarium fish fungal and bacterial diseases. Relating intensity of parasitism to an immunological marker, like globulins, would be a useful non-destructive way to indirectly assay prevalence and intensity of parasitism.

Competition

The presence or absence of competition has been too loosely considered for cavefish because too few recognize the distinction between direct and indirect competition and evidence needed to show the importance of each (Griffith and Poulson 1993). Indirect competition is also known as resource competition or exploitation competition and in fish operates mainly through differential efficiency of locating food; it need not involve even proximity of individuals to operate. Direct competition is also known as interference competition and in fish operates mainly by defense of a territory and maintaining exclusive access to local food resources; it must involve contact between individuals. There are three kinds of evidence needed to support a hypothesis that exploitation competition is ongoing: 1. Fish must reduce the abundance of a food resource; 2. Reduced food abundance must reduce food-finding; and 3. Reduced food-finding must result in reduced fecundity and/or survival. To demonstrate ongoing interference competition one criterion must be met: either 1. Fish must directly reduce each other's survival e.g. by fighting, or 2. Fish must indirectly reduce each other's survival or fecundity by reducing each other's foraging effectiveness. If this reduction is dependent on food density then it overlaps with criterion 3 for exploitation competition. Of course absence of evidence for ongoing competition does not exclude the importance of a 'ghost of competition past' as an agent of selection.

Astyanax fasciatus (Parzefall 2000) and *Rhamdia transiitoria* (Trajano 2001) are the only species where agonistic behavior (a kind of direct competition) has been compared in a troglobite and what is known to be its surface ancestor. The surface catfish shows clear preadaptations and its agonistic behaviors did not differ from its troglobiotic relative. However the surface characin is only marginally preadapted and researchers argue that its agonistic behavior in caves is reduced and different than for the surface fish where vision is so important. Parzefall writes that surface *Astyanax* defend small territories when they are not schooling or shoaling but it is hard to imagine how this would help an individual. It seems to me that food in a surface stream would not be extremely patchy or defendable. Eyed surface fish in the dark show no aggression but blind cavefish do show some aggression. However it is not clear to me that increasing 'defense' of a few cm larger areas as cave individuals decrease activity with food deprivation (Parzefall 2000) will allow them more access to food in a laboratory tank much less in a cave.

Bechler (pers. comm.) found clear intraspecific territorial behavior in 4 *Forbesichthys* or 4 *Typhlichthys* in the realistic scale of a 6.3 m long 4 m^2 artificial stream only where there were clearly defendable rock piles. But both species of *Amblyopsis* moved through all parts of the stream, were rarely in a rock pile, and only showed subtle and occasional agonistic behavior.

The general reduction in agonistic behavior in troglobitic fish (Parzefall 2000 and Bechler 1983) is consistent with my hypotheses that there are no defendable resources in caves that would favor interference/direct competition. But the evolution of increased foraging and metabolic efficiencies does not mean that resources cannot still be limiting for troglobites. In the case of one *Amblyopsis spelaea* population (Poulson 1969) over a number of years numbers fluctuated 2.3 fold, total mass 1.3 fold, and total metabolic demand only 1.15 fold. This natural experiment suggests to me that exploitation competition is regulating populations in a density dependent manner. And the increase in reproduction in another *Amblyopsis spelaea* population after collecting many large fish is consistent with this hypothesis (Poulson 1969). This was an inadvertent manipulative experiment.

Commensalism and Mutualism

Despite the fact that there can always be cascading indirect effects along food chains (Culver 1982), nobody has explored possibilities of commensalisms and mutualism affecting cavefish. These +/0 and +/+ biotic interactions may not be important for cavefish that are always at the top of the food chain. Top predators are the trophic level that is likely to affect other species. For example if a cavefish controlled the population of an isopod species that was a competitive dominant over another species then that would be an indirect mutualism; the enemy of my enemy is my friend. Of course there are rarely two species of cavefish in one cave and when there are they are in different habitats. This means there is little chance for mutualism or commensalisms between cavefish species.

VI. ALLOPATRY, SYMPATRY, AND SYNTOPY IN CAVEFISH

A. Terminology

I start by reviewing terms that describe the geographic relation among incipient species, newly evolved species, or distantly related species. In the classic model of speciation the process starts when a barrier to dispersal arises due to a vicariant event like a prolonged drought or glacial cycle. This splits a species range so that the separated populations (now allopatric) begin to evolve differences because their environments are different. If these incipient species, designated as subspecies if they are morphologically recognizable, are separated long enough to accumulate sufficient differences they become reproductively isolated and are now full species. Species that are separated geographically are allopatric. If the geographic barrier to dispersal disappears then the species can become sympatric. And if they occur in the same same cave they have become syntopic. If these syntopic species are not different enough they may compete or interbreed. If hybrids are viable then introgression may occur. This is the case for *Astyanax fasciatus* cave species today in eastern Mexico. If newly syntopic species are reproductively isolated they may still compete. And if competition is rigorous they may evolve greater microhabitat preferences. This is called character displacement or reinforcement and is one hypothesis to explain the syntopy of two troglobitic Amblyopsids in Mammoth Cave, Kentucky. This brings us to consideration of alternative hypotheses to explain the few cases of syntopy in cavefish.

B. Local Organic Enrichment in a Large Cave

If organic enrichment is natural and localized in a large cave system, as with a maternity colony of bats, then food may not be limiting and several related species may coexist in syntopy. Alternatively one of two troglobites could have a refuge from competition in an area of the same cave with much less food. Shelta Cave is a large system in northern Alabama that may exemplify both of these possibilities. Before the gray bat colony was lost there were five species of macroscopic troglobites: an Amblyopsid fish, an Atyid shrimp, and three crayfish (Cooper 1975). This cave has unusually stable abiotic conditions with very infrequent and localized abiotic input of allochthonous organic matter. Even with a small gray bat colony one of the cave crayfish species had an age at first reproduction estimated as 45 years with only 1-3 successful reproductions for adults that may live to 100 years!

C. Gradients of Food Supply in a Large Cave

In the huge Mammoth Cave system three Amblyopsid species occur and two troglobites seem to replace one another along an upstream to downstream gradient of increasing depth and decreasing food supply (Poulson 1992). Possible hypotheses to explain this syntopy are discussed in the chapter on The Amblyopsidae (Niemiller and Poulson, this volume). One hypothesis involves competition and the other is that the species had different habitat and food needs before they became sympatric and syntopic.

VII. PROBLEMS, MYSTERIES, AND CONTROVERSIES

A. Problems of Measuring Food Supply for Cavefish

First I echo Trajano's (2001) lament that it is rare for cavefish researchers to describe much less quantify habitat characteristics. I add that it is even rarer for them to quantify food supply. I suggest that even the best studies could be better by considering rigor, variability and predictability of allochthonous food input.

Since most of us argue that food limitation is a central agent of selection in caves, it is incumbent on all of us to better characterize food supply. We should at least try to compare amounts of food in caves with large vs small populations or densities of each cavefish species we study. One way to measure potential food input from outsIde is the 'openness' and depth of the karst – cave system (e.g. Noltie and Wicks 2001, Niemiller and Poulson, this volume). More directly there are a number of ways to quantify the amount and patchiness, of fine and coarse particulate organic matter in cave streams in space and over time (e.g. Poulson 1992). If there are rocks it is easy to measure the frequencies and densities macroscopic potential prey, like isopods and amphipods.

Of course the fish are surely better than we are at finding food and we can measure this success, if large populations justify sacrifice, by the frequency and volume of different sized food items in the gut. Using fish success as a measure of food supply and food limitation has the danger of some circularity if comparing two cavefish species since one expectation of evolved troglomorphy is more efficient location and capture of food. Put another way, the same cave could support different densities of a more and less troglomorphic species.

In many caves we need to consider the + tradeoff of food input and the – tradeoff of injury or washout from flooding. Pearson (this volume) has followed up on my observations of *Amblyopsis spelaea* from upstream to downstream of a system that floods regularly and has input of epigean prey from the surface (a + tradeoff). Pertinent to risk from flooding (a – tradeoff) is the increased volume and current speed in the downstream cave. These caves (Spring Mill State Park Indiana where Eigenmann studied in the early 1900s) are separated by short karst windows. My starting observation on one date was the change in abundance and size-frequency distribution going downstream (Poulson 1961, Figure 4). The numbers, over comparable stream lengths, were 86 to 36. The sizes ranged from 20-83 mm SL upstream to 45-87 downstream. Interpreted today in the context of meta-populations this suggests that the upper cave is a source and the lower cave section is a sink. One explanation is that those small fish that are washed out of the upper cave perish; in the upper cave small fish are in streamlets and backwaters that are not as frequent in the lower cave. There is certainly plenty of food in the lower cave as there are high densities of isopods and amphipods in the shallow rocky riffles upstream of pools where the fish live. Consistent with this high food supply in the lower cave, in one year Pearson found about 20 percent of the adult females carrying eggs in their branchial cavities on one date. But the apparent risks of fast currents and injury are such that there was no recruitment of immature fish the next year.

B.   What is the Primary Habitat for Cavefish?

Background

Even for species that have been found in many caves, like three of the four troglobites in the Amblyopsidae, there are few caves with large populations of fish. Put another way, in most localities we see only a few and often only one fish at a time. Before discussing alternative hypotheses to explain these patterns we need to review some observations about cave passages. Here I greatly simplify the details provided in Palmer's complete treatise on *Cave Geology* (2007).

Caves Enterable by Humans

Humans can enter few caves because most have no entrances and even in caves with entrances we often cannot access much of the upstream areas of the subterranean drainage basins. On the other hand there is not likely to be a 'woodwork' around enterable passages that we cannot census for fish. Above a critical dimension of a few cm cave passage enlargement occurs along a few routes that quickly 'capture' >95% of water volume. Even in cave streams with lots of loose rocks many capture-mark-recapture studies reviewed by Trajano (2001) show that visual census at one time finds a third to half of the real population. And in many of the best-studied Amblyopsid cavefish caves there are no obvious hiding places and so visual censuses miss a maximum of 10-20 percent of the population estimated by mark-recapture statistics (Pearson, this volume).

Hypotheses

So this brings us to two main alternative hypotheses about where most individuals of a troglobitic cavefish species occur: 1. In areas accessible to census; or 2. In areas not accessible to census. Either hypothesis has to explain the general patterns that only a small percent of all known localities for a troglobitic fish species have more than a few fish seen at one time.

Let me start with the best mark-recapture study of a cavefish population I have read. This is a multi-year near monthly study of an *Amblyopsis rosae* cave with at least 35 fish seen and as many as 29 fish caught on each trip (Brown and Johnson 2001). Logan Cave has a gray bat maternity colony (the only other *Amblyopsis rosae* cave with large populations is also a gray bat cave; see Figure 4 in Poulson 1961, Cave Springs Cave). Logan has about 1.6 km of accessible stream but only about 10 percent of fish are seen in the upper third of the stream which is not optimum habitat and another 10 percent are found in the lower third of the stream near the entrance. From their two-year study with 23 trips Brown and Johnson infer a regular movement of unmarked fish from the aquifer, but not into the aquifer. And they conclude that most fish are in the inaccessible aquifer. They suggest that this conclusion helps explain why a few fish at a time are found in springs or wells and even in caves. In each case it is rare to see as many as five fish at a time even over multiple visits to the same site. They infer that these few fish are at the edge of aquifers that we cannot access. They rationalize the obvious issue of food scarcity in the aquifer by observations by video camera in newly-dug wells of extensive mats of organic, bacteria like material.

I find Brown and Johnson's (2001) observations and deductions intriguing but not convincing evidence for most fish being in the inaccessible aquifer. The organic matter in the well could be due to local organic pollution enrichment in the shallow aquifer. In the accessible cave they do not preclude the hypothesis that new unmarked fish were in the rubble bottom of the stream. This is consistent with studies in the other population in a grey bat cave where 27-72 fish, all those catchable, were collected by a Tulane University researcher at three times 1-6 months apart (Poulson 1961, Figure 4, Cave Springs Cave). Also they infer, from size-frequency data, that the fish are reproducing in the accessible parts of Logan Cave. I agree with this last interpretation.

So why do I come to an opposite conclusion that caves, but not the water table aquifers, are the primary habitat for *Amblyopsis rosae* (and for all cavefish)? First, in terms of energetics the two bat caves should be the primary habitats because they have the highest food supplies. Furthermore ~ 80% of censused fish are seen in the stream area below the bat roosts. At a larger scale each bat cave is a source population, rather than a sink. It is recognized as a source by having all fish sizes represented with at least some very small fish every few years. This means that the population is reproducing and at least maintaining itself. It is likely to be a larger source of emigrants than of immigrants.

On the other hand, the many sink populations each have only one or a few fish seen at any one time. For any one locality and in aggregate their size distribution is skewed toward larger individuals and if there are any small individuals they are rare over time and among sites. Thus these populations are probably not reproducing and so must be maintained by immigration. Thus they are sinks.

Sources and Sinks

Population ecologists argue that both sources and sinks in a meta-population are important. Sink populations can become sources if some disaster wipes out large populations in a source cave. Thus the many sink populations spread the risk of population loss so that the meta-population always survives. For cavefish the individuals in the sink populations are not necessarily doing poorly since they have few competitors for a low food supply and have metabolic economies and excellent food-searching behaviors (Niemiller and Poulson, this volume). And even if they have no food they lose weight slowly and recover quickly when they find food with their efficient search patterns (Niemiller and Poulson, this volume). In fact they may rarely starve even in the aquifer since the condition of *Amblyopsis rosae* in the bat caves, in terms of fat deposits and gut contents, is only subtly greater than for fish collected from springs, wells, or small caves with no obvious allochthonous organic matter (Poulson 1961).

Another perspective on sink populations is that they include individuals that are potential colonizers of new habitats. It is probable that sink populations are much larger than apparent from the "windows" at which we see them, like springs or wells, but they are surely spread out over large areas. When they encounter an open cave stream with a good food supply their search patterns will keep them there (Niemiller and Poulson, this volume). They are the successful individuals in what biogeographers call 'sweepstakes dispersal'.

Pearson (this volume) addresses the issue of primary habitat for at least six populations of *Amblyopsis spelaea* where he or I have visually censused 60-100 fish on every visit over 40-50 years. His data and mine are consistent with the inference that individuals seen at the edge of the aquifer are outliers of sink populations with high dispersion of individuals. And we know for this species that it carries eggs and newly hatched young in its branchial chamber for almost six months so a 'pregnant' female has a long time to disperse to a new cave through the water table and found a new population.

A corollary of all of the above discussion is that calculations of cavefish population sizes and densities (Trajano 2001) mean little or nothing without a lot of context about stream characteristics and food supply. And home range and movements (Trajano 2001) are only meaningful within a species where one can relate the data to habitat structure and food supply.

C. Why do Some Highly Troglomorphic Fish Species Show Relict Distributions?

Given the troglomorphic adaptations of cavefish and the existence of meta-populations, it is a mystery as to why some species come to have relict distributions.

The most likely hypothesis to me is that they always had restricted distributions. This is only an extension of what we now know to be a common pattern of different populations of one epigean ancestor becoming independently isolated in different cave systems in a karst region. If the cave region/limestone formation is limited in extent then there never could have been a widespread meta-population. We have seen that cavefish can disperse far and well but only within a limestone area without subterranean barriers.

We have several models of widespread surface species that have incipient troglomorphy in restricted parts of a karst region without obvious barriers to dispersal. Two of these are not cavefish (*Gyrinophilus* salamanders, Niemiller and Fitzpatrick, pers. comm., and *Gammarus minus* amphipods, summary in Culver and Wilkens 2001) but they have been exhaustively studied and analyzed so the interpretations could be generally applicable. The amphipod has been studied in most detail. The small karst area where they have evolved troglomorphy has extensive caves with a lot of detrital food and barriers to dispersal between the spring and cave populations. So if some climatic even wipes out the surface spring populations the species complex will have a relict distribution. *Cottus* sculpins may be a fish example parallel to the amphipod. Incipient troglomorphic populations occur only in a small karst area of Missouri with extensive caves and lots of detrital wash-in (Burr *et al.* 2001). I suspect there may be more such examples from areas other than the United States when sufficient detailed surveys the distribution of surface–cavefish complexes are done.

The hypothesis just discussed to explain relict distributions, of a small area where isolation in caves can occur, does not seen to fit the most troglomorphic Amblyopsid, my namesake *Speoplatyrhinus poulsoni*. It is restricted to Key Cave in an intensively studied karst area of northern Alabama with many caves and no apparent barriers to subterranean dispersal (Kuhajda and Mayden 2001).

One hypothesis relates to the character of the cave. Key Cave has a maze structure typical of caves formed at the water table in flat bedded limestones. As such the deeper parts of the cave may have been flooded by impoundments in the adjacent Tennessee River so that suitable habitat has been restricted. But this does not explain why many nearby caves have *Typhlichthys* which has occurred as a single individual for at least seven years in Key Cave.

Another hypothesis is that *Speoplatyrhinus* was out-competed by what is now a widespread species *Typhlichthys subterraneus* (e.g. Woods and Inger 1957). I think this is unlikely based on my observation of density dependent population regulation in *Amblyopsis spelaea* (Poulson 1969) which indicates that effective exploitation competition is operating in a highly troglomorphic species. If this is general then less troglomorphic fish, like *Typhlichthys subterraneus*, should never be able to competitively exclude more evolved species, like *Speoplatyrhinus poulsoni*. The reason is that the more efficient species will always have a refuge in finding food. As the less efficient, but perhaps faster feeding, species reduces food to lower and lower abundances it may suffer exploitation competition but the more efficient troglobite will not yet be food limited.

Another hypothesis to explain relict distributions comes from a possible answer to another mystery i.e. how do troglophiles apparently outcompete troglobites when there is pollution by organic enrichment (not just in cavefish)? The parallel on a longer time scale is that a new species, originally troglophilic, might be able to 'invade' a cave region with a pre-existing troglobite if the food supply in caves increased at a time that surface climate resulted in isolation of the new species. We cannot test this paleogeographic hypothesis but we can do experiments relevant to the observation that organic enrichment can result in a troglophile replacing a troglobite.

I suggest a testable model of observed replacement of troglobites by troglophiles with organic enrichment due not to competition but rather to demographic swamping. This is analogous to a human minority group replacing the original ethnic or racial group simply due to higher reproductive rate. Such replacement would proceed slowly and eventual loss of the original fish species would be due to random processes. This demographic swamping could be speeded by harassment if the less efficient troglophile is more aggressive than the troglobite even if resources are not defendable. Though I believe that this demographic swamping mechanism operates generally with organic enrichment, I have no evidence that it explains the restriction of *Speoplatyrhinus* to a single cave. If the swamping hypothesis were correct I would expect caves around Key Cave to have much higher food supplies than Key Cave and large populations of *Typhlichthys*.

D. Dating Isolation in Caves

Molecular Clocks

The accuracy of molecular clocks is controversial and in cavefish may be complicated by increasing generation times with evolution of troglomorphy. Another problem is finding cave independent traits for which mutations are neutral even in early stages of isolation in caves. We may have to wait for complete genome sequencing to find single nucleotide changes that are not part of genes.

Neutral Mutations

Despite problems using molecular clocks, our best candidates for cave independent traits associated with rudimentation in caves may give the relative timing of isolation in caves. To check such inferences we need to also use at least geo-morphological evidence (e.g uppermost – oldest passage sediments in Mammoth Cave dated back to 3.5 my by cosmogenic Al and Be) and palynological evidence (pond bottom cores dating back several hundred thousand years) to match timing of divergence inferred from genetic evidence with times of change in landscapes and climates. A major problem is that we do not have geomorphological or isotope data about climates and land forms that go far enough back in time, to the early Pliocene or even Miocene times, that molecular data suggest are times of Amblyopsid lineage divergence (Niemiller, pers. comm.)

E. Phylogeography and Parallel Evolution of Cavefish

Species Complexes

It is becoming increasingly clear from molecular data that many or most species of trogolobite are actually species complexes that represent independent isolations in caves (e.g. Trajano 2001). Until the relevant molecular studies are done (e.g. chapters in this volume by Borowsky and by Niemiller and Poulson) Trajano has suggested (2001) that we use geologically and hydrologically defined subterranean drainage basins as criteria for occurrence of sister species.

A related area of evolutionary research on cavefish is inferring phylogenies and current biogeographic history within and among species or species complexes using continually improving computer algorithms to analyze molecular data. This approach is emphasized in this volume by Borowsky and by Niemiller and Poulson. I hope experts will enlighten us about the relative efficacy of different genetic markers, e.g. mitochondrial vs nuclear genes and single nucleotide polymorphisms vs RAPDs vs microsatellite variation for answering questions at different temporal and spatial scales. It is not clear to me that any of these markers can tell us even the relative times that different troglobites have been isolated in caves.

However molecular markers may help us avoid the problem of distinguishing cave dependent characters subject to convergent and parallel evolution, from cave independent characters (see the benchmark paper by Christiansen 1961). Unfortunately no one has yet studied any demonstrably cave independent character(s) in enough depth to infer absolute evolutionary time in caves for any cavefish. In this context no morphological characters should be neglected in constructing phylogenies even if they reflect the action of many quantitative genes that we cannot identify with present techniques (but see Protas *et al.* 2008).

Phylogeography of the Amblyopsidae

One of the 'Mysteries of the cavefish' we will explore for the Amblyopsidae (Niemiller and Poulson, this volume) is how or whether successive glacial cycles resulted in isolation of different sympatric or different allopatric surface ancestors in different places resulting in very different current geographic distributions. Culver (pers. comm.) seems to favor the view that the differences in adaptation among Amblyopsid troglobites (Poulson 1963, 1985) may be due to differences in cave type and food supply in the areas where each of allopatric ancestors were isolated. I agree that some of the differences among troglobites can be explained by differences in geology and cave character among cave regions (see also the section above on **Phenotypic vs Genotypic Troglomorphy**).

I have also discussed how the relative time of isolation, based on overall eye and pigment rudimentation, can be used to infer different evolutionary time in caves (Poulson 1985). This supposition is supported by the great differences in millions of years of separation of Amblyopsid lineages as deduced from molecular data in which *Amblyopsis rosae* is the oldest lineage (Niemiller and Poulson, this volume). Niemiller's ongoing molecular genetic studies will give us some answers as to the relative ages of species divergence but cannot tell us even the relative durations of time since isolation in caves. What we can do is look at different species of Amblyopsids and other cavefish groups and ask whether the present abiotic and biotic agents of selection are different enough to expect more troglomorphy in some species or in some karst regions.

F. Mechanisms for Evolutionary Isolation of Cavefish

Hypotheses

Though we may not be able to date the time or even the relative times of isolation we can use natural experiments to evaluate alternative hypotheses for the mechanism of isolation. Many authors have discussed hypotheses about how epigean species came to be cave species but Holsinger (2000) provides the best review I have read of the two main competing processes (models for isolation).

Climate Effects

For cavefish the climate relict model aka Pleistocene effect model seems to me more applicable than the adaptive shift model aka habitat shift model.

Drying and warming during Pleistocene cycles has apparently been just as important to evolutionary isolation in caves in parts of the tropics, South America at least (Trajano 2001), as in the temperate zones of the world.

In North America the troglophilic spring cavefish *Forbesichthys agassizi* may or may not be a good model for the transition as you will read in our chapter on the Amblyopsidae (Niemiller and Poulson, this volume). In Brazil, despite likely loss of surface streams with drying, during Pleistocene times (Trajano 2001) there is and probably were much higher food supplies in caves than in periglacial temperate areas. With warm temperatures and high rainfalls overall I was the first to argue in detail that caves form faster and have more entrances, including karst windows, than in temperate zones. In addition many if not most tropical caves have bat populations (Poulson and Lavoie 2000, Poulson 2005). So perhaps it is not surprising that tropical troglobitic fish, many of which have extant ancestors that are demonstrably preadapted (e.g. Trajano 2001 and Wilkens 2005) have not been shown to have the extreme metabolic economies or sensory hypertrophies documented for temperate troglobitic fish, especially the Amblyopsidae.

Habitat Shift

I suggest that this is a rare mode of isolation for cavefish. Examples may be cases where the actual or putative surface ancestor is a diurnal, schooling fish with large eyes. Possible examples are *Caecobarbus geertsi* and *Astyanax fasciatus*. And even in these cases there were some preadaptations with some sensory systems useful in darkness. In an earlier section (***Astyanax fasciatus* vs Amblyopsidae as Models**) I have discussed the catastrophic stream capture that all agree was the mode of isolation for *Astyanax fasciatus*. This is in no way analogous to the active invasion of lava tubes by ancestors of cave Arthropods shown by Howarth's (1981) studies.

VIII. PROTOCOL FOR STUDY OF A NEW OR ENDANGERED SPECIES

A. Rationale

I provide the following outline both as a summary of some parts of this chapter and as a suggestion for priority of study for a newly discovered species, an endangered species, or fish in caves where study is difficult (e.g. remote and need for SCUBA). In sequence I suggest field observations, non-destructive studies of captured individuals in the cave, behavioral observations in an aquarium in situ, physiological and behavioral studies of live individuals held temporarily in the laboratory, and detailed studies of preserved specimens.

B. Field Observations

Both descriptions of the cave and its fauna and estimates of size-frequency distribution of the fish are critical. The likelihood of flooding, amounts of allochthonous organic matter, and substrate types in riffles, pools, or ponds are combined with measures of live food (see Poulson 1992 for detailed methods). The nature of the karst may also be important (e.g. Noltie and Wicks 2001). And the estimated size-frequency distribution is a powerful clue as to whether the cave is a source or sink and whether the species is long- or short-lived.

C. Field Data on Captured Individuals

I start with observations of behavior as you try to capture a fish; remember that troglomorphy is often associated with loss of 'fright' reactions. Once a fish is caught measure standard length and combine this with volume displacement to estimate condition factor. Assess the amount and location of visible fat deposits and external parasites or lesions. Then take a scale sample for possible determination of minimum age from scale marks that show slowing of growth. Also take a fin clip for genetic studies. With care these studies can be done without anesthesia (e.g. Pearson, this volume).

D. In situ Aquarium Observations

Start with a hand-carried mini aquarium in which you watch a fish after taking data on the fish in hand. Note how calmly or frenetically the fish swims and how fast it calms down. Once calm, time the frequency and estimate the depth of respiratory movements; this may give a clue to metabolic rate. In a larger aquarium on the stream bank, with a rock for structure and possible hiding place, take notes on swimming and exploratory behavior. Does the fish swim in the open water as wll as along the sides or bottom? Is it highly thigmotactic? Does it investigate or pause under the rock? If there are live prey available in the cave put some in the aquarium and watch possible searching behavior and prey capture details.

E. Laboratory Aquarium and Respirometer Studies

Short-term Studies of Days to a Week

Use video (e.g. Bechler 1983 and Sheryl Coombs' website) to better quantify swimming, exploring, searching, and prey capture behaviors and efficiencies with different prey types and prey densities (e.g. Niemiller and Poulson, this volume). Then use a torus respirometer with IR light activity sensor to relate metabolic rate to swimming activity. This gives a measure of adaptation to chronic food scarcity.

Longer-term Studies of Weeks to Months

If you have several individuals then keep them in separate aquaria with a rock and then study agonistic behavior of home and intruder fish (Bechler 1983). Repeat the experiments with well-fed and fish deprived of food for a week. In fish derived of food, follow weight loss and condition factor for a month, or longer if weight loss is slow; this gives a measure of any adaptations to periodic food scarcity. At the end of the starvation period provide food ad libitum and measure the recovery of weight and condition factor along with any obvious fat storage.

F. Study of Preserved Specimens

Even endangered species will have specimens in collections upon which the species description was based. Urge the museum curators to allow detailed studies and even partial dissections to get the maximum data possible. There are some data that can only come from preserved specimens.

Data without Dissection

Describe external sense organs and quantify their number and degree of development (e.g. Poulson 1961, 1963 and 1985). If a size series is available also determine their positve or negative allometric coefficients. Do the same for fin and head lengths. Determine the weight length relation, relative head size, and branching of fin rays as clues to possible neoteny (e.g. Cooper and Kuehne 1974). Quantify the density, sizes, and character of pigmented melanophores and their allometric coefficient.

Data from Dissections

At a minimum open the body cavity and measure fat deposits, gono-somatic index, and mature ova number and maximum size. If allowed, determine frequency and volume of detritus and prey items in the stomach and intestine and relate this to data on food supply in the cave where the specimens were collected. And, if allowed, carefully open the cranial cavity to measure and draw brain parts, olfactory rosette, any visible eye remnant, and semi-circular canals and otoliths (e.g. Poulson 1963 and Cooper and Kuehne 1974).

Acknowledgements

I start by thanking the editors of this volume, especially 'Leo' or Eleonora Trajano, both for inviting me to write this introductory chapter and for organizing this book in my honor. I was surprised and am still touched. But the acme of my biospeleological career was the naming of the fish I 'predicted' as *Speoplatyrhinus poulsoni* by two of my closest colleagues (cOOp and Bob Kuehne 1974) aided and abetted by the good humor and photos of Tom Barr. My namesake has even made the Burger King kids' collectible cards on endangered species. Next I thank all of my students and colleagues over the past ~50 years who have been sounding boards for my ideas about cave biology and cavefishes as they have changed and matured. Especially influential have been Tom Barr, Dave Bechler, Richard Borowsky, Ken Christiansen, cOOp Cooper, Dave Culver, Dave

Griffith, John Holsinger, Bill Jeffery, Tom Kane (deceased), Kathy Lavoie, Bill Pearson, Leo Trajano, and Horst Wilkens. Next, I thank in advance any and all who build on my contributions by considering my suggestions and testing my hypotheses. I especially throw down the gauntlet to my new young colleague Matt Niemiller. And finally I admonish all readers to practice my mantra of using multiple hypotheses and doing long-term studies as exemplified in one of my limericks.

Tom always uses multiple hypotheses

To avoid those that smack of sophistries

To Tom long studies aren't dippity and he eschews serendipity

The right times and right places aren't wizardries

References

Bechler, D.L. 1983. The evolution of agonistic behavior in Amblyopsid fishes. *Behavioral Ecology and Sociobiology* 12: 35-42.

Borowsky, R. 2008. Restoring sight in blind cavefish. *Current Biology* 18: R23-R24.

Brown, J.Z. and J.E. Johnson. 2001. Population biology and growth of Ozark cavefishes in Logan Cave National Wildlife Refuge, Arkansas. *Environmental Biology of Fishes* 62: 161-169.

Burr, B.M., G.L. Adams, J.K. Krejca, R.J. Paul and M.L. Warren, Jr. 2001. Troglomorphic sculpins of the *Cottus carolinae* species group in Perry County, Missouri: Distribution, external morphology, and conservation status. *Environmental Biology of Fishes* 62: 279-286.

Chapin, F.S. III, A. Kellar and F. Pugnaire. 1993. Evolution of suites of traits in response to environmental stress. *American Naturalist* 142 (Supplement): 61-84.

Christiansen, K. 1961. Convergence and parallelism in cave Entomobryinae. *Evolution* 15: 288-301.

Cooper, J.E. 1975. Ecological and behavioral studies in Shelta Cave, Alabama with emphasis on decapod crustaceans. PhD Dissertation, Department of Zoology, University of Kentucky, Lexington. 364 pp.

Cooper, J.E. and R.A. Kuehne. 1974. *Speoplatyrhinus poulsoni*, a new genus and species of subterranean fish from Alabama. *Copeia* 1974: 486-493.

Culver, D.C. 1982. *Cave Life: Evolution and Ecology*. Harvard University Press, Cambridge.

Culver, D.C. 1985 (ed.). *Regressive Evolution, Bulletin of the National Speleological Society* 47: 70-162.

Culver, D.C. and T.L. Poulson. 1971. Oxygen consumption and activity in closely related Amphipod populations from cave and surface habitats. *American Midland Naturalist* 85: 74-84.

Culver, D.C. and H. Wilkens. 2000. Critical review of the relevant theories of the evolution of subterranean animals. In: *Ecosystems of the World*, Volume 30,

Subterranean Ecosystems, H. Wilkens, D.C. Culver and W.F. Humphreys (eds.), Elsevier, Amsterdam, pp. 381-398.

Culver, D.C. and W.B. White (eds.). 2005. *Encyclopedia of Caves*. Elsevier, Amsterdam.

Culver, D.C., T.C. Kane and D.F. Fong. 1995. *Adaptation and Natural Selection in Caves: The Evolution of* Gammarus minus. Harvard University Press, Cambridge.

Griffith, D.M. and T.L. Poulson. 1993. Mechanisms and consequences of intraspecific competition in a Carabid cave beetle. *Ecology* 74: 1373-1383.

Heuts, M.J. 1951. Ecology, variation, and adaptation of the blind African cave fish *Caecobarbus geertsi* Blgr. *Annales Musee Royale Zoologie Belgique* 82: 155-230.

Hill, L.G. 1966. Studies of the biology of the spring cavefish, *Chologaster agassizi* Putnam. PhD Dissertation, Department of Biology, University of Louisville. 104 pp.

Holsinger, J.R. 2000. Ecological derivation, colonization, and speciation. In: *Ecosystems of the World*, Volume 30, *Subterranean Ecosystems*, H. Wilkens, D.C. Culver and W.F. Humphreys (eds.), Elsevier, Oxford, pp. 399-415.

Howarth, F.G. 1980. The zoogeography of specialized cave animals: A bioclimatic model. *Evolution* 34: 394-406.

Huppop, K. 1985. The role of metabolism in the evolution of cave animals. *The National Speleological Society Bulletin* 47: 136-146.

Huppop, K. 2000. How do cave animals cope with the food scarcity in caves? In: *Ecosystems of the World*, Volume 30, *Subterranean Ecosystems*, H. Wilkens, D.C. Culver and W.F. Humphreys (eds.), Elsevier, Oxford, pp. 159-188.

Jeffery, W.R. 2005. Adaptive evolution of eye degeneration in the Mexican blind cavefish. *Journal of Heredity* 96: 185-196.

Keith, J.H. and T.L. Poulson. Broken-back syndrome in *Amblyopsis spelaea*, Cave Research Foundation 1979 Annual Report: 45-48.

Kuhajda, B.R. and R.L. Mayden. 2001. Status of the federally endangered Alabama cavefish *Speoplatyrhinus poulsoni* (Amblyopsidae), in Key Cave and surrounding caves, Alabama. *Environmental Biology of Fishes* 62: 215-222.

Mohr, C.E. and T.L. Poulson. 1967. *The Life of the Cave*. McGraw-Hill, New York, 232 pp.

Montgomery, J.C., S. Coombs and C.F. Baker. 2001. The mechanosensory lateral line system of the hypogean form of *Astyanax fasciatus*. *Environmental Biology of Fishes* 62: 87-96.

Niemiller, M.L. and B.M. Fitzpatrick. Phyologenetics of the southern cavefish, *Typhlichthys subterraneus*: Implications for conservation and management. *Proceedings of the 2007 National Cave Management Symposium*. (In Press)

Noltie, D.B. and C.B. Wicks. 2001. How hydrogeology has shaped the ecology of Missouri's Ozark cavefish, *Amblyopsis rosae*, and southern cavefish, *Typhlichthys subterraneus*: Insights on the sightless from understanding the underground. *Environmental Biology of Fishes* 62: 171-194.

Palmer, A.N. 2007. *Cave Geology*. Cave Books, Dayton, Ohio, 454 pp.

Parzefall, J. 2000. Ecological role of aggressiveness in the dark. In: *Ecosystems of the World*, Volume 30, *Subterranean Ecosystems*, H. Wilkens, D.C. Culver and W.F. Humphreys (eds.), Elsevier, Amsterdam, pp. 221-228.

Porter, M.L., K. Dittmar and M. Perez-Losada. 2007. How long does evolution of troglomorphic form take? Estimating divergence times in *Astyanax mexicanus*. *Acta Carsologica, Time in Karst, Postojna*, pp. 173-182.

Poulson, T.L. 1960. Cave adaptation in Amblyopsid fishes. PhD Dissertation, Department of Zoology, University of Michigan, Ann Arbor. University Microfilms 61-2787.

Poulson, T.L. 1963. Cave adaptation in Amblyopsid fishes. *American Midland Naturalist* 70: 257-290.

Poulson, T.L. 1964. Animals in aquatic environments: Animals in caves. In: *Handbook of Physiology: Environment*. D.B. Dill (ed.), American Physiological Society, Washington, pp. 749-771.

Poulson, T.L. 1969. Population size, density, and regulation in cave fishes. *Actes of the 4th International Congress of Speleology*, Ljubljana, Yugoslavia 4-5: 189-192.

Poulson, T.L. 1985. Evolutionary reduction by neutral mutations: Plausibility arguments and data from Amblyopsid fishes and Linyphiid spiders. *Bulletin of the National Speleological Society* 47: 109-117.

Poulson, T.L. 1992. The Mammoth Cave ecosystem. In: *The Natural History of Biospeleology*, A. Camacho (ed.), Museo Nacional de Ciencias Naturales, Madrid, pp. 569-611.

Poulson, T.L. 2001. Adaptations of cave fishes with some comparisons to deep-sea fishes. *Environmental Biology of Fishes* 62: 345-364.

Poulson, T.L. 2005. Food Sources. In: *Encyclopedia of Caves*, D.C. Culver and W.B. White (eds.), Elsevier, Amsterdam, pp. 255-263.

Poulson, T.L. and W.B. White. 1969. The cave environment. *Science* 165: 971-981.

Poulson, T.L. and K.H. Lavoie. 2000. The trophic basis of subsurface ecosystems. In: *Ecosystems of the World*, Volume 30, *Subterranean Ecosystems*, H. Wilkens, D.C. Culver and W.F. Humphreys (eds.), Elsevier, Amsterdam, pp. 118-136.

Protas, M., I. Tabansky, M. Conrad, J.B. Gross, O. Vidal, C.J. Tabin and R. Borowsky. 2008. Multi-trait evolution in a cave fish, *Astyanax mexicanus*. *Evolution and Development* 10: 196-205.

Proudlove, G.S. 2006. *Subterranean Fishes of the World*. International Society for Subterranean Biology, Moulis.

Romero, A. 2001 (ed.). *The Biology of Hypogean Fishes*. Kluwer Academic Publishers, Dordrecht, 376 pp.

Sexton, E.W. and A.R. Clark. 1936. A summary of the work on the amphipod *Gammarus chevreuxi*, Sexton, carried out at the Plymouth Laboratory (1912-1936). *Journal of the Marine Biological Association, United Kingdom* 21: 357-414.

Trajano, E. 2001. Ecology of subterranean fishes: An overview. *Environmental Biology of Fishes* 62: 133-160.

Weber, A. 2001. Fish and Amphibia. In: *Ecosystems of the World*, Volume 30, *Subterranean Ecosystems*, H. Wilkens, D.C. Culver and W.F. Humphreys (eds.), Elsevier, Amsterdam, pp. 109-132.

West-Eberhard, M.J. 2003. *Developmental Plasticity and Evolution*. Oxford University Press, Oxford, 816 pp.

Whittaker, F.H. and L.G. Hill. 1968. *Proteocephalus chologaster* sp. N. (Cestoda: Proteocephalidae) from the spring cavefish *Chologaster agassizi* Putnam (Pisces: Amblyopsidae) of Kentucky. *Proceedings of the Helminthological Society of Washington* 35: 15-18.

Whittaker, F.H. and S.J. Zober. 1978. *Proteocephalus poulsoni* sp. N (Cestoda: Proteocephalidae) from the northern cavefish *Amblyopsis spelaea* Dekay, 1842 (Pisces: Amblyopsidae) of Kentucky. *Folia Parasitologica* 25: 277-280.

Wilkens, H. 2007. Regresssive evolution: Ontogeny and genetics of cavefish eye rudimentation. *Biological Journal of the Linnean Society* 92: 287-296.

Wilkens, H., D.C. Culver and W.F. Humphreys (eds.). 2005. *Ecosystems of the World, Volume 30, Subterranean Ecosystems*, Elsevier, Amsterdam, 791 pp.

Woods, L.P. and R.F. Inger. 1957. The cave, spring, and swamp fishes of the family Amblyopsidae of central and eastern United States. *American Midland Naturalist* 58: 232-256.

CHAPTER 2

Biodiversity and Distribution of the Subterranean Fishes of the World

Graham S. Proudlove

Department of Zoology, The Manchester Museum, The University of
Manchester, Manchester M13 9PL, UK
E-mail: g.proudlove@manchester.ac.uk

INTRODUCTION

The fishes are the largest vertebrate group with something over 28,000 described species. The majority (around 16,000 species) are marine but around 12,000 species are found only in the freshwaters found on continental and island landmasses. These freshwaters are found as steams, rivers, lakes and ponds and eventually all freshwater finds its way to the oceans in the global hydrological cycle. During their course from deposition as rainwater, to eventual arrival at the oceans, some freshwater penetrates the surface to flow through various types, and various depths, of subterranean habitats. These vary from the tiny interstices between sand grains, through caves enterable by man, to vast groundwater bodies at great depth below the surface. Evolution has been at work within these subterranean places and animals which have found themselves in the darkness have adapted and evolved. They have evolved a set of very distinctive characteristics typical of, and found only in, subterranean environments. These are found in all taxa, vertebrate and invertebrate, which are now obligate inhabitants of the subterranean (or hypogean) realm. Most obviously these animals loose their eyes and any melanin and carotenoid pigmentation. They very often appear white in colour, or even transparent. These obvious losses have lead to the oft-used phrase "blind white cave animals". In addition to these losses are a set a gains which are adaptations to life without light and most often very food-poor conditions. For example, the limbs of Crustacea are elongated, the lateral line of fishes is expanded, and the sensory setae of insects are lengthened. There are parallel behavioural, physiological and neurological alterations all aimed at survival in the hypogean. All of these changes are referred to as troglomorphic. The relative degree of troglomorphy is related directly to the time of evolution – the phylogenetic age – of the particular taxon in question (see Porter *et al.* 2007 and Trajano 2007 for discussion of this point).

Out of all 12,000 species of freshwater fish only a very few, currently just over 150, have successfully colonised the hypogean. The number known is climbing fast but it is probably not likely to pass 1000 species. The purpose of this opening chapter is to set out, in broad terms, the overall patterns of biodiversity in subterranean fishes and to examine their distribution at global (biogeographic) scale. It sets the scene for the more detailed chapters making the main body of the book.

There is a large literature (of more than 2500 publications) on all aspects of the biology of subterranean fishes. The most recent synthesis of this literature is Proudlove (2006) which provides detailed accounts for 104 species and includes the complete bibliography to 2004. Wilkens (2005) is a highly readable account covering all aspects of their biology. Similar, though briefer, accounts are provided by Proudlove (2004) and Romero (2004). Romero (2001) is a volume similar to this one with a set of contributed papers covering many aspects. The chapters by Weber (2000) and Weber *et al.* (1998) are both excellent syntheses. Though now very out of date, and out of print, the book by Thines (1969) is well worth seeing.

THE CHECKLIST

The following checklist contains taxonomic details of all known subterranean species. Most are formally described but a few are known but have yet to receive a formal description. The most recent species description (of *Trichomycterus uisae*) was published after the Tables in this chapter were compiled and is not included in these Tables. All authorship details were checked with the Catalogue of Fishes (Eschmeyer, W.N. (ed.) Catalog of Fishes, electronic version (updated 29 August 2008) www.calacademy. org/research/ichthyology/catalog/fishcatsearch.html).

The Development of Knowledge: How Many Subterranean Fishes are There?

The first subterranean fish species (*Amblyopsis spelaea* in the USA) was formally described in 1842. Since then a further 149 species (to December 2008) have been discovered. We can plot the rate at which they have been discovered (as cumulative totals) and then use the curve to extrapolate in order to see how many species may exist. The curves are shown in Fig. 1 (an arithmetic plot) and Fig. 2 (a logarithmic plot). Several features are evident in the Figures. First the gradient changes dramatically around 1981 (year 140 in the figures). This is almost certainly the result of more species being found because more caving expeditions were being mounted from this time onwards. China especially was almost unknown speleologically before the 1980s but since then very large mileages of passage have been explored and many cave fish species discovered. (Subterranean fishes are unusual in that most are found by non-biologists whereas this is not the case for nearly all other newly described fish species.)

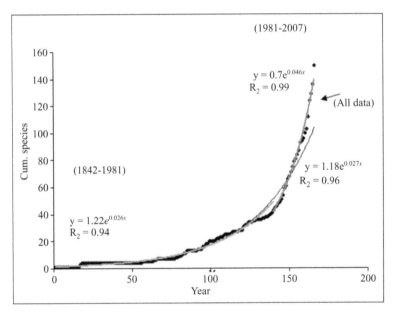

Fig. 1 An arithmetic plot of the cumulative number of subterranean fish species versus the date they were described. Year 0 = 1842 when the first subterranean fish was described.

Color image of this figure appears in the color plate section at the end of the book.

The fitted curve can be used to extrapolate in order to see how many species may exist in total. In order to see if the curve is a good fit to the data we can compare the expected total species for a certain year with the actual observed number. It is easy to se that the curve is a very good fit. We can then predict that there will be 166 species by 2010 (16 more than 2008) and 209 species by 2015 (59 more than 2008). Unfortunately there is

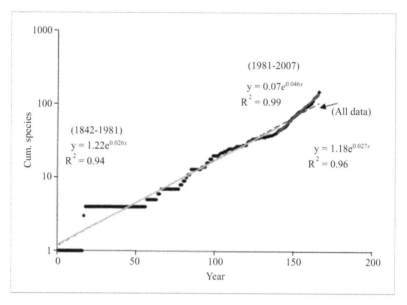

Fig. 2 A logarithmic plot of the cumulative number of subterranean fish species versus the date they were described. Year 0 = 1842 when the first subterranean fish was described.

Color image of this figure appears in the color plate section at the end of the book.

no evidence that the curve has reached an asymptote so it is not possible to predict the total global number of species.

	1981	1991	2006	2007	2010	2015
Expected	44	69	138	145	166	209
Observed	43	68	136	150	–	–

SYSTEMATIC DISTRIBUTION

Summaries of the systematic composition of subterranean species are provided in Table 1 (Orders) and Table 2 (Families). There is an obvious split into two groups: fishes within the clade Otophysi and those not in this clade. Seventy-nine per cent of all species are otophysans, with only 21% not being, despite there being many more non-Otophsyi. The explanation for this is that otophysans make up the majority of the primary freshwater fishes (Helfman *et al.* 2009) and it is mainly freshwater fishes that can get into the subterranean domain. Only the families Bythitidae, Gobiidae and Eleotridae have invaded caves from the sea. Table 3 lists all genera which have only one subterranean representative (38 genera and species, 60% of genera, 25% of species). There are 20 monotypic genera (32%) which are restricted to subterranean habitats (marked HR = hypogean restricted). There are eight monotypic genera which are currently known only from the type locality (marked TLO). Table 4 lists genera with two or more hypogean species and those where all species are hypogean restricted (HR), and known only from the type locality. The genera (25, 40%) and species (35, 23%) which are known only from subterranean sites (HR on the previous tables) are summarised in Table 5.

Table 1 The numbers of genera and species in each Order which contains hypogean fishes

Order	Genera	Species	Monotypic genera	Multitypic genera (species)	
Characiformes	2	3	1	0	
Cypriniformes	23	66	11	0	Otophysi
Siluriformes	19	49	3	3(7)	45 genera (71%)
Gymnotiformes	1	1	0	0	119 species (79%)
Persopsiformes	3	4	2	1(2)	
Ophidiiformes	3	8	1	1(6)	
Cyprinodontiformes	1	1	0	0	18 genera (29%)
Synbranchiformes	2	5	0	0	31 species (21%)
Scorpaeniformes	1	2	0	0	
Perciformes	8	11	2	1(3)	
10 orders	**63**	**150**	**20**	**6(18)**	

Table 2 The numbers of genera and species in each Family that contains hypogean fishes

Family	Genera	Species	Monotypic genera	Multitypic genera (species)	
Characidae	2	3	1	0	
Cyprinidae	12	37	8	0	
Nemacheilidae	8	26	1	0	
Balitoridae	1	1	1	0	
Cobitidae	2	2	1	0	Otophysi
Ictaluridae	3	4	2	1(2)	45 genera (71%)
Siluridae	1	3	0	0	119 species (79%)
Clariidae	3	4	1	1(2)	
Heptapteridae	3	13	0	0	
Callichthyidae	1	1	0	0	
Trichomycteridae	5	16	0	0	
Loricariidae	1	3	0	0	
Astroblepidae	1	2	0	0	

Table 2 Contd..

Table 2 Contd..

Siluriformes in. sed.	1	3	0	1(3)
Sternopygidae	1	1	0	0
Amblyopsidae	3	4	2	1(2)
Bythitidae	3	8	1	1(6)
Poeciliidae	1	1	0	0
Synbranchidae	2	5	0	0
Cottidae	1	2	0	0
Gobiidae	3	4	1	0
Eleotridae	5	7	1	1(3)
21 Families plus 1 in. sed.	**63**	**150**	**20**	**6(18)**

(Poeciliidae row note: 18 genera (29%), 31 species (21%))

Table 3 Genera with only a single known hypogean species. The figure in brackets is the total number of species in the family (from Nelson 2006). HR = hypogean restricted, TLO = known from the type locality only

Genus	Family	Notes
Aspidoras	Callichthyidae (177)	
Barbopsis	Cyprinidae (2420)	HR, monotypic
Bostrychus	Eleotridae (155)	
Caecobarbus	Cyprinidae (2420)	HR, monotypic
Caecocypris	Cyprinidae (2420)	HR, TLO, monotypic
Caecogobius	Gobiidae (1950)	HR, TLO, monotypic
Clarias	Clariidae (90)	
Copionodon	Trichomycteridae (201)	
Cryptotora	Balitoridae (170)	HR, monotypic
Eigenmannia	Sternopygidae (28)	
Eleotris	Eleotridae (155)	
Glossogobius	Gobiidae (1950)	
Glaphyropoma	Trichomycteridae (201)	
Indoreonectes	Nemacheilidae (420)	
Iranocypris	Cyprinidae (2420)	HR, TLO, monotypic
Milyeringa	Eleotridae (155)	HR, monotypic
Nemacheilus	Nemacheilidae (420)	
Neolissocheilus	Cyprinidae (2420)	
Ogilbia	Bythitidae (107)	
Oxeleotris	Eleotridae (155)	
Paralepidocephalus	Cobitidae (177)	
Phreatichthys	Cyprinidae (2420)	HR, monotypic
Poecilia	Poeciliidae (304)	
Poropuntius	Cyprinidae (2420)	

Table 3 Contd..

Table 3 Contd..

Protocobitis	Cobitidae (177)	HR, TLO, monotypic
Satan	Ictaluridae (46)	HR, monotypic
Silvinichthys	Trichomycteridae (201)	
Speoplatyrhinus	Amblyopsidae (6)	HR, TLO, monotypic
Stygichthys	Characidae (962)	HR, monotypic
Sundoreonectes	Nemacheilidae (420)	
Troglocobitis	Cobitidae (177)	HR, TLO, monotypic
Troglocyclocheilus	Cyprinidae (2420)	HR, TLO, monotypic
Trogloglanis	Ictaluridae (46)	HR, monotypic
Typhliasina	Bythitidae (107)	HR, monotypic
Typhlichthys	Amblyopsidae (6)	HR, monotypic
Tyhlobarbus	Cyprinidae (2420)	HR, TLO, monotypic
Tyhlogarra	Cyprinidae (2420)	HR, monotypic
Uegitglanis	Clariidae (90)	HR, monotypic
38 genera and species		**20 genera and species HR** **32% of genera, 13% of species**

Table 4 Genera with two or more hypogean species. The figure in brackets is the total number of species in the family (from Nelson 2006). HR = hypogean restricted, TLO = known from the type locality only

Genus	Family	No. of hypogean species	Notes
Amblyopsis	Amblyopsidae (6)	2	All HR
Ancistrus	Loricariidae (684)	3	
Astroblepus	Astroblepidae (54)	2	
Astyanax	Characidae (962)	2	
Cottus	Cottidae (275)	2	
Garra	Cyprinidae (2420)	2	
Horaglanis	Clariidae (90)	2	All HR
Ituglanis	Trichomycteridae (201)	5	
Luciogobius	Gobiidae (1950)	2	
Lucifuga	Bythitidae (107)	6	All HR
Monopterus	Synbranchidae (17)	3	
Ophisternon	Synbranchidae (17)	2	
Oreonectes	Nemacheilidae (420)	2	
Paracobitis	Nemacheilidae (420)	4	All TLO
Phreatobius	Siluriformes *in. sed.*	3	
Pimelodella	Heptapteridae (175)	2	
Prietella	Ictaluridae (46)	2	All HR
Pterocryptis	Siluridae (97)	3	All TLO
Rhamdia	Heptapteridae (175)	8	
Rhamdiopsis	Heptapteridae (175)	3	

Table 4 Contd..

Table 4 Contd..

Schistura	Namacheilidae (420)	8	
Sinocyclocheilus	Cyprinidae (2420)	26	
Trichomycterus	Trichomycteridae (201)	8	
Triplophysa	Nemacheilidae (420)	8	
Typhleotris	Eleotridae (155)	3	All HR
25 genera		**113 species**	

Table 5 The species and genera which are hypogean restricted

Family	Genus and species
Cyprinidae	Stygichthys typhlops
	Caecobarbus geertsii
	Barbopsis devecchii
	Phreatichthys andruzzii
	Iranocycpris typhlops
	Typhlogarra widdowsoni
	Caecocypris basimi
	Typhlobarbus nudiventris
	Troglocyclocheilus khammouanensis
Nemacheilidae	Troglocobitis starostini
Balitoridae	Cryptotora thamicola
Cobitidae	Protocobitis typhlops
Ictaluridae	Trogloglanis pattersoni
	Satan eurystomus
	Prietella phreatophila
	Prietella lundbergi
Clariidae	Uegitglanis zammaranoi
	Horaglanis krishnai
	Horaglanis alikunhii
Amblyopsidae	Amblyopsis spelaea
	Amblyopsis rosae
	Typhlichthys subterraneus
	Speoplatyrhinus poulsoni
Bythitidae	Lucifuga subterraneus
	Lucifuga dentatus
	Lucifuga spelaeotes
	Lucifuga simile
	Lucifuga teresinarum
	Lucifuga lucayana
	Typhliasina pearsei
Gobiidae	Caecogobius cryptophthalmus

Table 5 Contd..

Table 5 Contd..

Eleotridae	*Typhleotris madagascariensis*
	Typhleotris pauliani
	Typhleotris undescribed
	Milyeringa veritas
10 families	**25 genera and 35 species**

Table 6 The number of hypogean species of fishes in countries where they are known. It is highly significant that half of the total known species are from only three countries, and that over a quarter are from China. It is also significant that all Chinese species have been discovered since 1979

Country	Number of species	%	Proportion
China	42	28	
Brazil	23	15	50%
Mexico	11	7	
Thailand	9	6	
India	8	5	
USA	8	5	
Venezuela	4	3	
Cuba	4	3	
Madagascar	4	3	
Somalia	3	2	
Iran	2	1	
Iraq	2	1	
Oman	2	1	
Laos	2	1	
Vietnam	2	1	
Bolivia	2	1	
Colombia	2	1	
Bahamas	2	1	50%
Australia	2	1	
Japan	2	1	
Democratis Rep. Congo	1	0.5	
Malaysia	1	0.5	
Turkmenistan	1	0.5	
Namibia	1	0.5	
Trinidad	1	0.5	
Bolivia	1	0.5	
Argentina	1	0.5	
Ecuador	1	0.5	
Peru	1	0.5	

Table 6 Contd..

Table 6 Contd..

Galapagos	1	0.5	
Philippines	1	0.5	
Papua New Guinea	1	0.5	
Guam	1	0.5	
Indonesia	1	0.5	
34 countries	**150**		

Table 7 The genera, and number of species per genus, for the three countries with the largest number of known species. HR = hypogean restricted.

Country	Genera	No. of species	Notes
China	*Sinocyclocheilus*	26	
	Paracobitis	3	
	Oreonectes	2	
	Triplophysa	8	
	Protocobitis	1	HR
	Paralepidochephalus	1	
	Pterocryptis	1	
	7 genera	**42 species**	
Brazil	*Stygichthys*	1	HR
	Pimelodella	2	
	Rhamdia	1	
	Rhamdiopsis	3	
	Aspidoras	1	
	Trichomycterus	3	
	Ituglanis	5	
	Glaphyropoma	1	
	Copionodon	1	
	Ancistrus	2	
	Phreatobius	2	All HR
	Eigenmannia	1	
	12 genera	**23 species**	
Mexico	*Astyanax*	2	
	Prietella	2	All HR
	Rhamdia	4	
	Typhliasina	1	HR
	Poecilia	1	
	Ophisternon	1	
	6 genera	**11 species**	

GEOGRAPHIC DISTRIBUTION

Subterranean fishes are known from 34 countries. Three of these countries (China, Brazil and Mexico) have very large areas of karstic rocks, mainly in limestones but also in other lithologies such as quartzite, and many caves. They also have very active cave exploration and study. A combination of these factors means that these three countries account for half of all known subterranean fishes with China having 28%, Brazil 15% and Mexico 7% of all known species (Table 6). The remaining 50% are found in 31 countries, with 14 countries with one species, 10 with two species, one with three species, three with four species, two with eight species and one with nine species. The countries with three or more species (10 in all) account for 76% of all known species. In most of these countries cave exploration is ongoing, and advancing, and more species are expected in all of them. Table 7 lists the numbers of species in each of genera known from China, Brazil and Mexico. Most genera have only a few known subterranean species. Exceptions are *Sinocyclocheilus* with twenty-six and *Ituglanis* with five.

Acknowledgements

Figures 1 and 2 were developed for me by Professor Malcolm Elliott of the Freshwater Biological Association and I am most grateful for this very valuable assistance.

References

Helfman, G., B.B. Collette, D.E. Facey and B.W. Bowen. 2009. *The diversity of Fishes. Biology, Evolution and Ecology.* Wiley-Blackwell, Oxford, 2nd Edition.

Nelson, J.S. 2006. *Fishes of the World.* John Wiley & Sons, Inc., Hoboken. 4th Edition.

Porter, M.L., K. Dittmar and M. Perez-Losada. 2007. How long does evolution of the troglomorphic form take? Estimating divergence times in *Astyanax mexicanus*. In: *Time in Karst,* A. Kranjc, F. Gabrovsek, D.C. Culver and I.D. Sasowsky (eds.). Karst Water Institute Special Publication 12. Karst Waters Institute, Leesburg, pp. 173-182.

Proudlove, G.S. 2004. Pisces (Fishes). In: *Encyclopaedia of Caves and Karst Science.* J. Gunn (ed.). Fitzroy Dearborn, New York, pp. 593-595.

Proudlove, G.S. 2006. *Subterranean Fishes of the World. An Account of the Subterranean (Hypogean) Fishes Described up to 2003 with a Bibliography 1541–2004.* Includes Note added in press with details 2003-2006. International Society for Subterranean Biology, Moulis.

Romero, A. (ed.). 2001. The biology of hypogean fishes. *Environmental Biology of Fishes* 62: 1-364.

Romero, A. 2004. Pisces (Fishes): Amblyopsidae. In: *Encyclopaedia of Caves and Karst Science.* J. Gunn (ed.). Fitzroy Dearborn, New York, pp. 595-597.

Slechtova, V., J. Bohlen and H.H. Tan. 2007. Families of Cobitoidea (Teleostei: Cypriniformes) as revealed from nuclear genetic data and the position of the mysterious genera *Barbucca, Psilorhynchus, Serpenticobitis* and *Vaillantella. Molecular Phylogenetics and Evolution* 44: 1358-1365.

Tang, Q., H. Liu, , R.L. Mayden and B. Xiong. 2006. Comparison of evolutionary rates in the mitochondrial DNA cytochrome b gene and control region and their implications for phylogeny of the Cobitoidea (Teleostei: Cypriniformes). *Molecular Phylogenetics and Evolution* 39: 347-357.

Thines, G. 1969. *L'evolution regressive des poissons cavernicoles et abyssaux.* Masson et Cie, Paris.

Trajano, E. 2007. The challenge of estimating the age of subterranean lineages: Examples from Brazil. In: *Time in Karst,* A. Kranjc, F. Gabrovsek, D.C. Culver and I.D. Sasowsky (eds.). Karst Water Institute Special Publication 12. Karst Waters Institute, Leesburg, pp. 191-198.

Weber, A. 2000. Fish and amphibia. In: *Ecosystems of the World 30. Subterranean Ecosystems,* H. Wilkens, D.C. Culver and W.F. Humphreys (eds.). Elsevier, Amsterdam, pp. 109-132.

Weber, A., G.S. Proudlove and T.T. Nalbant. 1998. Morphology, systematic diversity, distribution and ecology of stygobitic fishes. In: *Encyclopaedia Biospeologica,* C. Juberthie and V. Decu (eds.). Societe de Biospeologie, Moulis and Bucarest, pp. 1179-1190.

Wilkens, H. 2005. Fish. In: *Encyclopedia of Caves,* D.C. Culver and W.B. White (eds.). Elsevier, Burlington, pp. 241-251.

CHECKLIST OF SUBTERRANEAN FISHES OF THE WORLD TO DECEMBER 2009

Checklist of subterranean fishes as discovered to December 2009. Generic placement and suprageneric classification follow Nelson (2006), with the following exceptions: Nemacheilidae, Balitoridae and Cobitidae follow Tang, Liu, Mayden and Xiong (2006) and Slechtova, Bohlen and Tan (2007); Amblyopsiformes follows Murray and Wilson (1999). Currently accepted name combinations are numbered and are followed by their original author and date and by revisers (if any). Other names which have been used (synonyms, previous combinations and misidentifications) are indented and follow immediately after the current name for the species. Author names and dates in square brackets are for species which have not been formally described, and identify the source with most information about that particular species.

Note: The analyses in this chapter were made on the 150 species known to me at the time the chapter was prepared. Since then further species have been described, or discovered in the literature. This checklist contains all troglobiotic fish species known at December 2009. The additional 14 species over those analysed in the text do not in any way alter the conclusions of the chapter.

Additional notation is as follows:

MG Monotypic subterranean-restricted genus.

MuG Subterranean-restricted genus with two or more species.

TLO Species known only from its type locality.

TMP Troglomorphic (and probably troglobiotic) hypogean population of a previously known, and named, epigean fish. These species exhibit some, or many, of the characteristics termed troglomorphy, i.e. they are depigmented and may have a reduction in eye size. Further work is required accurately to asses their status.

DRC = Democratic Republic of Congo (formerly Zaire), Madagasc. = Madagascar, PNG = Papua New Guinea, UAE = United Arab Emirates

Order Characiformes

Family Characidae

1. *Astyanax jordani* (Hubbs and Innes, 1936) Buckup, 2003   **México**
   Anoptichthys jordani Hubbs and Innes, 1936
   Anoptichthys antrobius Alvarez, 1946
   Anoptichthys hubbsi Alvarez, 1947

Astyanax fasciatus (Cuvier, 1819) cave form

Astyanax mexicanus (De Filippi, 1853) cave form

2. *Astyanax* undescribed species [Espinasa, Rivas and Espinosa Perez, 2001][TLO] — México

3. *Stygichthys typhlops* Brittan and Bohlke, 1965[MG] — Brazil

Order Cypriniformes

Family Cyprinidae

4. *Caecobarbus geertsii* Boulenger, 1921[MG] — DRC

5. *Phreatichthys andruzzii* Vinciguerra, 1924[MG] — Somalia

6. *Barbopsis devecchii* Di Caporiacco, 1926[MG] — Somalia

  Eilichthys microphthalmus Pellegrin, 1929

  Barbopsis stefaninii Gianferrari, 1930

  Zaccarinia stefaninii Gianferrari, 1934

7. *Iranocypris typhlops* Bruun and Kaiser, 1944[MG TLO] — Iran

8. *Typhlogarra widdowsoni* Trewavas, 1955[MG] — Iraq

9. *Caecocypris basimi* Banister and Bunni, 1980[MG TLO] — Iraq

10. *Garra barreimiae* Fowler and Steinitz, 1956[TMP] — Oman

11. *Garra barreimiae wurayahi* Khalaf, 2009[TMP] — UAE

12. *Garra dunsirei* Banister, 1987[TLO] — Oman

13. *Typhlobarbus nudiventris* Chu and Chen, 1982[MG TLO] — China

14. *Sinocyclocheilus guilinensis* Ji, 1985 — China

15. *Sinocyclocheilus anatirostris* Lin and Luo, 1986 — China

16. *Sinocyclocheilus bicornutus* Wang and Liao, 1997 — China

17. *Sinocyclocheilus anophthalmus* Chen, Chu, Luo and Wu 1988[TLO] — China

18. *Sinocyclocheilus cyphotergous* (Dai, 1988) Kottelat and Brehier, 1999[TLO] — China

  Gibbibarbus cyphotergous Dai, 1988

19. *Sinocyclocheilus microphthalmus* Li, 1989[TLO] — China

20. *Sinocyclocheilus angularis* Zheng and Wang, 1990[TLO] — China

21. *Sinocyclocheilus altishoulderus* (Li and Lan, 1992) Wang and Chen, 1998[TLO?] — China

  Anchicyclocheilus altishoulderus Li and Lan, 1992

22. *Sinocyclocheilus hyalinus* Chen and Yang, 1993[TLO] — China

23. *Sinocyclocheilus rhinocerous* Li and Tao, 1994[TLO?]   China
24. *Sinocyclocheilus furcodorsalis* Chen, Yang and Lan,
    1997[TLO?]   China
25. *Sinocyclocheilus guangxiensis* Zhou and Li, 1998   China
26. *Sinocyclocheilus albeoguttatus* Zhou and Li, 1998   China
27. *Sinocyclocheilus lingyunensis* Li, Xiao, Zan, Luo and
    Li, 2000[TLO]   China
28. *Sinocyclocheilus jiuxuensis* Li and Lan, 2003   China
29. *Sinocyclocheilus tianeensis* Li, Xiao and Luo, 2003   China
30. *Sinocyclocheilus tianlinensis* Zhou, Zhang and He, 2004   China
31. *Sinocyclocheilus tileihornes* Mao, Lu and Li, 2003   China
32. *Sinocyclocheilus hugeibarbus* Li and Ran, 2003   China
33. *Sinocyclocheilus xunlensis* Lan, Zhao and Zhang, 2004   China
34. *Sinocyclocheilus liboensis* Li, Chen and Ran, 2004   China
35. *Sinocyclocheilus donglanensis* Zhao, Watanabe and
    Zhang, 2006   China
36. *Sinocyclocheilus halfibindus* (Li and Lan, 1992)
    Wang and Chen, 1998[TLO?]   China
    Anchicyclocheilus halfibindus Li and Lan, 1992
37. *Sinocyclocheilus aquihornes* Li and Yang, 2007   China
38. *Sinocyclocheilus brevibarbatus* Zhao, Lan and
    Zhang, 2008[TLO]   China
39. *Sinocyclocheilus broadihornes* Li and Mao, 2007   China
40. *Sinocyclocheilus huaningensis* Li, 1998   China
41. *Sinocyclocheilus longifinus* Li, 1996   China
42. *Poropuntius speleops* (Roberts, 1991) Kottelat and
    Brehier, 1999   Thailand
    Barbus speleops Roberts, 1991
43. *Troglocyclocheilus khammouanensis* Kottelat and
    Brehier, 1999[MGTLO]   Laos
44. *Neolissochilus subterraneus* Vidthayanon and
    Kottelat, 2003[TLO]   Thailand
45. *Longanalus macrochirous* Li, Ran and Chen, 2006   China

Family Nemacheilidae

46. *Paracobitis smithi* (Greenwood, 1976) Nalbant and
    Bianco, 1998[TLO]   Iran

Nemacheilus smithi Greenwood, 1976

Noemacheilus smithi Greenwood, 1976

47. *Paracobitis posterodarsalus* Li, Ran and Chen, 2006 [TLO]   China

48. *Paracobitis maolanensis* Li, Ran and Chen, 2006 [TLO]   China

49. *Nemacheilus troglocataractus* Kottelat and Gery, 1989 [TLO]   Thailand

50. *Oreonectes anophthalmus* Zheng, 1981 [TLO]   China

51. *Oreonectes furcocaudalis* Zhu and Cao, 1987   China

52. *Oreonectes translucens* Zhang, Zhao and Zhang, 2006   China

53. *Oreonectes microphthalmus* Du, Chen and Yang, 2008   China

54. *Oreonectes retrodorsalis* Lan, Yang and Chen, 1995   China

55. *Triplophysa gejiuensis* (Chu and Chen, 1979) Chen, Yang and Xu, 1992 [TLO]   China

      Noemacheilus gejiuensis Chu and Chen, 1979

56. *Triplophysa xiangxiensis* (Yang, Yuan and Liao,1986) Chen, Yang and Xu, 1992 [TLO] *Noemacheilus xiangxiensis* Yang, Yuan and Liao, 1986   China

57. *Triplophysa yunnanensis* Yang, 1990 [TLO]   China

58. *Triplophysa shilinensis* Chen and Yang, 1992 [TLO]   China

59. *Triplophysa nandanensis* Lan, Yang and Chen, 1995   China

60. *Triplophysa aluensis* Li and Zhu, 2000   China

61. *Triplophysa longibarbata* (Chen, Yang, Sket and Aljancic, 1998) Du, Chen and Yang, 2008 [TLO]

    Paracobitis longibarbatus Chen, Yang, Sket and Aljancic, 1998   China

      Nemacheilus liboensis Ran, 2000 and Chen, 2000 *nomen nudum*

      Oreonectes liboensis Wang, 1991 *nomen nudum*

62. *Triplophysa tianeensis* Chen, Cui and Yang, 2004   China

63. *Triplophysa rosa* Chen and Yang, 2005   China

      Linemacheilus wulong Wu, He and Li, 2005 *nomen nudum*

      Heminoemacheilus wulongensis Wu, He and Li, 2007 *nomen nudum*

64. *Triplophysa qiubeiensis* Li, Yang, Chen, Tao, Qi and Han, 2008   China

65. *Triplophysa longipectoralis* Zheng, Du, Chen and Yang, 2009 [TLO]   China

66. *Triplophysa sinensis* [Chen, Li and Yang, 2008] China
67. *Schistura sijuensis* (Menon, 1987) Kottelat, 1990[TLO] India
      Noemacheilus sijuensis Menon, 1987
68. *Schistura oedipus* (Kottelat, 1988) Kottelat, 1990 Thailand
      Nemacheilus oedipus Kottelat, 1988
69. *Schistura jarutanini* Kottelat, 1990[TLO] Thailand
70. *Schistura kaysonei* Vidthayanon and Jaruthanin, 2002[TLO] Laos
71. *Schistura spiesi* Vidthayanon and Kottelat, 2003[TLO] Thailand
72. *Schistura deansmarti* Vidthayanon and Kottelat, 2003[TLO] Thailand
73. *Schistura spekuli* Kottelat, 2004 Vietnam
74. *Schistura papulifera* Kottelat, Harries and Proudlove, 2007 India
75. *Sundoreonectes tiomanensis* Kottelat, 1990[TLO] Malaysia
76. *Indoreonectes evezardi* (Day, 1872) Kottelat, 1990[TLO TMP] India
      Noemacheilus evezardi Day, 1872
77. *Troglocobitis starostini* (Parin, 1983)[MG TLO] Turkmenia
      Noemacheilus starostini Parin, 1983
78. *Heminoemacheilus hyalinus* Lan, Yang and Chen, 1996 China

Family Balitoridae

79. *Cryptotora thamicola* (Kottelat, 1988) Kottelat, 1998[MG] Thailand
      Homaloptera thamicola Kottelat, 1988

Family Cobitidae

80. *Protocobitis typhlops* Yang and Chen, 1993[MG TLO] China
81. *Protocobitis polylepis* Zhu, Lu, Yang and Zhang, 2008 China

Order Siluriformes

Family Ictaluridae

82. *Trogloglanis pattersoni* Eigenmann, 1919[MG] USA
83. *Satan eurystomus* Hubbs and Bailey, 1947[MG] USA
84. *Prietella phreatophila* Carranza, 1954[MuG] México
85. *Prietella lundbergi* Walsh and Gilbert, 1995[MuG] México

Family Siluridae

86. *Pterocryptis cucphuongensis* (Mai, 1978) Ng and
    Kottelat, 1998 [TLO]   Vietnam
       Silurus cucphuongensis Mai, 1978
87. *Pterocryptis buccata* Ng and Kottelat, 1998 [TLO]   Thailand
88. *Pterocryptis* undescribed species [Clarke pers.
    comm. 2001] [TLO]   China

Family Clariidae

89. *Uegitglanis zammaranoi* Gianferrari, 1923 [MG]   Somalia
90. *Clarias cavernicola* Trewavas, 1936 [TLO]   Namibia
91. *Horaglanis krishnai* Menon, 1950 [MuG]   India
92. *Horaglanis alikunhii* Babu and Nayar, 2004 [MuG]   India

Family Heptapteridae

93. *Pimelodella kronei* (Miranda Ribeiro, 1907) Pavan, 1946   Brazil
       Typhlobagrus kronei Miranda Ribeiro, 1907
       Pimelodella lateristriga var. *kronei* Haseman, 1911
       Caecorhamdella brasiliensis Borodin, 1927
94. *Pimelodella spelaea* Trajano, Reis and Bichuette, 2004   Brazil
95. *Rhamdia quelen urichi* (Norman, 1926) Mees, 1974 [TLO]   Trinidad
       Caecorhamdia urichi Norman, 1926
       Caecorhamdella urichi (Norman, 1926) Hubbs,
       1938 *lapsus calami*
96. *Rhamdia laticauda typhla* Greenfield, Greenfield, and
    Woods 1982 [TLO]   Belize
97. *Rhamdia reddelli* Miller, 1984   México
98. *Rhamdia zongolicensis* Wilkens, 1993 [TLO]   México
99. *Rhamdia macuspanensis* Weber and Wilkens, 1998 [TLO]   México
100. *Rhamdia laluchensis* Weber, Allegrucci and Sbordoni,
     2003 [TLO]   México
101. *Rhamdia guasarensis* DoNascimiento, Provenzano and
     Lundberg, 2004   Venezuela
102. *Rhamdia enfurnada* Bichuette and Trajano, 2005   Brazil
103. *Rhamdia* undescribed species [Cordeiro pers. comm.]   Brazil

104. *Rhamdiopsis* undescribed species 1
     [Trajano and Bichuette, 2007]   **Brazil**

105. *Rhamdiopsis* undescribed species 2
     [Trajano and Bichuette, 2007]   **Brazil**

106. *Rhamdiopsis* undescribed species 3
     [Trajano and Bichuette, 2007]   **Brazil**

Family Callichthyidae

107. *Aspidoras* sp.
     [Trajano and Bichuette, 2007]   **Brazil**

Family Trichomycteridae

108. *Trichomycterus chaberti* Durand, 1968 [TLO]   **Bolivia**

109. *Trichomycterus itacarambiensis* Trajano and
     De Pinna, 1996 [TLO]   **Brazil**

110. *Trichomycterus spelaeus* DoNascimiento, Villarreal
     and Provenzano, 2001 [TLO]   **Venezuela**

111. *Trichomycterus sandovali* Ardila Rodriguez, 2006   **Colombia**

112. *Trichomycterus santanderensis* Castellanos-Morales,
     2007 [TLO]   **Colombia**

113. *Trichomycterus uisae* Castellanos-Morales, 2008 [TLO]   **Colombia**

114. *Trichomycterus* undescribed species 1
     [Trajano and Bichuette, 2007]   **Brazil**

115. *Trichomycterus* undescribed species 2
     [Trajano and Bichuette, 2007]   **Brazil**

116. *Trichomycterus* undescribed species 3 [Schultz, 1949] [TLO]   **Venezuela**

117. *Ituglanis passensis* Fernandez and Bichuette, 2002 [TLO]   **Brazil**

118. *Ituglanis bambui* Bichuette and Trajano, 2004   **Brazil**

119. *Ituglanis epikarsticus* Bichuette and Trajano, 2004   **Brazil**

120. *Ituglanis ramiroi* Bichuette and Trajano, 2004   **Brazil**

121. *Ituglanis mambai* Bichuette and Trajano, 2008   **Brazil**

122. *Ituglanis* undescribed species
     [Trajano pers. comm. 11.2009]   **Brazil**

123. *Glaphyropoma spinosum* Bichuette, de Pinna
     and Trajano, 2008 [TLO]   **Brazil**

124. *Copionodon* undescribed species
     [Trajano and Bichuette, 2007]   **Brazil**
125. *Silvinichthys bortayro* Fernandez and De Pinna, 2005   **Argentina**

Family Loricariidae

126. *Ancistrus cryptophthalmus* Reis, 1987   **Brazil**
127. *Ancistrus galani* Perez and Viloria, 1994 [TLO]   **Venezuela**
128. *Ancistrus formoso* Sabino and Trajano, 1997   **Brazil**

Family Astroblepidae

129. *Astroblepus pholeter* Collette, 1962 [TLO]   **Ecuador**
130. *Astroblepus riberae* Cardona and Guerao, 1994 [TLO]   **Peru**

Siluriformes incertae sedis

131. *Phreatobius cisternarum* Goeldi, 1905 [MuG]   **Brazil**
132. *Phreatobius dracunculus* Shibatta, Muriel-Cunha
     and De Pinna, 2007 [MuG]   **Brazil**
133. *Phreatobius sanguijuela* Fernandez, Saucedo,
     Carvajal and Schaefer, 2007 [MuG]   **Bolivia**

Order Gymnotiformes

Family Sternopygidae

134. *Eigenmannia vicentespelaea* Triques, 1996 [TLO]   **Brazil**

Order Amblyopsiformes

Family Amblyopsidae

135. *Amblyopsis spelaea* DeKay, 1842 [MuG]   **USA**
136. *Amblyopsis rosae* (Eigenmann, 1898) Woods and
     Inger, 1957 [MuG]   **USA**

         Typhlichthys rosae Eigenmann, 1898

         Troglichthys rosae (Eigenmann, 1898)
         Eigenmann, 1899

         Amblyopsis rosae whitae Romero, 1998
         nomen nudum

137. *Typhlichthys subterraneus* Girard, 1859 [MG]   **USA**
      Typhlichthys osborni Eigenmann, 1905
      Typhlichthys wyandotte Eigenmann, 1905
      Typhlichthys eigenmanni Charlton, 1933
      Typhlichthys eigenmanni Hubbs, 1938 *nomen nudum*

138. *Speoplatyrhinus poulsoni* Cooper and Kuehne,
    1974 [MGTLO]   **USA**

Order Ophidiiformes

Family Bythitidae

139. *Lucifuga subterranea* Poey, 1858 [MuG]   **Cuba**
      Lucifuga subterraneus Poey, 1858
      Lucifuga teresinarum Diaz Perez, 1988

140. *Lucifuga dentata* Poey, 1858 [MuG]   **Cuba**
      Lucifuga dentatus Poey, 1858
      Stygicola dentatus Gill, 1863

141. *Lucifuga simile* Nalbant, 1981 [MuG]   **Cuba**

142. *Lucifuga spelaeotes* Cohen and Robins, 1970 [MuG]   **Bahamas**

143. *Lucifuga lucayana* Moller, Schwarzhans, Iliffe and
    Nielsen, 2006 [MuG]   **Bahamas**

144. *Typhliasina pearsei* (Hubbs, 1938) Whitley, 1951 [MG]   **México**
      Typhlias pearsei Hubbs, 1938
      Ogilbia pearsei (Hubbs, 1938) Cohen and
      Nielson, 1978

145. *Ogilbia galapagosensis* (Poll and Leleup, 1965)
    Cohen and Nielson, 1978   **Galapagos**
      Caecogilbia galapagosensis Poll and Leleup, 1965

Order Cyprinodontiformes

Family Poeciliidae

146. *Poecilia mexicana* Steindachner, 1863 [TMP]   **México**
      Poecilia sphenops Valenciennes in Cuvier and
      Valenciennes, 1846
      Mollienisia sphenops (Valenciennes in Cuvier and
      Valenciennes, 1846) Regan, 1913

Order Synbranchiformes

Family Synbranchidae

147. *Ophisternon infernale* (Hubbs, 1938) Rosen and
     Greenwood, 1976   México
        Pluto infernalis Hubbs, 1938
        Furmastix infernalis (Hubbs, 1938) Whitley, 1957
        Synbranchus infernalis (Hubbs, 1938) Mees, 1962

148. *Ophisternon candidum* (Mees, 1962) Rosen and
     Greenwood, 1976   Australia
        Anomatophasma candidum Mees, 1962

149. *Monopterus eapeni* Talwar in Talwar and Jhingran, 1992[TLO]   India
        Monopterus indicus Eapen, 1963
        Monopterus "indicus" (Eapen, 1963) Rosen
        and Greenwood, 1976

150. *Monopterus roseni* Bailey and Gans, 1998[TLO]   India
151. *Monopterus digressus* Gopi, 2002[TLO]   India

Order Scorpaeniformes

Family Cottidae

152. *Cottus carolinae* species group
     [Burr, Adams, Krejca and Warren, 2001][TMP]   USA
153. *Cottus "bairdii-cognatus"* complex
     [Espinasa and Jeffery, 2003][TLO TMP]   USA

Order Perciformes

Family Gobiidae

154. *Luciogobius pallidus* Regan, 1940   Japan
155. *Luciogobius albus* Regan, 1940   Japan
156. *Caecogobius cryptophthalmus* Berti and Ercolini,
     1991[MG TLO]   Philippines
157. *Glossogobius ankaranensis* Banister, 1994[TLO]   Madagasc
158. *Bostrychus microphthalmus* Hoese and Kottelat, 2005   Indonesia

Family Eleotridae

159. *Typhleotris madagascariensis* Petit, 1933 [MuG]   **Madagasc.**

160. *Typhleotris pauliani* Arnoult, 1959 [MuG]   **Madagasc.**

161. *Typhleotris* undescribed species
    [Sparks and Stiassny, 2004] [MuG]   **Madagasc.**

162. *Milyeringa veritas* Whitley, 1945 [MG]   **Australia**

163. *Oxyeleotris caeca* Allen, 1996 [TLO]   **PNG**

164. *Eleotris* undescribed species
    [Tibbatts pers. Comm. 2006]   **Guam**

CHAPTER **3**

Conservation of Subterranean Fishes

Maria Elina Bichuette[1] and Eleonora Trajano[2]
[1]Departamento de Ecologia e Biologia Evolutiva, Universidade Federal
de São Carlos, São Carlos, SP, Brazil. E-mail: bichuette@uol.com.br
[2]Departamento de Zoologia, Instituto de Biociências, Universidade de
São Paulo, São Paulo, SP, Brazil. E-mail: etrajano@usp.br

INTRODUCTION

Subterranean ecosystems pose special problems for conservation due to their intrinsic fragility and the distinctive features of subterranean communities, including a high degree of endemism and morphological, ecological and behavioral specialization. Such fragility is a consequence of the relatively low biological diversity of subterranean ecosystems and their general dependence on nutrients imported from the surface, making them strongly influenced by disturbances in epigean habitats. Furthermore, many troglobites have small geographic ranges and low population densities, resulting in small population sizes, K-selected life histories, thus population turnover is slow and their ability to recover from population losses is decreased compared to epigean related taxa, and they may be highly susceptible to environmental fluctuations (Trajano 2000).

Around 98% of unfrozen freshwater in the world is subterranean and Juberthie (2004) claims at least 50% is inhabited by troglobitic and troglophilic species. These aquatic habitats include karst aquifers (including the epikarst zone), phreatic aquifers in valleys in many kinds of sediments and rocks, interstitial and hyporheic habitats, and base-level streams. Troglobitic fishes have been found in all these kinds of habitats (see, for instance, Trajano and Bichuette, this volume).

1. THREATS AND VULNERABILITY OF AQUATIC SUBTERRANEAN HABITATS AND SPECIES

Aquatic subterranean habitats are often disturbed by human actions, such as habitat degradation, hydrological manipulations, environmental pollution, overexploitation and impacts of introduced fauna (Proudlove, 2001). Main threats to subterranean fishes throughout the world are discussed and exemplified below (see also Sullivan 1956, Poulson 1968, Culver 1986, Proudlove 2001, Juberthie 2004, Elliott 2005).

1.1 Habitat Destruction: Quarrying and Open Pit Mining

The ultimate threat for subterranean faunas as a whole, as well as for epigean communities, is the removal of rock layers for limestone and ore (iron, gold etc.) exploitation. Removal of gravel from stream-beds may destroy the interstitial and stream habitats as well. As far as we know, there is no reported instance of extinction of troglobitic fishes caused by this extreme threat, but it is possible that populations have already been destroyed before being noticed. This is particularly true for species living in less obvious, non-cave habitats, whose existence has been recently acknowledged, such as troglomorphic catfishes, genus *Phreatobius*, from underground waters in the Amazon basin.

1.2 Habitat Changes: Use of Water

1.2.1 Dams and Reservoirs

Building of dams and reservoirs for hydro-electrical plants and water storage for human use has profound effects over the structure and physical-chemical characteristic of aquatic habitats, changing from lotic to lentic conditions: velocity of water current decreases, water depth increases and soft bottom predominates, dissolved oxygen may drop to quite low levels. Due to the elevation of the regional water table, caves in the surrounding areas may also be affected. Increased sedimentation may be harmful for aquatic organisms with gills, like fishes, making it difficult for rock-dwellers to find proper substrates, low oxygen levels will exclude many organisms, and so on. As a consequence, stream-dwellers may disappear.

The construction of dams in the Green River in Kentucky, U.S.A., raised the water level in the Mammoth Cave, and this may have affected the amblyopsid *Amblyopsis spelaea*, either because of silt deposition on previously sand bottomed areas and/or due to the reduction of particulate organic matter, a major food source for aquatic communities (Proudlove 2001).

1.2.2 Over-pumping of Subterranean Aquifers

Withdrawal of large amounts of water for human use, for domestic, agricultural or industrial purposes, is one of the most serious threats for aquatic troglobites throughout the world. When recharge rates become insufficient for replenishing the aquifer, the water table gradually lowers, water quantity and quality declines and habitat for subterranean organisms is lost. In the extreme, it becomes an instance of habitat destruction (see Section 1.1).

Proudlove (2001) mentions several cases of actual or potential habitat loss for troglobitic fishes due to over-pumping, including the North American ictalurid catfishes *Satan eurystomus* and *Trogloglanis pattersoni*, endemic to the San Antonio pools of the Edwards aquifer, one of the main water sources in Texas, U.S.A., and *Prietella phreatophila*, from Mexico; the African clariids *Clarias cavernicola*, from Namibia, and *Uegitglanis zammaranoi*, from Somalia, as well as the cyprinids *Barbopsis devecchii* and *Phreatichthys andruzzi*; the Australian *Ophisternon candidum* (Synbranchidae) and *Milyeringa veritas* (Eleotridae).

Over-pumping also threatens non-troglomorphic fish in caves. In Devil's Sinkhole, Nevada, U.S.A., the Devil's Hole pupfish, *Cyprinodon diabolis*, is endangered by over-pumping of the regional aquifer, which has reduced this cave fish to a tiny population at the bottom of a sinkhole (Elliott 1981). Other North American pupfishes, totally dependent on karst aquifers, are also endangered, as is the case of *Cyprinodon elegans*, from Comanche Springs, and *C. bovines*, from Leon Springs. (Elliott, op. cit.).

Moreira and colaborators (In Press) called the attention for the highly specialized Brazilian phreatic fish, *Stygichthys typhlops*, from eastern Brazil (Jaíba karst area). Information provided by town inhabitants, well-drilling operators, local farm owners and workers indicates that *S. typhlops* was quite common and widespread in the Jaíba area, at least up to the early 1990's, occurring in springs, shallow hand dug wells, and also coming out during drilling of artesian wells. Almost all of these reports referred to sightings at least ten years ago. Most of the wells referred to are now dry, and in one case where the well was not dry it was clogged with clay preventing (or severely decreasing) the influx of water. In the past twenty years, the Jaíba region has undergone extensive irrigation projects for agriculture, mostly drawing from subterranean water. In the first half of the 1980's, the intensity and frequency of utilization of this resource was low, probably one-tenth the recharge capacity of the aquifer (Silva 1984). However, there has been a marked decrease in water table levels, with several wells and springs reportedly dried out in the past few years. This has promoted a shift from traditional wells or open springs to artesian wells as source of water in recent years.

This situation poses a major concern for the future of *S. typhlops*. The lowering of the water level may result in significant reduction of available habitat and reduce the input of nutrients from the surface, what could lead to a rapid population decline. In 2004, *S. typhlops* was included in the Brazilian official list of threatened species (MMA 2004), which hopefully will help preserving this species.

Further lowering of two meters in the water level will result in drying of the two wells where we collected specimens. Unless deeper access to the fish habitat is found, collecting of additional material for future studies may become impossible in a few years.

1.3 Pollution

Pollution includes eutrophication and contamination from agricultural and industrial runoff (e.g. fertilizers, pesticides and heavy metals – Proudlove, 2001). Underground waters generally are very sensitive to pollution because pollutants may rapidly spread into subterranean aquifers due to their limited filter capacity. Hence, the detection and removal of pollutants from groundwater is especially difficult.

1.3.1 *Organic Pollution: Eutrophication due to Garbage Dumping, Inadequate Sewage Treatment and Introduction of Fauna*

Organic pollution of subterranean habitats may be caused by legal and illegal thrash and garbage dumping and infiltration of wastewater from faulty sewage (Culver 1986, Bolner and Tardy 1989). Making waste "disappear" by dumping it into caves and sinkholes has been a widespread practice, and it is still practiced in many places throughout the world. This illusion is especially dangerous due to the facility with which the pollutants disperse and their general invisible nature after dispersed.

In many studies (e.g. Cruickshank *et al.* 1980, Varela *et al.* 1980), an evaluation of water quality from samples collected in caves showed chemical and bacteriological contamination affecting cave fish populations. Proudlove (2001) mentions cases of contamination by high levels of nitrogen compounds affecting subterranean fishes. The proliferation of bacteria may also affect the density of some aquatic organisms, causing an imbalance on their population parameters. The skin of fishes is protected by mucus and it is relatively thin, representing a main entrance for pathogens, and any injury on fish bodies allows the proliferation of diseases, for example, the *Lucifuga* species from Cuba.

Specialized troglobitic fishes exhibit a better feeding performance when density of prey is low than troglophilic and epigean close relatives, the opposite being true when prey density is high. This has been shown for the amblyopsids *Typhlichthys subterraneus, Amblyopsis rosae* and *A. spelaea*, when compared to *Forbesichthys agassizi* and *C. cornuta* (Poulson 1963). As a consequence, eutrophication favors the invasion of naturally food-limited subterranean habitats by epigean species, which may displace troglobites by successfully competing with them.

The impacts caused by the introduction, either accidental or deliberate, of alien animals may be considered a special case of organic pollution. Proudlove (2001) cites some examples of epigean fishes introduced in cave areas, where they may threat troglobitic species due to competition (*Clarias gariepinus* versus *C. cavernicola*, in Namibia) and/or predation (some marine species versus *Lucifuga* spp., in Cuba), or hybridization (*Astyanax mexicanus* versus troglobitic *Astyanax*, Mexico).

In Brazil, carps (*Cyprinus carpio*) introduced in the epigean drainage of Intervales State Park, São Paulo State, invaded caves and are presently threatening, probably due to competition and/or habitat change (siltation), cave decapods, genus *Aegla*. In the neighbor Alto Ribeira Tourist State Park, these decapods are apparently endangered by the expansion of freshwater shrimps, genus *Macrobrachium* (E. Trajano *et al.*, pers. obs.).

1.3.2 Chemical Pollution

Chemical pollution by heavy metals, fertilizers and pesticides is a major environmental problem worldwide. Vesper (2005) summarizes the problem of contamination of cave waters by heavy metals. There are probably many aquatic troglobites threatened by chemical pollution throughout the world, but few cases have been well-documented.

The blind catfish, *Pimelodella kronei*, occurs in several caves at both sides of the Betari stream, in the Upper Ribeira karst area, SE Brazil: the Areias cave system (Areias de Cima – type-locality, by far with the largest known population – Areias de Baixo and Ressurgências das Areias da Água Quente), Ressurgência das Bombas and Córrego Seco Cave, in at least partially isolated aquifers at the right margin of the Betari stream; and the Alambari cave system (Alambari de Cima and Gurutuva caves), at the left margin. The populations in the two sides of the Betari are separated by an insoluble, diabase dyke and may represent distinct, cryptic species. During the 1980's, the Alambari system population practically disappeared, probably due to pollution caused by uncontrolled mining activities uspstream from the cave system (Trajano 1997, 2000). The Alambari de Cima cave population did not recover till now, more than two decades after its sharp decline. Apparently, other aquatic troglobites also disappeared or almost, including undescribed decapods, genus *Aegla*, and planarians. This well illustrates how a single pollution event may have catastrophic consequences for the fragile subterranean fauna.

1.4 Human Visitation

Excessive human visitation is one of the major causes of impact for subterranean faunas. It is the result of the considerable development of speleology as sports and adventure, overcrowding many caves. This activity can cause a significant impact on river margins, trampled by visitors and changing the bottom of streams and rivers through siltation. Considering that troglobitic fishes use chemical and mechanical reception to detect food, reproductive partners and to avoid any threat, alteration of water physical conditions can cause a serious damage to these animals. For example, the Brazilian cave catfish *Ituglanis bambui* has a relatively large population but it occurs in a vadose tributary at Angélica Cave, central Brazil (Fig. 1). This kind of habitat is extremely fragile; the bottom is formed by silt and clay and the main visitation in this cave cross the small tributary. For another species of *Ituglanis* from central Brazil, the concern is high – *I. epikarsticus* is extremely rare in the cave habitat, and the number of observed specimens conspicuously decreased since the 1970's. So far, it has only been found in a set of rimstone pools situated right in the pathway for visitors reaching the far end of the cave (Fig. 2). Thus, there is reason for major concern about the future of this highly endemic species. A strict control of visitation is urgently needed to assure its protection, until evidence appears that the species is safe in non-cave habitats.

Fig. 1 Locality of the Brazilian cave catfish *Ituglanis bambui* – a vadose tributary at Angélica Cave, central Brazil (Photo: Maria Elina Bichuette).

Other negative consequences of visitation are pollution of water as result of toxic products left in caves (such as carbide and batteries) and accumulation of non-biodegradable products, topoclimatic changes caused, for instance, by hot light sources and the opening of artificial passages during cave exploration and management; introduction of alien species and materials, such as metal, cement and treated wood; trampling of cavernicoles and soil compacting of sediment banks, the easiest pathway inside caves; destruction of specific habits, like rimstone pools formed

Fig. 2 Locality of the Brazilian cave catfish *I. epikarsticus* – a set of rimstone pools situated at São Mateus Cave, central Brazil (Photo: Flavio Dias Passos).

by infiltration of epikarstic aquifers; and direct disturbance (lights, loud noises, handling of specimens, introduction of pathogens) (Trajano 2000).

1.5 Deforestation

In many regions, karst depressions have been used for cultivation since its soil is commonly deep and fertile, differently from karst slopes where the soil is generally shallow (Ford and Williams 1989). The removal of the natural vegetation cover causes changes in karst hydrological regimes, particularly by disruption storage and release mechanisms in the regolith. Fire associated with hazard reduction in agricultural areas can cause sealing of the epikarst, reduction infiltration capacity, so that surface waters may carry high sediment loads, impacting on groundwater quality (Holland 1994). The main damage is the siltation of the water bodies, in some cases filling and obstructing the caves, and destroying an entire habitat.

Tree roots hold the soil in the fissures of the rock and prevent denudation and subsidence of the terrain. Besides buffering erosion, natural vegetation around karst areas is source of input of allochthonous food items for the subterranean biota.

In Madagascar, the gobiid *Glossogobius ankaranensis* may be threatened by increasing the level of catastrophic floods in some areas and siltation in others (Proudlove 2001). The same may be true for several other troglobitic fishes.

1.6 Overcollecting

The harmful effects of overcollecting and handling cave animals were an early concern of biospeleologists. Sullivan (1956) cautioned against excessive collecting of specimens, and the National Speleological Society – NSS (U.S.A.) adopted a conservation policy that discouraged collecting by amateurs or for any commercial use.

Overcollections may threaten small populations with low reproductive rates, such as the crayfish, *Orconectes sheltae*, in Shelta Cave, Alabama, U.S.A., which may not become sexually mature until reaching the age of 40+ years (Cooper 1975). In general, larger troglobites, such as cavefish, salamanders and crayfish, are long-lived, have small population sizes, and reproduce slowly; therefore collecting should be highly restrained. Smaller, more abundant troglobites, such as some amphipods, isopods, millipedes and beetles, may withstand intensive collections. However, some small species are exceedingly rare. Culver (1982) emphasized the need of restraining collections by admitting that he had caused a severe decline in populations of several cave isopods.

Overcollecting was suspected to have caused the decline of the cavefish *Amblyopsis spelaea* in Mammoth Cave, which were sold as curios in the

19th Century (Poulson 1968). The Mexican characins *Astyanax fasciatus* and the African cyprinid *Caecobarbus geertsii* have been subject to collection for aquarism. The former bred very easy in captivity. Apparently, the later cannot bred in captivity and probably all the individuals sold were from wild stocks. In 1993, 1,500 individuals were reported as imports by the U.S.A. (Proudlove 2001). Culver (1986) considered *Amblyopsis rosae* and *Prietella phreatophila* threatened by overcollecting.

In Brazil, the *Pimelodella kronei* catfish population from the type-locality, Areais de Cima Cave, was subject to over-collections by an amateur during the mid-1970's, whose effects were visible ten years later (Trajano and Britski 1992, Trajano 2000). Tourist visitation was forbidden and, although this prohibition was not vigorously enforced, the population recovered after 15-20 years (E. Trajano, pers. obs.). It is interesting to note that this is the maximum lifespan estimated for *P. kronei* (Trajano 1991). As a preliminary approach, we suggest that the minimum time for a troglobitic population recovering should be two times the average estimated lifespan.

1.7 Climate Change

In general, stream-dwelling troglobitic fish depend, at least in part, on floods carrying organic matter into the subterranean biotope. In regions subject to highly seasonal climates, flash floods typically occur at the peak of the rainy season (January and February in Brazilian karst areas, in the Cerrado and Caatiga domains), and there is no significant precipitation during all the dry season (May to September). Population studies have shown that the dry season may represent a highly stressful period for cavefish due to the progressive depletion of food resources, with many individuals loosing weight and some even shrinking, as observed for the catfishes *Trichomycterus itacarambiensis* (Trajano 1997) and *Pimelodella spelaea* (Trajano *et al.* 2004). As indicated by the relatively low degree of troglomorphism (reduction of eyes and pigmentation), these species may have been isolated for a relatively short time in caves, lacking specializations to cope with a pronounced, long-term shortage of nutrients, and living in a delicate energetic equilibrium.

A decrease in precipitation throughout the last two decades was reported by local inhabitants of Itacarambi region (locality for *T. itacarambiensis*), with flash floods becoming progressively rarer. This is coincident with a significant decrease in the catfish density observed during visual censuses in 2006 and 2007 (Trajano *et al.* 2009) at the resurgence end of the 5,000 m long cave where the only known population is found. On the other hand, the population density at the 3,000 m upstream was similar to that recorded in 1994 (Trajano 1997). Such decrease may be due to a deficiency in restoration of nutrient sources far from the sinkhole. Because the Olhos d'Água stream is normally a small water course, with a short carrying capacity, such restoration should be done mainly by flash floods, which occur every year or nearly so.

A similar situation was recently observed for *P. spelaea*: after a particular long dry season, with a two-month delay in the beginning of the rainy period, the population studied in April 2008 has apparently decreased in relation to previous years and the few individuals captured were very thin (Trajano *et al.*, unpubl. data).

It is still unclear whether these phenomena are part of a natural cycle of climatic fluctuations, illustrating an evolutionary event (bottleneck effect), or a consequence of progressive climatic changes due to human activities, perhaps within the global warming context. This being the case, measures for protection of these interesting cave species would be the same necessary for the whole planet.

2. ENDANGERED TROGLOBITIC FISHES AROUND THE WORLD

There are 164 species of troglobitic fishes in the world (Proudlove, this volume), several still undescribed. The formal description of new species may be a slow work, especially in the case of those belonging to taxa ill-defined or in need of revision, or for which close relatives are not clearly recognized (the so-called "phylogenetic relicts"). For this reason many species are not included in the Red Books published by the International Union for the Conservation of Nature and Natural Resources (IUCN), which categorizes the threat levels and produces the lists of nominal endangered species. These lists are very helpful in the protection and proposal of conservationist actions. The first list of cave fishes under the categories proposed by the IUCN in 1994 was published in 1996 (IUCN 1996). After this, an updated list was published ten years later in a specialized publication (Proudlove 2006).

There is another difficulty when revisions of the list are elaborated. Some data about the life history, ecology and behaviour are essential to propose species like in an endangered category, and for many species, very little has been studied and are few references in the literature. However, a number of species has been extensively studied and the literature is large.

The rarity of the blindfish *Amblyopsis spelaea* in Mammoth Cave and its absence from adjacent areas to the north led to speculation that it was either introduced or decimated during the long period when blindfish were sold as curios. Poulson (1968) examined historical and scientific records and found that most early records from Mammoth were for *A. spelaea*, not *Typhlichthys subterraneus*, which also inhabits the cave. *A. spelaea* was the dominant species in the Echo and Roaring River areas around 1890, and it is still common in Roaring River. The present rarity of *A. spelaea* is probably related to silting and flooding associated with deforestation, forest fires, and Lock and Dam (Poulson 1968).

3. ENDANGERED AQUATIC TROGLOBITES: HOW TO PROTECT THEM?

Caves are windows to the subterranean realm, but the actual subterranean habitats occupied by troglobitic species are usually much larger, encompassing inaccessible spaces. In order to guarantee the protection of these species, it would be necessary to estimate the habitat extension needed for a minimum sized effective population. Especially for species living in phreatic waters, it is expected that such minimum habitat size will be larger than the water bodies appearing in large spaces accessible to humans, i.e., caves. In order to be really effective, protection of aquatic troglobites should focus on cave systems and/or aquifers. Unfortunately, conservation policies throughout the world generally focus on caves, and not on the entire underground environment.

A major problem for the protection of many troglobitic fishes with multiples populations is the lack of detailed genetic and morphological allowing for the recognition of the taxonomic status of each of these populations. It has been observed that the degree of isolation, indicated by differentiation, may vary between populations of a same nominal species (see, for instance, Reis *et al.* 2006), and the possibility of replacement of loss in a population by immigrants from cospecific ones will vary accordingly. In cases with good evidence of isolation, even partial, the isolated sets of populations should be the taxonomic units for conservation purposes. As well, undescribed species should be included in red lists when the available evidence point to important menaces threatening their survival. In view of their generally high degree of endemism, an acceptable descriptive reference for troglobites would be its locality.

The *Pimelodella kronei* population occurring in the right margin of the Betari stream (see Section 1.3.2) was included in the Red List of Threatened Fauna in the State of São Paulo (Secretaria do Meio Ambiente 1998) separately from the populations found in the left margin (*P. kronei sensu stricto*). It received the status of critically endangered, whereas the latter were considered as vulnerable, assuring the proper protection for each population. Unfortunately, the Federal environmental agency decided not to include, in the Brazilian Red List, undescribed species. As a consequence, many Brazilian troglobitic species, some in urgent need of protection actions such as the above mentioned blind catfishes and several invertebrates, were left aside.

As a general plan for an effective protection of subterranean species, we suggest the following steps (not necessarily chronological):

- field, taxonomic and genetic studies focusing on distribution, taxonomic status and connectiveness of populations, population dynamics (total and effective population sizes and densities, home ranges) and life history; complementary laboratory research on behavior and physiology are also important to establish the species vulnerability to disturbance;
- when actual or potential threats are detected, conservation actions directed to the species/population ecology and biology, including but not restricted to habitat protection;
- long-term monitoring of disturbed populations, lasting for at least two generations-time after conservation actions are taken;
- education aiming to developing public awareness of the beauty and importance of subterranean organisms and ecosystems, and self-consciousness of individual, group and institutional responsibility for environmental protection.

4. RESTORATION PROJECTS

Cave restoration projects are beneficial in removing harmful materials, trash, and graffiti. Some precautions are needed to avoid stressing cave communities that may have colonized organic materials, particularly old woodpiles that may be a heaven for invertebrates. Elliott (1982) and Hubbard (1995) emphasized that wood should be examined by a biologist during cleanup projects or left alone. Such woodpiles may have attracted a large population of invertebrates over decades of time, and to remove it suddenly may, in effect, trap out a significant portion of the population in that area of the cave. Oftentimes such wood can be removed gradually but a small residue should be kept in order to provide habitat for the remaining fauna (Elliott 1997a).

An important concept is to give some caves or microhabitats time to "rest". In fact, it has been observed that a moderate, but continuous visitation has more harmful effects to cave ecosystems than peaks of intensive visitation intercalated with relatively long (weeks to months) periods without or with few visitations. For example, there is a proposal of seasonal visitation in some rooms of the Kartchner Caverns, in the State of Arizona (U.S.A.). There is a Big Room area in this cave which is the summer home of the cave bat, *Myotis velifer;* the colonies reach to 1,000-2,000 bats and, in order to protect these bats, cave development work ceases in this room between May and mid-September (Toomey 2006). As mentioned above, it is herein suggested that the minimum time for a troglobitic fish population recovering should be two times the average lifespan estimated for this population.

5. FUTURE PERSPECTIVES

The ultimate survival of a cave or karst community depends on the proper protection and management of the cave and the surrounding terrain. In the case of some endemic, terrestrial cave invertebrates, a one-hectare preserved area unit may be sufficient as long as nutrient and water inputs are not altered and disturbances are minimal. However, often we must reach beyond the immediate karst area to contributing watersheds or, in the case of bats, to alternate roost habitats. With migratory bats, these considerations must be on a continental scale to insure success. Many conservation units established for cave protection have been set aside only to protect the entrance area, and not the entire watershed for the cave system. Many recognize the need for ecosystem management, which considers the whole karst area. The trend by governments has been toward increasing aspirations for karst ecosystem management, but in practice it is very difficult financially and politically. In the Austin, area (Texas), even though seven terrestrial troglobites have been listed as endangered, the federal government has not acquired any area to preserve them, even though it has spent millions of dollars to purchase a national wildlife refuge nearby for the benefit of two endangered bird species. Sixty-five areas that were planned to preserve caves will have to await funding through a local development fee collected by the county and city (Elliott 1997c).

Some believe that caves cannot be managed, and that they should be left alone (except for occasional caving trips). Such a passive conservation ideal cannot work in developed areas where drainage patterns are changed, native vegetation and faunas are gradually perturbed, nutrient and moisture inputs are altered, and pests invade. In managing cave preserved areas, it is important to be able to measure the results from time to time. Therefore, baseline surveys, census surveys, and written management plans are essential.

Karst groundwater issues are so economically and politically important that they move beyond the influence of the technical experts who best understand them (Elliott 1994, 1996b). In Texas, the regulation of overpumping of the Edwards Aquifer, one of the most important karst aquifers in North America, has been gradually improved through a succession of state rulings and legislations that were contradicted by federal court decisions. It is quite possible that the San Marcos and Comal

Springs will go dry within the next decades, endangering many species and the human economy (Elliott 1993, 1996a).

It has become axiomatic that baseline faunal surveys with simultaneous monitoring of ecological factors such as temperature, humidity, moisture, air movements, and nutrient inputs, are essential for understanding changes in cave faunas (Poulson and Kane 1977, Elliott and Reddell 1989, Perkins 1990, Elliott 1994, Northup and Welbourn 1995, Lewis 1996). Subterranean faunas change seasonally and also in response to climate changes at larger temporal scales. Initial faunal surveys can be accomplished in several ways, including hand collecting, baiting, Berlese extraction, and pitfall trapping. Once a faunal list is compiled, monitoring censuses in different seasons and years increases the range and value of the data set. Large fluctuations in some trogloxenic populations can occur on an annual time scale; for instance, cave cricket populations in some Central Texas caves dropped by half after a drought year in 1996 (Elliott, unpublished data). Historical data from the Mammoth Cave ecosystem over 100 years have been useful in understanding the current ecology of the cave. Comparative studies of different types of cave communities give us a better idea of how to manage a particular cave, such as a show cave that has been perturbed for many years.

References

Bolner, K. and J. Tardy. 1989. Bacteriological and chemical investigations of dripping waters in the caves of Budapest. In: *International Symposium on Physical, Chemical and Hydrological Research of Karst*, Slovenská Speleologická Spločnos, Liptovský Mikuláš. pp. 102-111.

Cooper, J.E. 1975. Ecological and behavioral studies in Shelta Cave, Alabama, with emphasis on decapods crustaceans. Ph.D. dissertation, University of Kentucky, USA.

Cruickshank, G., J. Aguirre, D. Kraemer and E. Craviota. 1980. Bacterial contamination of the limestone aquifer beneath Merida, Mexico. In: *Aquifer Contamination and Protection*, R.E. Jackson (ed.). Project 8.3 of the International Hydrological Programme, UNESCO, Paris. pp. 341-345.

Culver, D.C. 1982. *Cave life: evolution and ecology*. Harvard University Press, Cambridge, 189p.

Culver, D.C. 1986. Cave faunas. In: *Conservation Biology: The Science of Scarcity and Diversity*, M. Soule (ed.). Sinauer Associates, Sunderland. pp. 427-442.

Dickson, G.W., L.A. Briese and J.P. Giesy, Jr. 1979. Tissue metal concentrations in two crayfish species cohabiting a Tennessee cave stream. *Oecologia* 44(1): 8-12.

Elliott, W.R. 1981. Damming up the caves. *Caving International* 10: 38-41.

Elliott, W.R. 1982. An introduction to biospeleology. In J. Hassemer (ed.), *Caving Basics*. 1987. National Speleological Society, Huntsville, pp.100-108.

Elliott, W.R. 1993. Cave fauna conservation in Texas. In D.L. Foster (ed.), *1991 Proceedings of the National Cave Management Symposium, Bowling Green, Kentucky,* pp. 323-337.

Elliott, W.R. 1994. Biodiversity and conservation of North American cave faunas: an overview. In B. Mixon (ed.), *1994 NSS Convention Program, Abstracts, National Speleological Society, Huntsville,* p. 48.

Elliott, W.R. 1996a. The Barton Spring salamander. *American Caves,* 9(1): 14-15.

Elliott, W.R. 1996b. The evolution of cave gating – how the philosophy and technology have changed. *American Caves,* 9(2): 9-15.

Elliott, W.R. 1997a. A survey of ecologically disturbed areas in Carlsbad Cavern, New Mexico. *Report to Carlsbad Caverns National Park,* 10p.

Elliott, W.R. 1997b. Biological, geological and engineering aspects of Mayor Elliott Cave, Georgetown, Texas. *Report to Paul Price Associates and the City of Georgetown, Texas,* 8p.

Elliott, W.R. 1997c. The caves of the Balcones Canyonlands Conservation Plan, Travis County, Texas. *Report to Travis County, Texas, Transportation and Natural Resources Department, Balcones Canyolands Conservation Plan,* 156p.

Elliott, W.R. 2005. Protecting caves and cave life. In: *Encyclopedia of Caves,* D.C. Culver and W.B. White (eds.). Academic Press Elsevier. pp. 458-467.

Elliott, W.R. and J.R. Reddell. 1989. The status and range of five endangered arthropods from caves in the Austin, Texas, Region. *Austin Regional Habitat Conservation Plan,* 100p.

Ford, D.C. and P.W. Williams. 1989. *Karst Geomorphology and Hydrology.* Unwin Hyman, London.

IUCN, 1996. *1996 IUCN Red list of Threatened Animals.* IUCN, Gland. 368 pp.

Holland, E. 1994. The effects of fire on soluble rock landscapes. *Helictite* 32(1): 3-9.

Hubbard, D.A. 1995. Cave conservation and cave clean-ups: not always one in the same. *National Speleological Society News,* 53:31.

Juberthie, C. 2004. Conservation of subterranean habitats and species. In: *Ecosystems of the World 30, Subterranean Ecosystems.* H. Wilkens, D.C. Culver and W.F. Humphreys (eds.). Academic Press, Elsevier. pp. 691-700.

Lewis, J.J. 1996. Cave bioinventory as a management tool. In G.T. Rea (ed.), *1995 Proceedings of the National Cave Management Symposium. Spring Mill State Park, Mitchell, Indiana,* pp. 228-236.

MMA——Ministério do Meio Ambiente. (2004). Lista Nacional das Espécies de Invertebrados Aquáticos e Peixes Ameaçados de Extinção. Instrução Normativa n° 5, 21st May of 2004. Brasília, Brazil.

Moreira, C.R., M.E. Bichuette, O.T. Oyakawa, M.C.C. de Pinna and E. Trajano. Rediscovery of *Stygichthys typhlops* Brittan and Böhlke, 1965, a subterranean characiform fish from Brazil. *Journal of Fish Biology* (In Press).

Northup, D.E. and W.C. Welbourn. 1995. Conservation of Invertebrates and Microorganisms in the Cave Environment. In Dale L. Pate (ed) *1993 Proceedings of the National Cave Management Symposium, Carlsbad,* pp. 292-301.

Poulson, T.L. 1963. Cave adaptation in Amblyopsid fishes. *American Midland Naturalist*, 70(2): 257-290.

Poulson, T.L. 1968. Aquatic cave communities. *Cave Research Foundation Annual Report*: 16-18.

Poulson, T.L. and T.C. Kane. 1977. Ecological diversity and stability. In T. Aley and D. Rhodes (eds.), 1976 Proceedings of the national Cave management Symposium, Mountain View, Arkansas, pp.18-21.

Proudlove, G.S. 2001. The conservation status of hypogean fishes. In Romero, A. (ed.). *The biology of hypogean fishes*. Kluwer Academic Publ., Dordrecht, pp. 201-213.

Proudlove, G.S. 2006. *Subterranean Fishes of the World. An Account of the Subterranean (Hypogean) Fishes Described up to 2003 with a Bibliography 1541-2004*. International Society for Subterranean Biology, Moulis, 300 pp.

Silva, A.B. 1984. *Análise morfoestrutural, hidrogeológica e hidroquímica no estudo do aqüífero cárstico do Jaíba, Norte de Minas Gerais*. Unpubl. Ph.D. Thesis, Universidade de São Paulo, São Paulo.

Sullivan, G.N. 1956. Biological aspects of cave conservation. *The News, National Speleological Society* 14(3): 30-31.

Reis, R.E., E. Trajano and E. Hingst-Zaher. 2006. Shape variation in surface and cave populations of the armoured catfish *Ancistrus* (Siluriformes: Loricariidae) from the São Domingos karst area, Upper Tocantins River, Brazil. *Journal of Fish Biology* 68: 414-429.

Toomey, R.S. 2006. Defining limits and protocol at Karchner Caverns. In: *Cave Conservation and Restoration*, V. Hildreth-Werker and J.C. Werker (eds.). National Speleological Society, Huntsville, pp. 141-145.

Trajano, E. 1997. Synopsis of Brazilian troglomorphic fishes. *Mémoires de Biospéologie*, 24: 119-126.

Trajano, E. 2000. Cave faunas in the Atlantic tropical rain forest: composition, ecology and conservation. *Biotropica*, 32, (4b): 882-893.

Trajano, E. and H.A. Britski. 1992. *Pimelodella kronei* (Ribeiro, 1907) e seu sinônimo *Caecorhamdella brasiliensis* Borodin, 1927: Morfologia externa, taxonomia e evolução (Teleostomi, Siluriformes). *Boletim de Zoologia, São Paulo* 12: 53-89.

Trajano, E., R.E. Reis and M.E. Bichuette. 2004. *Pimelodella spelaea*, a new cave catfish from Central Brazil, with data on ecology and evolutionary considerations (Siluriformes: Heptapteridae). *Copeia* 2004(2): 315-325.

Trajano, E., S. Secutti and M.E. Bichuette. 2009. Population decline in a Brazilian cave catfish, *Trichomycterus itacarambiensis* Trajano & Pinna, 1986 (Siluriformes): reduced flashflood as a probable cause. *Speleobiology Notes* 1: 24-27.

Varela, E., O. Barros and S. Jimenez. 1980. The effects of sugar-waste disposal on a karstic limestone aquifer. In: *Aquifer Contamination and Protection*, R.E. Jackson (ed.). Project 8.3 of the International Hydrological Programme, UNESCO, Paris. pp. 329-339.

Vesper, D.J. 2005. Contamination of cave waters by heavy metals. In: *Encyclopedia of Caves*. D.C. Culver and W.B. White (eds.). Academic Press, Elsevier. pp. 127-131.

CHAPTER 4

Behavioral Patterns in Subterranean Fishes

Jakob Parzefall[1] and Eleonora Trajano[2]

[1]University of Hamburg, Biozentrum Grindel, Germany.
E-mail: parzefall@zoologie.uni-hamburg.de
[2]Departamento de Zoologia, Instituto de Biociências da Universidade de
São Paulo, Brazil. E-mail: etrajano@usp.br

I. INTRODUCTION

The ecological conditions in the subterranean environment are characterized by two main factors: permanent darkness, thus absence of green plants and of photoperiods, and a more or less constant temperature. Subterranean habitats in different parts of the world have been colonized successfully by teleostean fishes from 20 families (Proudlove, this volume). In the majority of these families, epigean (surface) members are known to be active at dusk and/or night, so it is not surprising that those species with a preference for activity in darkness colonize subterranean habitats such as caves. But among the ancestors of troglobitic fish we also note members of the families Characidae (tetras), Cobitidae (loaches), Balitoridae (loaches), Cyprinidae (carp-fish) and Poeciliidae (toothcarps), which include many species with clear diurnal activity.

Mainly for these species the question arises what enables certain families to survive and to reproduce in caves. One possibility would be that members of these families show behavioral preadaptations to life in darkness. Preadaptations (or exaptations) may be defined as character states conferring performance advantage in a given selective regime, but which have been selected in another, independent previous regime (Trajano 2005). In addition, or associated with the predominantly nocturnal activity, preadaptations in epigean fishes include enhanced mechano and/or chemosensorial systems, as in the amblyopsids and siluriforms respectively, and an opportunistic, generalist feeding.

Nevertheless, in order to distinguish preadaptations, which are plesiomorphic states for troglobites, from autopormorphies resulting from the differentiation in the subterranean habitats, it is necessary to establish the direction of transformation series into cladograms, by applying the comparative method. For behavior, both the outgroup comparison and the ontogenetic methods may be used, but they require phylogenetic trees and ontogenetic series respectively. Unfortunately, there are few ontogenetic series available for cavefish, because relatively few species reproduce in laboratory and collection of juveniles directly in the natural habitat is rare, and even fewer behavioral data for the younger stages. As well, phylogenies including troglobitic fish species are scanty.

The evolutionary interpretation of character states observed in subterranean organisms is particularly difficult for the so-called phylogenetic relicts, i.e., species without known epigean close relatives, as is the case with the Brazilian phreatic characiform, *Stygichthys typhlops*. At first considered a characid, recent taxonomic studies pointed to a more basal position within the characiforms, but its exact familial position was not established yet. As well, the phylogenetic positions of the four recognized Brazilian species of troglobitic *Rhamdiopsis* (see Trajano and Bichuette, this volume) within the genus and that of the phreatic African catfish, *Uegitglanis zamaranoi*, within the Clariidae, were not established. Consequently, hypotheses on direction of evolutionary change for character states observed in this highly modified species are questionable.

At the moment, the most parsimonious solution is to compare the studied troglobitic species with the geographically closest epigean congeneric species, in the assumption that this would be a less modified descendant of a common ancestor with the differentiated cave species (the hypothesis of extinction of the epigean sister-species followed by dispersion of less related species to a closer geographic area requires more evolutionary steps). For the outgroup comparison method, a third species is required. In the absence of phylogenies, available data on any other epigean congeners has been used (e.g. Trajano and Bockmann 1999).

In this chapter, we compare behavior patterns of troglobitic fishes with their closest epigean relatives, aiming to detect apomorphic character states resulting from specialization to the subterranean way of life, and discuss common patterns among troglobitic species, in search of homoplastic traits indicating convergent adaptation to this peculiar life due to similar selective regimes. We examine whether differences observed in relation to epigean relatives are phenotypically plastic responses to cave life, or whether the observed behavioral changes are genetically fixed.

II. BEHAVIORAL PATTERNS IN SUBTERRANEAN FISHES

The ethology of subterranean fishes is a comparatively well developed discipline of Subterranean Biology in great part thanks to the pioneering work of the Belgium researcher G. Thinés and collaborators, who published a most impressive experimental work focusing on different species from the 1950's and 1980's, and the German geneticist C. Kosswig, who carried out the first studies on the genetics of behavioral traits, from the late 1940's to the early 1960's. In the United States, on the early 1960's T. Poulson published the first comprehensive comparative study on cave fishes, which included behavioral aspects. Kosswig's students, conducted by J. Parzefall and H. Wilkens, continued his work with special focus on Mexican cavefish. This legacy was augmented by Italian researchers leaded by R. Berti from the 1970's on and by Brazilian (E. Trajano and collaborators) and Indian researchers (J. Biswas, A. Pati and collaborators) from the late 1980's on.

As a consequence, several species, mainly from Mexico, Brazil, India and northern Africa, have been studied in greater or less detail. Nevertheless, behavioral studies on subterranean fishes still present a great taxonomic heterogeneity. A few species have been intensively studied, especially the Mexican tetra characins, genus *Astyanax*, whereas important groups with many troglobitic derivatives, such as the cyprinids, are scarcely known with regard to their behavior in particular and biology in general. Siluriformes (catfishes), another important fish group in subterranean habitats, are in the mid term: several Brazilian species belonging to three families (Heptapteridae, Trichomycteridae and Loricariidae) have been studied in detail, but most species from other countries call for investigation. Therefore, general patterns are not quite clear.

The reasons why Mexican blind tetras are so intensively studied are mainly practical, because these fish are easily maintained and bred in laboratory, being an excellent material for genetic studies. Nevertheless, they do not provide a good paradigm because, being one among only three known troglobitic characiforms (in contrast with 78 cypriniforms and 52 siluriforms – Proudlove, this volume), the Mexican blind *Astyanax* represent a taxonomic exception. In fact, their putative epigean sister-species, *A. mexicanus*, active at dusk and with a chemically oriented reproductive behavior, is an exception among the generally diurnal, visually oriented characids. As well, several behavioral traits of the blind tetras do not have clear equivalents in many other subterranean fishes.

Another difficulty arises from the fact that congeneric troglobitic organisms may present a mosaic distribution of character states related to an exclusively subterranean life, indicating not only a rather independent differentiation but also different evolutionary rates affecting these characters. A mosaic distribution of character states (although not presented as such), including behavioral traits, was first noticed for amblyopsids (Poulson 1985). This small North American family, which encompasses an exclusively epigean, a troglophilic and four troglobitic species, provides an excellent material for comparative research (see Niemiller and Poulson, this volume).

Well documented cases of a mosaic distribution of character states in fishes include Brazilian armored catfish, genus *Ancistrus* (Bessa and Trajano 2002), and Thai cave balitorids, genus *Schistura* (Trajano and Borowsky 2003). The same is true for several confamilial taxa such as the neotropical heptapterid catfishes (Trajano and Bockmann 1999). Besides morphological traits, such mosaics affect behavioral characters as diverse as vertical distribution, cryptobiotic (hiding) habits, reaction to light and agonistic interactions. The factors underlying this phenomenon are still poorly understood because, at a coarse-grained scale, the selective regimes under which these species are seem not to differ significantly.

Among the troglobitic fishes treated below, the Mexican cave *Astyanax* were compared to the epigean *A. mexicanus*, the cave form of *Poecilia mexicana*, characterized by variable eyes and pigmentation, to the normally eyed and pigmented epigean form, Thai cave balitorids with epigean species belonging to the same and different genera, troglobitic amblyopsids with epigean and troglophilic. Among the Brazilian catfishes, the heptapterids *Pimelodella kronei* was compared to its putative epigean sister-species, *P. transitoria*, and to other congeneric catfish living in other basins (including the troglobitic *P. spelaea*), *Rhamdia enfurnada* to the epigean widely distributed nominal species, *Rhamdia quelen*, and the troglobitic trichomycterids and loricariids with any epigean congeners for which behavioral data were available.

These studies included mainly laboratory research, complemented by naturalistic observations in the habitat in several cases, including most of the Brazilian species so far investigated. They showed that, although it is not clear how general are the detected patterns for troglobitic fishes, some phenomena are quite common and would represent patterns at least for some taxa. In fact, among the aspects discussed below, there are many cases of loss or reduction of behavioral traits, similar to the reduction of eyes and pigmentation, well know and characteristic of troglobites in general. Such reductions would represent instances of regressive characters, i.e., characters where the derived states correspond to a decrease in the size, intensity and/or complexity of structures, behaviors or physiological processes (Trajano 2005), a process frequently attributed to the loss of function or, in other words, to the relaxation of stabilizing selective pressures (Wilkens 1992).

II.1 Spontaneous Behavior

Most epigean siluriforms and balitorids, two taxa with many subterranean representatives, are bottom-dwellers. Except for highly specialized benthonic taxa such as the armored catfishes (Loricariidae), a tendency to increase the midwater activity, in parallel with a decrease in cryptobiotic habits, has been observed in the studied troglobitic catfishes and also in balitorids from Thailand.

Both in caves and in laboratory, the catfish *Pimelodella kronei* (Heptapteridae), from southern Brazil, is frequently observed swimming in the midwater and near the surface. In contrast, *Pimelodella transitoria* is a typical bottom-dweller, very shy, elusive and photophobic catfish. In a quantitative laboratory study comparing these species, with a few exceptions the blind catfish spent at least half the observation time calmly swimming, and most specimens stayed hidden under dens less than one third that time; in contrast, the epigean catfish remained almost all time inside their hiding places (Trajano 1989).

Not quantified observations point to a similar adaptation in the highly troglomorphic heptapterids of the *Nemuroglanis* subclade (*sensu* Trajano and Bockmann 1999), *Rhamdiopsis* sp. from Chapada Diamantina (cited as "new heptapterine" in Trajano 2003) and *Rhamdiopsis* sp. from Campo Formoso (formerly cited as *Taunayia* sp.), and the trichomycterid *Trichomycterus* sp. from Serra da Bodoquena (see Trajano and Bichuette this volume), and, to a lesser degree, in the Rhamdiini heptapterids, *Rhamdia enfurnada* (Bichuette and Trajano 2005) and *Rhamdia macuspanensis* (Weber and Wilkens 1998), the trichomycterid *Trichomycterus itacarambiensis* and a new population of *Aspidoras*, the first known troglomorphic callichthyid catfish (Trajano and Secutti, pers. obs.). As expected, the morphologically less specialized *Pimelodella spelaea* and *Rhamdiopsis* sp. from Cordisburgo (Trajano and Bichuette, *op. cit.*) retained the basically bethonic habits.

The increased midwater activity probably enhances the chances of finding food under conditions of low density of prey, especially if associated to low predation rates and low competition at midwater. As a matter of fact, the proportion of allochthonous (terrestrial) food items in the diet of *Pimelodella kronei* is higher than the observed for its eyed relative, *P. transitoria* (Trajano 1989). In the case with *P. kronei*, midwater + bottom activity may be a plesiomorphic character state, since it was also observed for other *Pimelodella* species, including *P. transitoria* (Trajano and Bockmann 1999). As well, constant swimming in the water column in addition to the bottom may be a plesiomorphic state for Trichomycterinae catfishes. Thus, high activity levels with frequent midwater swimming may be a plesiomorphic trait for cave trichomycterids. On the other hand, midwater activity in the two above mentioned *Rhamdiopsis* species is better interpreted as an autapomorphic adaptation of these species to the cave habitat (Trajano and Bockmann 1999). Apparently, this is also the case with the cave *Aspidoras* catfish.

Unlike their epigean relatives, several Brazilian troglobitic catfishes rarely hide. These are possible examples of behavioral regression related to cave life, which affects cryptobiotic habits. Such evolutionary reduction is consistent with the hypothesis that cryptobiotic habits are part of a strategy against visually-oriented predators in the surface habitat. Hiding is not effective against non-visual, chemo-oriented predators, and other strategies are required against predators able to feed in caves, such as nocturnal predaceous fishes. Therefore, selection for maintenance of hiding habits may be relaxed and the character regress, as much as eyes and pigmentation would regress in the absence of light (Trajano 2003).

A mosaic distribution of behavioral traits was reported for Brazilian loricariids, genus *Ancistrus*. The morphologically specialized *Ancistrus formoso* is conspicuously cryptobiotic (although less than epigean *Ancistrus* spp.), spending almost 70% of the time hidden, whereas an inter-population variation is observed in the less specialized *Ancistrus cryptophthalmus*, in parallel with the degree of eye reduction (but not with the pigmentation reduction – Reis *et al.* 2006), from 60% of the time hidden in catfish from Angélica Cave to less than 15% in specimens from Passa Três Cave (Bessa and Trajano 2002).

For some Brazilian troglobitic fishes, potentially important predators were found in the cave habitat, as is the case with *Rhamdiopsis* sp. from Campo Formoso (Trajano and Bockmann 2000) and *Ancistrus cryptophthalmus* (Trajano and Bichuette, unpubl. data). This may explain the strong avoidance of mechanical stimuli in the former, which presents enlarged cephalic lateral line canals. Probably due to the absence of important predation rates, other cave fishes, such as *P. kronei, R. enfurnada, Rhamdiopsis* sp. from Chapada Diamantina and *T. itacarambiensis*, do not show strong phobic reactions to stimuli as those from water movements and sudden illumination.

A mosaic of behavioral character states was also observed for four troglobitic balitorids from Thailand, *Schistura oedipus, S. jaruthanini, S. spiesi* and *Nemacheilus troglocataractus* (which would also be a *Schistura* – R. Borowsky, pers. comm.), studied in laboratory (Trajano and Borowsky 2003). The epigean balitorids, *S. mahnert, S. similis* and *Acanthocobitis zonalternans*, were studied under the same standardized conditions, allowing for the application of the out-group comparison method. The epigean balitorids were typical bottom-dwellers, with occasional incursions onto the midwater. *S. jaruthanini*, the least troglomorphic cave balitorid we studied, presented a significant activity at midwater in addition to the bottom, a specialized behavior convergent with those observed in several Brazilian troglobitic catfishes. In contrast, the highly troglomorphic species *N. troglocataractus* and *S. oedipus* retained the predominant bottom activity. *Schistura spiesi* is intermediate in this respect. Cryptobiotic habits varied among cave balitorids, with no pattern in relation to those of the epigean ones: *S. oedipus* was the most cryptobiotic, and *S. jaruthanini*, the least.

II.2 Feeding Behavior

Suitable food sources and food quantity vary among different subterranean habitats, and also from cave to cave. In general, subterranean animals depend upon food brought in from outside the cave and tend to be opportunistic carnivores or omnivores. Even though there are exceptions represented by caves with abundant food (Langecker *et al.* 1996), most subterranean habitats, especially the phreatic ones, have less food available than surface habitats. Food sources can be widely distributed or concentrated in patches and are often unpredictable in space and time. Therefore, food finding abilities are often improved and additional storage tissues help cave fishes to survive prolonged starvation periods.

The blind cave form of *Astyanax mexicanus* from the Pachon cave was studied in competition experiments in darkness. In the case of the cave form, the focal fish detected 80% of small pieces of meat that were distributed on the bottom of an aquarium, whereas the epigean fish managed to find only 20% of food items (Hüppop 1987). Schemmel (1980) found that the improved food finding ability in the cave form of *A. mexicanus* is due to a behavioral adaptation: the cave fish swim and feed in a typical angle of about 45° to the substrate (Fig. 1), whereas the epigean fish swim vertically while feeding. Studies using population hybrids confirmed that this difference is based on genetic factors. In addition, the taste buds, which are restricted to the mouth region in the epigean fish, are distributed over the lower jaw and cover also ventral areas of the head in the cave form. A similar feeding mode is exhibited by the Texan blind catfish *Trogloglanis pattersoni* (Ictaluridae), which have a very specialized mouth used for the chemical localization of food on the bottom (Langecker and Longley 1993).

When investigating the feeding behavior of amblyopsid fishes, Poulson (1963) obtained results similar to those reported for cave *Astyanax*: the blind *Amblyopsis spelaea* detected a small prey item hours before the

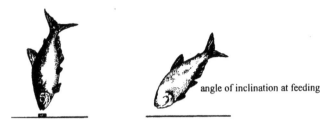

Fig. 1 The position in cave and epigean form of *Astyanax mexicanus* to pick food from the bottom (after Schemmel 1980).

troglophilic *Forbesichthys agassizi*. *A. spelaea* ate all ten *Daphnia* presented in the test aquarium before *C. agassizi* had eaten half of them (Niemiller and Poulson, this volume; Poulson and White 1969).

On the other hand, no important difference in the food finding ability per se was observed for *P. kronei* when compared to *P. transitoria* – the higher searching times recorded in laboratory for the latter are due to its hesitation to leave the den, once outside both species are equally efficient in food taking, even under dim light. This supports the idea that the epigean catfish are highly preadapted to colonize subterranean habitats and few adaptations were necessary for a successful cave life in *P. kronei*.

With regard to the feeding tactics, in addition to the crepuscular to nocturnal preying of bottom animals and grubbing (excavating while moving) usual for epigean bottom-dwelling catfish, the troglobitic catfishes with increased midwater activity are also surface pickers. For instance, *P. kronei* catfish are frequently observed swimming with the open maxillary barbels sweeping the water surface and promptly catching prey near or at the water surface. In the natural habitat, allochthonous items are more important for this species than for *P. transitoria* (Trajano 1989).

In subterranean amblyopsids, density dependent cannibalism seems to account for most of the mortality of the young (Poulson 1963). This would be an important mechanism of population control, in view of the food scarcity and lack of predators observed in the cave habitat for these fish (Poulson, *op. cit.*). Cannibalism was also observed for the Brazilian fishes, *Rhamdiopsis* sp. from Toca do Gonçalo and *Stygichthys typhlops*. Both are highly specialized troglobites living in phreatic habitats apparently subject to severe food shortage, and conspecifics could represent an important source of food. *Rhamdiopsis* sp. catfish are able to ingest very large prey, even larger than themselves, an adaptation enabling a quite wide feeding niche. A specialized behavior was observed for the ingestion of large organisms, including conspecifics, which are swallowed posterior part first and digestion starts with the prey only partially swallowed (Trajano and Bockmann 2002).

II.3 Biological Rhythms

The cyclic activity of organisms is internally controlled by oscillators, which are organized in internal timing systems ("biological clocks"). These oscillators produce endogenous rhythms which are synchronized by external cyclic stimuli, the *zeitgebers* (time givers) or forcing signals, resulting in the observed rhythmic behaviors. Cycles with a period close to 24 h are circadian rhythms, those with period close to four weeks are circalunar, those close to 12 months are circannual and so on. The most powerful *zeitgebers* for activity, including the locomotor and feeding behaviors, are the daily cycles of light and temperature. The seasonal variation in the length of photoperiods and monthly temperatures are important *zeitgebers* of reproductive cycles, migrations etc.

Troglobitic species, especially those evolving in very constant environments, provide good opportunities to test the two main hypotheses on the evolution of circadian rhythms, the internal versus external, ecological selection (Lamprecht and Weber 1992). Having evolved in the absence of 24-h *zeitgebers*, it is predicted that, if external factors provide the main selectives forces for circadian rhythms, these could regress at some degree in many troglobites like the observed for eyes, melanic pigmentation and other characters related to light. On the other hand, if internal order were of prime importance for the maintenance of circadian rhythms, these would not be lost in troglobites (Volpato and Trajano 2006).

Several studies pointed to the loss or weakening of circadian components of activity in troglobites as diverse as beetles, crustaceans and fishes, suggesting that such rhythmicity is not necessary for the maintenance of internal temporal order, thus corroborating the hypothesis of external selection as a major selective force for the appearance and maintenance of endogenous circadian rhythms, at least for locomotor activity. Such pattern is quite clear for the several studied subterranean fishes. In general, these species were studied by monitoring their locomotor activity in laboratory, both under constant (free-run) conditions, usually darkness, and LD cycles, using an infrared photocells setup connected to a computer system which automatically records the number of beam crossings along the time. All these troglobitic species belong to typically nocturnal taxa and, when known and available, the epigean sister-species were also studied for comparison.

In *Astyanax mexicanus*, free-running activity rhythms were detected, as expected for an epigean species. Light/dark (LD) cycles with different periods (12:12h, 6:6h, 4:4h, 16:8h etc.) acted as *zeitgebers*, entraining the locomotor activity, and residual oscillations (post-oscillations) were observed after the transition from LD to DD, during one or a few cycles. This suggests the existence of an endogenous circadian oscillator, which becomes obvious in free-running conditions. On the other hand, in the highly specialized troglobite, *"A. antrobius"* (from El Pachón Cave), although activity was entrained by all applied LDs, the signal energies were lower than in the tests with epigean fishes, the rhythms of total activity disappeared immediately after the transition from LD to DD (no residual oscillations), and in no LD with a period frequency differing from 24-h a circadian rhythm could be observed in addition to the entrained frequency. It is concluded that the internal clock of *"A. antrobius"* was simplified in relation to its epigean ancestor: the passive system has developed into an extremely passive one, incapable of synchronizing, thus the circadian oscillator was subject to regression, but it was not completely lost (Erckens and Martin 1982). The hybrid population, *"Astyanax jordani"*, from La Chica Cave, seems to be intermediate in this aspect, because one or two residuals oscillations were observed after a transition from LD (12:12) to DD (Erckens and Weber 1976). The persistence of a circadian clock in this cave fish was pointed out by Cordiner and Morgan (1987), who recorded free-running circadian activity rhythms (not shown by Erckens and Weber 1976), often masked by apparently random, infradian oscillations.

Several Brazilian catfishes, with different degress of troglomorphy (see Trajano and Bichuette, this volume) had their locomotor rhythmicity investigated (Trajano and Menna-Barreto 1995, 1996, 2000, Trajano *et al.* 2005, unpubl. data). In most cases, individual variation concerning the presence of free-running circadian rhythms and the frequency and periodicity of other rhythms was reported. In the less specialized, stream-dwelling troglobitic *Pimelodella kronei*, *P. spelaea*, *Rhamdia enfurnada*, *Rhamdiopsis* sp. from Cordisburgo and *Trichomycterus itacarambiensis*, at least 2/3 of the monitored specimens did present significant circadian rhythms under constant darkness. It is notheworthy that all the studied epigean catfishes, *P. transitoria* and *Taunayia bifasciata*, exhibited strong, significant free-running circadian components of locomotor activity.

On the other hand, in highly troglomorphic Brazilian subterranean fishes, the proportion of individuals with free-running circadian rhythms was much lower, as in *Rhamdiopsis* sp. from Chapada Diamantina, or none, as is the case with *Stygichthys typhlops* and *Rhamdiopsis* sp. from Toca do Gonçalo. The two *Rhamdiopsis* species were also studied under LD cycles, 12:12 h. In both species, activity was entrained by these cycles, but no residual oscillations were observed, indicating a possible masking effect.

Activity rhythms in six species of balitorids from Thailand, two epigean *Schistura* species and four troglobitic species (*Schistura oedipus*, *S. jaruthanini*, *S. spies* and *Nemacheilus troglocataractus*) were investigated in laboratory. All species exhibited an approximate 24 hour rhythms, except *S. oedipus*, which had a longer cycle, 38-40 hours (Borowsky and Trajano 2003).

The Indian balitorid, *Nemacheilus evezardi*, was subject to several studies on circadian and circannual rhythmicity, with focus on locomotor activity, air gulping behavior and phototactic behavior (Pradhan *et al*. 1989, Biswas *et al*. 1990, Biswas 1991). The epigean "form" *N. evezardi* is basically a dusk active bottom, with maximum locomotor activity at the early part of the dark phase, concealing themselves under stones during most of the day. This is another example of epigean fish preadapted to the subterranean life.

Biswas *et al*. (1991) studied the locomotor activity and surfacing frequency of *N. evezardi* during the pre-spawning, spawning and post-spawning phases, comparing epigean (kept under natural LD cycles) and hypogean individuals (kept under DD). Significant rhythms were detected for total activity in both epigean and hypogean loaches during the pre-spawning phase, but only for the epigean fishes during the spawning and post-spawning phases. Significant circadian rhythms of surfacing behavior related to air-gulping were observed during the pre-spawning and spawning phases only. In hypogean loaches, the levels of total and surfacing activity were lower in the pre-spawning and spawning phases, suddenly increasing during the post-spawning phase, whereas no change in total activity was noticed for the epigean loaches. These results demonstrate the influence of the reproductive condition over the expression of circadian rhythmicity, which differs among cave and epigean populations.

In conclusion, the comparison between troglobitic fishes and their epigean close relatives (troglobitic *Astyanax* versus *A. mexicanus*, *P. kronei* versus *P. transitoria*; troglobitic *Rhamdiopsis* versus *Taunayia bifasciata*, *S. oedipus* versus epigean and other cave balitorids, *N. evezardi* cave form versus epigean form) analyzed under the same conditions and using the same protocols, provides good evidence for the hypothesis of an evolutionary regression of time-control mechanisms in troglobitic species, either affecting the oscillator(s) itself (themselves) or due to an uncoupling between the oscillators and at least one of their related functions; in the case, the locomotor activity. Regression of retina and, at least for some of the studied species, possibly also of the photoreceptor of the pineal organ, where circadian oscillators of fishes would be located, may be involved in the disorganization of the circadian system verified in several troglobitic species (Volpato and Trajano 2006).

Besides differences in general locomotor activity, cavefishes do not reproduce as regularly as their epigean relatives: e.g., amblyopsid cavefishes reproduce only when food is abundant in the cave (Poulson 1963). In various cave populations of *Astanax mexicanus* we could observe offspring or very young fish only occasionally during 25 years of field work (Parzefall, pers. obs.).

II.4 Reaction to Light

Reaction to light is understandably one of the earliest and most intensively studied behaviors in subterranean fishes. So far, the phototactic behavior has been studied in troglobitic fishes from ten families: Characidae (genus *Astyanax*), Cyprinidae, Balitoridae, Heptapteridae (formerly in Pimelodidae), Trichomycteridae, Loricariidae, Clariidae, Amblyopsidae, Bythitidae (formerly in Brotulidae) and Poeciliidae (Parzefall 1998, Gerhard and Trajano 1997, Bessa and Trajano 2002, Trajano and Borowsky 2003).

Many cavefishes have epigean relatives that are photophobic, as is the case with the catfishes. A decreased photonegative response in relation to the highly photophobic epigean relative, *P. transitoria*, was described for *P. kronei*, with a considerable individual variation. A tendency towards a weak photonegative or even a photopositive response was also recorded for other catfishes such as the Somalian clariid, *Uegitglanis zammaranoi*, and the Brazilian trichomycterid, *Trichomycterus itacarambiensis*. For a discussion on the reaction to light in siluriforms, see Trajano (2003).

It is noteworthy that the mosaic distribution in morphological and behavioral characteristics observed in the Brazilian *Ancistrus* catfish includes the reaction to light (Trajano and Bessa 2002): adult *A. cryptophthalmus* catfish from Passa Três Cave are indifferent, the reaction to light in *Ancistrus formoso* varies individually, and *A. cryptophthalmus* catfish from Angélica and Bezerra caves are basically photophobic, although less than the epigean *Ancistrus* from the same and other regions too. Juvenile from Passa Três, with eyes still not regressed, are photonegative, indicating that the absence of reaction to light in choice-chamber experiments exhibited by the adults, with regressed eyes sunken into the head (Secutti and Trajano 2009), may be a consequence of poor light sensitivity; nevertheless, the possibility of an ontogenetic behavioral change may not be ruled out.

In members of the families Amblyopsidae, Bythitidae, Characidae, Claridae, Cyprinidae, photonegative behavior with respect to light intensities above 30 lux has been described. Only the presumed phylogenetically young cave form of the poeciliid *Poecilia mexicana* showed a photopositive reaction at three different light intensities (Parzefall *et al.* 2007, Plath and Tobler, this volume).

The most detailed study has been conducted in different populations of the characid *Astyanax mexicanus* by Langecker (1992). Langecker could demonstrate that this behavior in the epigean fish depends on light intensity. Juveniles showed a slightly positive response under all tested conditions (Fig. 2), while adult fish were photonegative. Similar results were found in juvenile and adult cave *Garra barreimiae* (Timmermann and Plath, unpublished data). The photonegative behavior in surface dwelling *Astyanax* was distinctly stronger in blinded fish. Blinded fish that had been pineal-ectomized showed a weaker response, suggesting an important role of the pineal organ for light perception and phototactic behavior. In contrast to the epigean fish, the removal of the pineal organ had no significant effect on the phototactic behavior of the cavefish. Thus, it could be argued that differences in the phototactic behavior between surface and cavefish would be caused by a reduction process in parts of the pineal organ. Ultrastructurally, however, surprisingly little change in this organ was detected (Langecker 1990). In addition, the degree of differentiation of the photoreceptor cells indicates that they are probably still light sensitive (Langecker 1990). Therefore, it seems that the different phototactic responses of surface and cave *Astyanax* represent a reduction process on the behavioral level. The same mechanisms may also be operating in the case of several other cavefishes, but data are still lacking for most systems.

Thai balitorids were also investigated with focus on their reaction to light in a series of tests under different light intensities and in different phases of 12:12 h light-dark cycles (Trajano and Borowsky, unpubl. data).

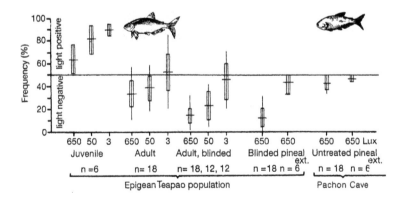

Fig. 2 Preference for the lighted area in a light-dark choice experiment of *Astyanax mexicanus*. Mean and variability of eyed, blinded and pineal ectomized epigean fish and blind Pachon cave fish under different light conditions (after Langecker 1989, 1992).

In general *Schistura* loaches, both epigean *(S. moiensis, S. similis* and *S. mahnert)* and troglobitic *(S. oedipus, S. jaruthanini* and *S. spiesi)*, exhibited photonegative responses in choice-chamber experiments, although non-significant responses were observed in some tests. As well, individual variation was observed in the three studied *N. troglocataractus* fish. Except for *S. jaruthanini*, the troglobitic species showed a weaker photophobia in the night phase of the 12:12 LD cycle; whether these differences between the phototactic behavior in the light and in the night phases are endogenous or due to masking is an open question. On the other hand, *Acanthocobitis* specimens, studied as another balitorid outgroup, did not show a photonegative behavior, indicating a diurnal activity (Trajano and Borowsky, unpubl. data). This is consistent with field observations on epigean balitorids in Thailand (Trajano *et al.* 2002), and points to a possible nocturnal activity as a specialized behavior in *Schistura*. Such preadaptation to the subterranean life could explain the diversity of troglobitic *Schistura* species.

II.5   Dorsal Light Reaction

With some exceptions, nearly all fish swim with their back oriented towards the light. Overall, the vertical orientation of the dorso-ventral axis is triggered by the direction of light and gravity. Illumination from different angles relative to an upright position of the test fish result in a deviation of this normal position and can be measured experimentally as the angle of inclination from $0°$ (= vertical orientation).

To date this reaction has been studied in a comparative approach, including surface and cave forms of the same species, in *Astyanax mexicanus* only. Using hybrids between cave and epigean populations as well as the phylogentically young Micos cave population with functional eyes, Langecker (1990) found that the angle of inclination has changed in the cave fish (Fig. 3). The dorsal light reaction is almost completely reduced in the cave fish, and absence of a response did not depend on the degree of eye development (Langecker 1990). Using crossbred fish, the regression was found to be genetically based and to underlie polygenic inheritance involving at least three genetic factors (Langecker 1993)

II.6   Chemical Communication and Alarm Reaction

Chemical signals are good candidates for social communication in the absence of light. In fact, most catfishes are especially dependent on chemical orientation and several studies pointed to their high sensitivity to chemical cues, showing its importance in feeding behavior and social

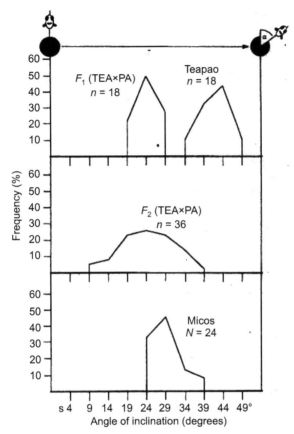

Fig. 3 Dorsal light reaction of *Astyanax mexicanus*. Frequency distribution in the epigean population, F_1 and F_2 hybrids with the blind Pachon Cave fish, and in the Micos Cave fish. All specimens tested had functional eyes (data from Langecker 1990).

communication. Several authors studied the role of chemoreception in the detection at distance of food and conspecific substances in troglobitic vertebrates. Mexican blind tetras, *Astyanax* sp., and cyprinids, *Caecobarbus geertsi* and *Phreatichthys andruzzii*, tested in choice-chamber setup, showed generalized attraction to water conditioned by conspecifics (Trajano 2003). In a food-poor environment with a low density of conspecifics, it would be advantageous for non-aggressive troglobitic fish to approach any stimuli that could signalize the presence of food or potential mates.

The situation is different for aggressive troglobites such as *Pimelodella kronei*, which showed a more complex behavior when compared to *P. transitoria* (Hoenen and Trajano 1995). While all tested eyed catfishes showed consistent repulsion to chemical cues from conspecifics and from blind catfishes as well, individual responses of *P. kronei* greatly varied in accordance with sex and size of the catfish: large males tended to be attracted and small fish (less than 12.0 mm SL) tended to be repelled by conspecifics. This indicates that social interactions are more complex in *P. kronei* than in *P. transitoria* – the latter simply avoid other fish in most situations, whereas blind catfish reactions vary according to the circumstances (Hoenen and Trajano, op. cit.).

The generalized repulsion of *Pimelodella transitoria* to other catfishes is possibly a correlate of its cryptobiotic habits and part of a generalized phobic behavior, including photophobia. It may be regarded as a mechanism allowing individual spacing in order to minimize direct competition for hiding places. If this is characteristic of the genus (or higher groups), then *P. kronei* behavior would represent a specialization to the subterranean life, through an increase in the frequency of positive responses to the presence of conspecifics. As in *Astyanax* sp. and other troglobitic fishes, positive responses to conspecifics cues would enhance the chance of finding food and reproductive partners. However, tendency to aggregate increase the intraspecific competition at food sites and, thus, the probability of agonistic interactions. Therefore, small individuals, whose chance of meeting a large one and being chased away is higher, would tend to avoid conspecifics in order to prevent disadvantageous interactions.

The alarm or fright reaction, i.e., the response to specific alarm substances, appears to be confined to the ostariophysan fishes (Pfeiffer 1963). Alarm pheromones are released from club cells in the epidermis when damaged (e.g., by a predator), and lead to a distinct set of behavioral responses: the alarmed conspecifics react by seeking cover, closer shoaling, rapid swimming or 'freezing' behavior (i.e. they stop moving).

In a detailed comparative study with different populations of epigean and hypogean *Astyanax mexicanus* and their hybrids (Fricke 1988, Fricke and Parzefall 1989), it could be demonstrated that fish from all populations studied stop feeding when *schreckstoff* is introduced into the tank. In the cave fish, some of the behavior patterns found in surface-dwelling populations seem to be reduced: cover seeking on the bottom, rapid swimming (zigzags) and freezing are still found in the cave fish. All individual cave fish tested avoided the site where the alarm substance was released (Fig. 4). In addition, a higher swimming activity was registered. It seems that the alarm behavior in *Astyanax mexicanus* was modified as an adaptation to life in darkness, i.e., in the absence of visual communication: the use of visual signals to inform conspecifics about the predation event is reduced. However, the avoidance reaction *per se* is still existent, even though avian

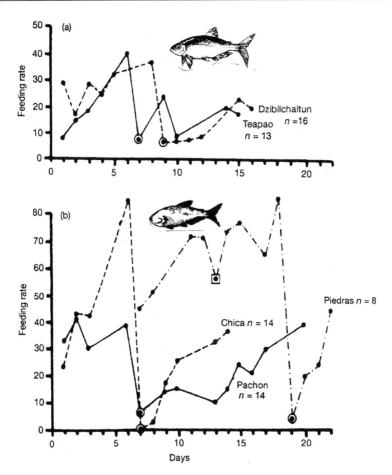

Fig. 4 Feeding rate at the water surface in two epigean and three cave populations of *Astyanax mexicanus*. The dot marks introduction of food, the circle with dot introduction of alarm substance, the square with dot introduction of skin extract of *Pterophyllum scalare* a species without alarm substance (after Fricke 1988).

and piscine predation on cave fishes seems to be an exception (see Plath and Tobler, this volume). The higher shoaling activity can be interpreted as a general response to predation threat (Plath and Schlupp, 2008), and may also lead the alarmed animals away from the predation area.

Up to now it was not possible to release this behavior in the field (Langecker and Fricke, unpubl. data). Thinés and Legrain (1973) described for the blind cave cyprinid *Caecobarbus geertsi* that alarm substance causes foraging behavior to be orientated towards the bottom of the aquarium. It seems tempting to speculate that these cavefish could experience predation from above, namely by fish-eating bats, but this result could also indicate the evolutionary persistence of a now functionless behavior.

II.7 Social Groups

For three species, *Astyanax mexicanus* (Characidae), *Garra barreimiae* (Cyprinidae) and *Poecilia mexicana* (Poeciliidae), social groups have been reported as the common form of social organization in the epigean forms. Such groups of fish that remain together for social reasons are termed by Pitcher (1983) as shoals. Following this definition, the epigean fish of these species are mostly shoaling, and sometimes, when disturbed by predators, they switch to schooling, which includes synchronized moving and escape maneuvers (Parzefall 1979, 1983). Shoaling is strongly reduced in all cave fish examined so far.

A detailed study has been made with the cave population of *P. mexicana* to answer the question whether this lack of shoaling behavior is just due to the absence of visual orientation, or whether a genetically based reduction on the behavior level can be found. Two epigean populations of *P. mexicana* have been compared with the extreme cave form from chamber XIII of the Cueva del Azufre (see Plath and Tobler, this volume) and with their hybrids (Parzefall 1993a). In light, the epigean fish from the Río Teapao (PST) were strongly attracted by a stimulus shoal, but the tendency to follow the shoal of conspecifics is already reduced in the epigean population living near the cave entrance (El Azufre, PSO). This creek has a milky appearance due to suspended sulfur, which may indeed provide some protection from predation. Furthermore, increased food competition in sulfidic habitats may explain this reduction (Plath and Schlupp 2008). The cave fish showed similarly low shoaling (Fig. 5). In both populations there was strikingly more variability than in the epigean fish. However, in darkness no preference for the stimulus shoal has been registered in any population examined. The F_1 hybrids exhibited the same level of shoaling as the epigean parental generation and significantly differed from the cave fish. The F_2 generation differed significantly from the putative epigean ancestor in the mean shoaling tendency, as well as in the high variability. There is only a significant difference in the cavefish backcross from the epigean and the F_1 generation. These data demonstrate a genetically fixed reduction which takes already place in the milky Arroyo del Azufre outside of the cave.

Also in a comparable study in different populations of *Astyanax mexicanus* (Characidae), a genetically based reduction of the schooling/shoaling in the cave forms of that species was found (Parzefall 1993a).

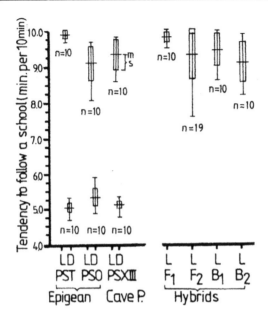

Fig. 5 Tendency to follow a school of subadult *Poecilia mexicana* in min/10 min. Test of single specimen in light (L) and darkness (D) of two epigean, one cave population and hybrids (after Parzefall 1993a).

Timmermann *et al.* (2004) provide another example for the reduction of shoaling behavior during the evolution of a cave fish from a previously surface dwelling species. The surface form of *Garra barreimiae* (Cyprinidae) showed a strong shoaling tendency even in darkness, but the cave form showed only a weak preference to associate with a stimulus shoal.

In the subterranean species studied by Berti and Thinés (1980) and Jankowska and Thinés (1982), shoaling/schooling was also found to be absent: *Caecobarbus geertsi* (Cyprinidae), *Barbobsis devecchi* (Cyprinidae) and *Uegitglanis zammeranoi* (Claridae). All these cave-dwellers show a random spatial distribution in their habitats as well as in aquaria. However, in the latter cases comparative data for the epigean relatives are needed to evaluate if shoaling was indeed reduced. For example, in several catfishes, also the epigean forms do not shoal.

After mapping behavioral character states into a phylogeny proposed for the Heptapteridae, Trajano and Bockmann (1999) hypothesized that the solitary habit observed for the Rhamdiini *Pimelodella kronei* is an autapomorphy related to the cave life, probably as a response to an increased intraspecific competition for food, since the epigean *P. transitoria* catfishes aggregate while feeding. Nevertheless, the same solitary habits exhibited by non-Rhamdiini heptapterids such as the *Rhamdiopsis* spp. from Chapada Diamantina and from Toca do Gonçalo, would be plesiomorphic traits inherited from a small sized, probably pedomorphic, stream-dwelling ancestor. Therefore, for the latter the solitary habits are not an adaptation for the food poor subterranean environment. This well illustrates the need for phylogenetic schemes in order to interpret character states observed in troglobites into a evolutionary context.

II.8   Reproductive Behavior

Cavefishes require behavioral adaptations that allow for effective mate finding and subsequent fertilization in the absence of any visual orientation.

In species living at high population density, males usually have an easy game to find a female. The livebearing poeciliid fish *Poecilia mexicana* lives in high population densities in its cave habitat. Like in other species of "shortfin mollies", already in the epigean form of *P. mexicana* any visual sexual display is lacking (Parzefall 1969, Plath *et al.* 2007). The mating behavior of *P. mexicana* males thus appears preadapted to cave life. The males check conspecific females by nipping at the enlarged genital region, the so-called genital pad (Parzefall 1969). Mature females that are in the receptive stage of their monthly sexual cycle advertise their receptivity chemically by substances emanating from their genital pad (Parzefall 1973). A female accepting a male stops fleeing and allows the male to copulate. Normally bigger males are preferred based on visual cues. Only in the cave fish, females are able to perform mate choice in darkness (Fig. 6), which is probably accomplished by switching from using the visual system to using the lateral line and possibly also chemical cues (Parzefall 2001, Plath *et al.* 2004). For the female choice behavior we found some other differences between cave and epigean populations in *P. mexicana*: only the females from the cave population exhibited a preference for well nourished males even under completely dark conditions (Plath *et al.* 2005). Overall, the sexual activity of males is also reduced in cave living *P. mexicana* (Plath *et al.* 2003). These behavioral changes are discussed in an ecological context by Plath and Tobler (this volume).

Species with a lower population density in the cave habitats like the characid fish *Astyanax mexicanus* (Fraipont 1992, Wilkens 1972) attract conspecifics by chemical signals emitted into the water. In addition, sexually active females mark the substrate by a chemical substance produced in the genital region and restrict their movement to a small area, which may help males localize the females. The swimming activity of the male increases after contact with such a marked substrate or with a female with ripe

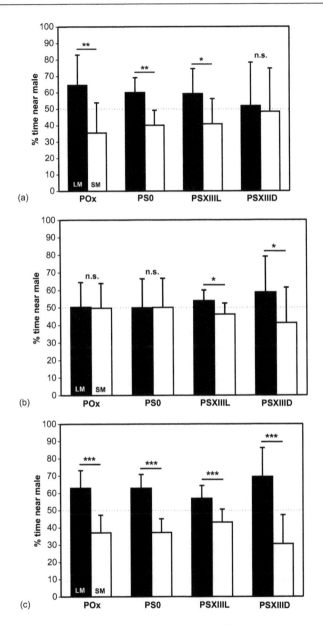

Fig. 6 Choice tests with females of *P. mexicana*. Epigean and cavefishes had the possibility to associate with a large (LM, black bars) or a small male (SM, white bars). Surface population POx, cave entrance population PS0 and cave population, reared in light PSXIIIL, or in darkness PSXIIID, * p < 0,05, ** p < 0,01, n.s. p > 0,05. Paired t-test. a) stimulus males in wire-mesh cylinders in light, b) in darkness, c) in clear Plexiglas cylinders in light (after Plath *et al.* 2004).

oocytes. The male is searching for the sexually active female. Upon finding a female, it swims parallel to the female and tries to achieve contact using fine hooks on his anal fin in order to spawn in the free water (Wilkens 1972). In a comparative study of six cave and one epigean population, the female preference for large male body size in *A. mexicanus* has been tested under different light conditions. Females from the surface form as well as females from the phylogentically young and eyed Micos cave form showed a strong preference for large males. When only non-visual cues were presented in darkness, a preference for large males was found in the case of two of the cave populations (Micos and Yerbaniz populations; Plath *et al.*, 2006 b). This shows that not all cavefishes necessarily switch to using non-visual cues for mate choice, and that the evolution of mating preferences may be different among different cave systems.

As already pointed out, most troglobitic fish are descendents of nocturnal species, thus few modifications are expected to affect their reproductive behavior. Unfortunately, few troglobitic fishes reproduce in laboratory and, with the exception of the above discussed studies, detailed descriptions of reproductive are mostly lacking. Among Brazilian subterranean fish, spontaneous reproduction of three species occurred in laboratory, the trichomycterids *Ituglanis passensis* and *Trichomycterus itacarambiensis*, and the loricariid, *Ancistrus cryptophthalmus* from Passa Três Cave, which bred three times in consecutive years. For the first two, the reproduction was noticed only after the juveniles were independent (but still with non-regressed eyes, supporting the notion that the ontogenetic eye regression is a rule for troglobitic fish), but the whole reproductive events could be recorded for the latter (Secutti and Trajano 2009). No differences in reproductive behavior were observed between *A. cryptophthalmus* and epigean *Ancistrus* catfish, which are typically rocky guarders presenting parental care usually displayed by the male, including territorialism and tending of eggs and newly hatched larvae. These would be preadaptive traits retained by the cave species.

II.9 Agonistic Behavior

Aggressive behavior consists of different patterns: threatening postures and attacks, which may be followed by fights if the attacked individual does not take flight. The threatening animal in general enlarges his body shape (e.g., by spreading its fins), shows striking, dark colors (in surface animals) or sends impressive calls. Then the attacker typically attempts to bite, strike or beat the opponent. To prevent an injury, fights can be ritualized to techniques which allow measuring how strong an opponent is. The weaker partner then has the chance to exhibit submissive behaviors to bring the agonistic interactions to an end without being injured. Both types of behavior patterns are combined under the term agonistic behavior here. Agonistic behavior is used in a variety of contexts. One of them is **competition for food.** Epigean *Astyanax mexicanus* (both males and females) defend small feeding territories of ten to twenty centimeters in diameter, depending on body size, by fin-spreading, snake-swimming and ramming attempts. In escalated fights, biting and circling can be observed. Submission is signaled by a head-up-position. In darkness the epigean fish do not perform these behavior patterns (Burchards *et al.* 1985, Hausberg 1995). In the cave form, territories are smaller, and only signals effective in close body contact are reported: biting, circling and tail-beating have been observed, but no fin-spreading, snake-swimming and ramming. Hausberg (1995) argues that the aggressive behavior of the cave fish is of a new type that has evolved as an adaptation to life in a lightless habitat after the reduction of the visually mediated aggressive behavior shown by the epigean ancestor.

The maintenance, or even enhancement of agonistic interactions, seems to be common for troglobitic fishes. Ercolini *et al.* (1981) studied the aggressive behavior in the clariid catfish, *Uegitglanis zammaranoi*, based on pairing individuals in a resident × intruder scheme. Using the same laboratory method, complemented with observations in the natural habitat, Trajano (1991) compared the agonistic behavior of *P. kronei* and its epigean sister-species, *P. transitoria*. Both *U. zammaranoi* and *P. kronei* present a dominant-subordinate relationship established through aggressive interactions usually starting immediately after unknown catfish are paired and expressed as the possession of the bottom by the dominant fish, which banishes the subordinate to the upper levels of the aquarium. At least in *P. kronei*, the dominance is primarily based on size, previous residence being a secondary determinant. Agonistic interactions were also observed in caves, with smaller catfish being chased away by the larger ones. Likewise, some of the aggressive acts described for *U. zammaranoi* in laboratory were also observed in the wells where the specimens have been captured.

In laboratory, *Pimelodella transitoria* shows more or less the same type of aggressive behavior as *P. kronei* (Fig. 7), but centered on dens. In epigean habitats, however, the eyed catfish are seen sharing hiding places and frequently aggregate during foraging, which is concentrated at dusk and dawn (Gerhard 1999). It is possible that the exclusively solitary habits of *P. kronei* were accompanied by an enhancement of aggressiveness towards conspecifics.

A rich behavioral repertoire was observed for these species, and behavioral components common for *U. zammaranoi* and *P. kronei*, also

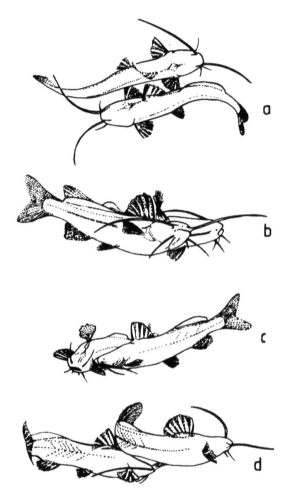

Fig. 7 Aggressive behavior in *Pimelodella kronei*: zigzagging in head-to-tail (a) and head-to-head position (b), ventral shove (c) and fin biting (d) (after Trajano 1991).

reported for epigean catfish such as *Ictalurus* spp., included "zig-zagging", "snout/head, caudal and lateral shoves", "bites", "patrolling", "chasing", "mutual chasing", "flight", "avoidance", "ascent/subordinate swimming". Categories exclusive to the former include "mouth-locking" and "gapping", and to the latter, "arched posture". This would correspond to a general pattern for aggressive siluriforms. A simplified version of this repertoire have been recorded for *Ancistrus cryptophthalmus* catfish, which has been seen defending food sources (edible leaves and commercial food in laboratory, grazing stones in the cave habitat) from conspecifics using lateral strokes, aided by extended opercular spines (odontodes), which are particularly developed in *Ancistrus* catfish; some chasing of smaller individuals by larger ones is also observed, mutual chasing and attempts to thrust from beneath the opponent, corresponding to "ventral shoves" are rare.

Less frequent but more intense agonistic interactions are exhibited by *R. enfurnada* catfish: in laboratory, when two catfish meet, usually the larger one chases away the smaller in a quick pursuit, or the smaller simply avoids the other one. However, in some occasions, intense fights occur, leading to the death of one fish in less than 24 h. The juveniles are more tolerant to each other proximity (Bichuette and Trajano 2005).

Agonistic interactions were observed for all studied balitorids (Trajano and Borowsky 2003), but they were more frequent and intense among the cave species, especially in *N. troglocataractus*, which presented an apparent territorial behavior. The latter are quite intolerant to the proximity of conspecifics, displaying a rich repertoire of agonistic interactions which include lateral and ventral shoves, lateral zig-zagging, bites, pursuits (including circular pursuits) and body elevation (Trajano and Borowsky, unpubl. data).

Competition for mates can result in defending breeding territories that only females are allowed to enter. Another possibility is that the males compete within a social group of conspecifics and eventually establish a rank order, which is often correlated with the physical strength of the group members, such or just their body size. In such a case only the high-ranking male has unlimited access to the females.

In epigean *Poecilia mexicana*, the dominant male responds to an attractive female by following behavior and dark body coloration. In the field, the pair separates from the shoal and the male defends a **territory** around the female, where the male is repeatedly nipping and attempts to copulate. During aggressive encounters with more or less equal sized males, small males adopting a female-like body coloration try to sneak copulations (Parzefall 1973, 1979). Males in the cave population of *P. mexicana* do not defend territories. In laboratory studies with epigean fish and cavefish (which have functional eyes), a quantitative genetically based reduction of the aggressive behavior was demonstrated (Parzefall 1974, 1979, 1993a). The reduction of the aggressive pattern was found to be highly variable within the population. Some of the cavefish tested in light did not perform any aggressive pattern and it seemed that they did not "understand" attacks by other (surface) males. Instead, they were answering aggression by nipping and copulation attempts against the aggressor. Attractive females can be seen followed by several males without any aggression (Parzefall, unpubl. data). There is no other example in cave-dwellers up to now with a quantitative reduction of the whole aggressive repertoire and a reduction of the rank order.

No aggressive interactions clearly recognizable as such have been observed for Brazilian troglobitic trichomycterids but, considering that trichomycterid catfish in general are not aggressive, this is probably a plesiomorphic trait of the troglobitic species. Occasional chasing of smaller individuals by the larger ones was observed in *Rhamdiopsis* sp. from Chapada Diamantina. On the other hand, it is not clear whether chasing in *Rhamdiopsis* sp. from Toca do Gonçalo kept together are agonistic interactions or cannibalistic attempts (Trajano and Bockmann 2000). In any case, the absence of robust comparable data for epigean relatives hampers further evolutionary interpretations of these behaviors.

A case for reduction of aggressive behavior in troglobitic fishes was made by Bechler (1983) for amblyopsids, based on parings in laboratory. The troglophilic *Forbesichthys agassizi* exhibit the whole set of observed agonistic acts, which encompassed both aggressive (tail-beat, head-butt, attack, bite, chase and jaw lock) and submissive (freeze, escape), while the exclusively epigean *C. cornuta* exhibited none. *F. agassizi* repertoire was also present in *Amblyopsis spelaea*, but some components were missing in *Typhlichthys subterraneus* (jaw-lock was not observed) and *A. rosae* (only tail-beat observed among the aggressive acts). It is noteworthy that in species having similar overall repertoires, frequency, duration and intensity of different acts vary between species. Hence, among cave species the higher specialized amblyopsids, *Typhlichthys subterraneus* and *Amblyopsis rosae*, are engaged in simpler, less intense bouts, which were considerably shorter in length. A first step in the process of reduction of agonistic acts could be a decreasing in intensity and duration of the interactions.

Considering *F. agassizi* as a representative of the behavior present in the ancestor(s) of these troglobitic species, Bechler (*op. cit.*) concluded for a reduction of the aggressiveness in *T. subterraneus* and, in a more advanced form, in *A. rosae*, hypothesizing that such regression was correlated with a reduction in the metabolic rates of these species – as the metabolic needs decreased, these fish would be less food-limited, therefore intraspecific competition would decrease. This is a very interesting and compelling hypothesis, although the lack of generally accepted phylogenies for the family (see Niemiller and Poulson, this volume) difficultates any proposal of direction of evolutionary character changes in the troglobitic species.

III. HOW TO EXPLAIN BEHAVIORAL CHANGES IN CAVEFISHES? A GENETIC APPROACH

The studies of changes in the behavior of cave-dwellers reported here reveal that a complex trigger system releases the right behavioral response on the basis of chemical and tactile stimuli. Many animals can therefore survive in complete darkness by using non-visual signals. In some cases the existing behavior like aggression in *Astyanax mexicanus* has changed to a more efficient behavioral system. In all cases these behavioral innovations seem to be derived from behavior patterns already existent in the surface forms, and to date no completely new character has been found in cavefishes.

For the traits presented in this chapter, the question arises whether the behavioral changes can be explained by selection or on the basis of the neutral mutation theory as developed by Kosswig (1948, 1963). Kosswig hypothesized that the absence of stabilizing selection in darkness allows for an accumulation of selectively neutral mutations, leading to increased variance in non-selected traits, such as eye structure. Over time the variability then decreases within a population because of genetic drift. In phylogenetically old cave animals this may lead to homozygosity of alleles that do not allow for the expression of a given trait, and thus, diminished variability for the reduced patterns (Wilkens 1988). Other authors favor selective explanations involving selection for increased metabolic economy (Poulson 1963) or indirect effects of pleiotropy (Barr 1968, Jeffery 2005). A more detailed review of the relevant theories of the evolution of subterranean animals has been given by Culver and Wilkens (2000). There is only a general consensus that differentiation of cave populations, especially with regard to regressive characters, cannot occur when there is substantial gene flow between the cave and surface populations.

It seems that for the reduction of eyes, pigmentation, the circadian clock, the dorsal light reaction, phototactic behavior and – in part – the lack of schooling/shoaling, the neutral mutation theory is a good explanation (but see Plath and Tobler, this volume, for an effect of food shortage on the reduction of shoaling). For other characters presented here, directional selection seems a plausible explanation.

IV. GENERAL CONCLUSIONS

The majority of troglobitic fishes belong to epigean taxa which are active at dusk and/or night, photophobic, cryptobiotic (with strong hiding habits), most of them bottom-dwellers. The analysis of behavioral character states in these troglobitic species into an evolutionary context reveals several apomorphies which can be related to their exclusively subterranean life:

- Weaker cryptobiotic habits, probably as a consequence of the absence of visually oriented predation;
- Increased midwater activity observed in catfishes and balitorids, probably as an adaptation to enhance the chances of finding food, in the absence of ecologically important predators and midwater-active competitors;
- Weaker internal time-control mechanisms at the circadian scale, possibly due to an evolutionary regression affecting the oscillator(s) itself (themselves) or to an uncoupling between the oscillators and at least one of their related functions, such as the locomotor activity;
- Decreased photonegative response to light;
- with some exceptions, tendency towards solitary habits and moderate to high aggressiveness, probably related to a high intraspecific competition for food and potential mates, which may represent the maintenance of the character state already present in the epigean ancestor or a behavioral specialization to the subterranean life.

Part of the above-mentioned behavioral patterns would represent adaptive (constructive) character states, as seems to be the case with the increased midwater activity, the loss of generalized phobic reactions and the enhanced aggressiveness, and part would be regressive traits resulting from evolutionary processes similar to those affecting eyes and melanic pigmentation, as in the case with the weaker cryptobiotic habits, photophobia and internal time-control mechanism, when compared to epigean relatives.

Only a minority has epigean ancestors with diurnal activity. After colonizing subterranean habitats, they developed striking behavioral changes like the reduction of schooling, dorsal light reaction and alarm behavior. For the reproductive behavior, chemical and tactile releasers, already working together with optical ones in the epigean ancestors, assume a important position. It is striking that within these families up to now no species has been found having a optical display which was able to colonize the dark habitat successfully.

Acknowledgements

The first author (JP) would like to thank the DFG (German research board) for financial support since many years, the Mexican government for providing research and collecting permits over many years and M. Hänel (Zoological Institute, Hamburg) for drawing the figures. Many thanks also to Martin Plath for comments on the manuscript. The junior author (ET) is greatly indebted to the continuous financial support from the Fundação de Amparo à Pesquisa do Estado de São – FAPESP and Conselho Nacional de Desenvolvimento Científico e Tecnológico – CNPQ.

References

Barr, T.C. 1968. The ecology and the evolution of troglobites. *Evoutionary Biology* 2: 35-102.

Bechler, D.L. 1983. The evolution of agonistic behavior in amblyopsid fishes. *Behavioral Ecology and Sociobiology* 12: 35-42.

Berti, R. and G. Thinés. 1980. Influences of chemical signals on the topographic orientation of the cavefish *Caecobarbus geertsi* (Pisces, Cyprinidae). *Experientia* 36: 1384-1385.

Bessa, E. and E. Trajano. 2002. Light reaction and cryptobiotic habits in armoured catfishes, genus *Ancistrus*, from caves in Central and Northwest Brazil (Siluriformes: Loricariidae). *Mémoires de Biospéologie* 28: 29-37.

Bichuette, M.E. and E. Trajano. 2005. A new cave species of *Rhamdia* (Siluriformes: Heptaperidae) from Serra do Ramalho, northeastern Brazil, with notes on ecology and behavior. *Neotropical Ichthyology* 3(4): 587-595.

Biswas, J. 1991. Annual modulation of diel activity rhythm of the dusk active loach *Nemacheilus evezardi* (Day): A correlation between day length and circadian parameters. *Proceedings of the Indian National Science Academy* B, 5: 339-346.

Biswas, J., A.K. Pati, K. Pradhan and R.S. Kanoje. 1990. Comparative aspects of reproductive phase dependent adjustments in behavioural circadian rhythms of epigean and hypogean fish. *Comparative Physiology and Ecology* 15(4): 134-139.

Borowsky, R. and E. Trajano. 2003. Biological rhythms in cavefishes from Thailand. In: 2003 Joint Meeting of Ichthyologists and Herpetologists, ASIH.

Burchards, H., A. Dölle and J. Parzefall. 1985. Aggressive behaviour of an epigean population of *Astyanax mexicanus* (Characidae, Pisces) and some observations of three subterranean populations. *Behavioral Processes* 11: 225-235.

Cordiner, S. and E. Morgan. 1987. An endogenous circadian rhythm in the swimming activity of the blind Mexican cavefish. In: *Chronobiology and Chronomedicine*. G. Hildebrandt, R. Moog and F. Raschke (eds.). Verlag Peter Lang, Frankfurt am Main. pp. 177-181.

Culver, D.C. 1982. *Cave Life. Evolution and Ecology*. Harvard University Press, Cambridge, MA, USA and London.

Culver, D.C. and H. Wilkens. 2000. Critical review of the relevant theories of the evolution of subterranean animals. In: *Ecosystems of the World*, Volume 30, *Subterranean Ecosystems*, H. Wilkens, D.C. Culver and W.F. Humphreys (eds.). Elsevier, Amsterdam pp. 381-398.

Erckens, W. and F. Weber. 1976. Rudiments of an ability for time measurement in the cavernicolous fish *Anoptichthys jordani* Hubbs and Innes (Pisces, Characidae). *Experientia* 32: 1297-1299.

Erckens, W. and W. Martin. 1982. Exogenous and endogenous control of swimming activity in *Astyanax mexicanus* (Characidae, Pisces) by direct light response and by a circadian oscillator. II. Features and time controlled behaviour of a cave population and their comparison to an epigean ancestral form. *Zeitschrift für. Naturforschung* C 37: 1266-1273.

Ercolini, A., R. Berti and A. Cianfanelli. 1981. Aggressive behaviour in *Uegitglanis zammaranoi* Gianferrari (Clariid, Siluriformes), anophthalmic phreatic fish from Somalia. *Monitore Zoologico Italiano* n.s. 14: 39-56.

Espinasa, L, P. Rivas-Manzano and H. Espinosa-Perez. 2001. A new blind cave fish population of genus *Astyanax*: Geography, morphology and behavior. *Environmental Biology of Fishes* 62: 339-344.

de Fraipont, M. 1992. Réponse d'*Astyanax mexicanus* aux stimulation chimiques provenant de groupes de congénères à différents stades du développement. *Mémoires de Biospéologie* 19: 209-214.

Fricke, D. 1988. Reaction to alarm substance in cave populations of *Astyanax fasciatus* (Characidae, Pisces). *Ethology* 76: 305-308.

Fricke, D. and J. Parzefall. 1989. Alarm reaction, aggression and schooling in cave and river populations of *Astyanax fasciatus* and their hybrids. *Mémoires de Biospéologie* 26: 177-182.

Gerhard, P. 1999. Ecologia de populações e comportamento de quatro espécies de bagres Heptapterinae (Teleostei: Siluriformes) em riachos do Alto Vale do rio Ribeira (Iporanga, São Paulo). M.Sc. Thesis, Instituto de Biociências, São Paulo.

Hausberg, C. 1995. Das Aggressionsverhalten von *Astyanax fasciatus* (Characidae, Teleostei): Zur Ontogenie, Genetik und Evolution der epigäischen und hypogäischen Form. Ph.D. thesis, University Hamburg.

Heuts, M.J. 1951. Ecology, variation and adaptation of the blind fish *Caecobarbus geertsi* Blgr. *Annales Société Royale Zoologique Belgique:* pp. 155-230.

Heuts, M.J. 1953. Regressive evolution in cave animals. *Symposium Society of Experimental Biology* 7: 290-309.

Hoenen, S.M.M. and E. Trajano. 1995. Locomotor responses of *Pimelodella* spp. from Southeastern Brazil to chemical signals of other individuals (Teleostei: Siluriformes: Pimelodidae). *Mémoires de Biospéologie* 22: 57-63.

Hüppop, K. 1987. Food finding ability in cave fish (*Astyanax fasciatus*). *International Journal of Speleology* 16: 59-66.

Hüppop, K. 1988. *Phänomene und Bedeutung der Energieersparnis bei dem Höhlenfisch Astyanax fasciatus* (Characidae). Ph.D. thesis, University Hamburg.

Hüppop, K. and H. Wirkens. 1991. Bigger eggs in subterranean *Astyanax fasciatus* (Characidae, Pisces). *Zeitschrift für Zoologische Systematik und Evolutionsforschung* 29: 280-288.

Jankowska, M. and G. Thinés. 1982. Etude comparative de la densité de groupes de poissons cavernicoles et épigés (Characidae, Cyprinidae, Claridae). *Behavioral Processes* 7: 281-294.

Jeffery, W.R. 2005. Evolution of eye degeneration in cavefish: the return of pleiotropy. *Subterranean Biology* 3: 1-11.

Junge, P. 1993. Egg size and larval development of epigean and cave forms of *Rhamdia guatemalensis* (Pimelodidae, Pisces). In: *Trends in Ichthyology*. H. Schröder, J. Bauer and M. Schartl (eds.). GSF-Bericht 7/92, pp. 165-173.

Kosswig, C. 1948. Genetische Beiträge zur Präadaptationstheorie. *Review of Faculty of Science (Istanbul) Series* B 5: 176-209.

Kosswig, C. 1963. Genetische Analyse konstruktiver und degenerative Evolutionsprozesse. *Zeitschrift für Zoologische Systematik und Evolutionsforschung* 1: 205-239.

Lamprecht, G. and F. Weber. 1992. Spontaneous locomotion behavior in cavernicolous animals: The regression of the endogenous circadian system. In: *The Natural History of Biospeleology*, A.I. Camacho (ed.). Monografias Museo., Nacional de Ciencias Naturales, pp. 295-326.

Langecker, T.G. 1989. Studies on the light reaction of epigean and cave populations of *Astyanax fasciatus* (Characidae, Pisces). *Mémoires de Biospéologie* 16: 169-176.

Langecker, T.G. 1990. *Der Einfluss des Lichts bei der Evolution von Höhlenfischen*. Ph.D. thesis, University Hamburg.

Langecker, T.G. 1992. Light sensitivity of cave vertebrates. Behavioral and morphological aspects. In: *The Natural History of Biospeleology*, A.I. Camacho (ed.). Monografias Museo Nacional de Ciencias Naturales, pp. 295-326.

Langecker, T.G. 1993. Genetic analysis of the dorsal light reaction in epigean and cave dwelling *Astyanax fasciatus* (Teleostei, Characidae). *Ethology, Ecology and Evolution* 3: 357-346.

Langecker, T.G. and G. Longley. 1993. Morphological adaptations of the Texas blind catfishes *Trogloglanis pattersoni* and *Satan eurystomus* (Siluriformes, Ictaluridae) to their underground environment. *Copeia* 1993: 976-986.

Langecker, T.G., H. Wilkens and J. Parzefall. 1996. Studies on the trophic structure of an energy-rich Mexican cave (Cueva del Azufre) containing sulphurous water. *Mémoires de Biospéologie* 23: 121-127.

Niemiller, M.L. and T.L. Poulson. Subterranean fishes of North America: Amblyopsidae. This volume.

Parzefall, J. 1969. Zur vergleichenden Ethologie verschiedener Mollienesia- Arten einschließlich einer Höhlenform von *Mollienesia sphenops*. *Behaviour* 33: 1-37.

Parzefall, J. 1973. Attraction and sexual cycle of Poeciliidae. In: *Genetics and Mutagenesis of Fish*, J.H. Schröder (ed.). Springer, Berlin, pp. 177-183.

Parzefall, J. 1974. Rückbildung aggressiver Verhaltensweisen bei einer Höhlenform von *Poecilia sphenops* (Pisces, Poeciliidae). *Zeitschrift für Tierpsychologie* 35: 66-84.

Parzefall, J. 1979. Zur Genetik und biologischen Bedeutung des Aggressions- verhaltens von *Poecilia sphenops* (Pisces, Poecilidae). Untersuchungen an

Bastarden ober- und unterirdisch lebenden Populationen. *Zeitschrift für Tierpsychologie* 50: 399-422.

Parzefall, J. 1983. Field observation in epigean and cave population of the Mexican characid *Astyanax mexicanus* (Pisces, Characidae). *Mémoires de Biospéologie* 10: 171-176.

Parzefall, J. 1993a. Schooling behaviour in population—hybrids of Astyanax fasciatus and Poecilia mexicana (Pisces, Characidae, Poeciliidae). In: Trends in Ichthyology. H. Schröder, J. Bauer and M. Schartl (eds.). GSF-Bericht 7/92, pp. 297-303.

Parzefall, J. 1993b. Behavioural ecology of cave dwelling fishes. In: *The Behaviour of Teleost Fishes*. T.J. Pitcher (ed.). Chapman and Hall, London, pp. 573-606.

Parzefall, J. 1998. Behavioural adaptation of cavefishes. In: *Encyclopaedia Biospeologica*, v. II, C. Juberthie and V. Decu (eds.). Société de Biospéologie, Moulis, pp. 1191-1200.

Parzefall, J. 2001. A review of morphological and behavioural changes in the cave molly, *Poecilia mexicana*, from Tabasco, Mexico. *Environmental Biology of Fishes* 62: 263-275.

Parzefall, J., C. Kraus, M. Tobler and M. Plath. 2007. Photophilic behaviour in surface- and cave-dwelling Atlantic mollies, *Poecilia mexicana* (Poeciliidae). *Journal of Fish Biology* 71: 1225-1231.

Pati, A. 2001. Temporal organization in locomotor activity of hypogean loach, *Nemacheilus evezardi,* and its epigean ancestor. *Environmental Biology of Fishes* 62: 119-129.

Peiffer, W. 1963. Vergleichende Untersuchung über die Schreckreaktion und den Schreckstoff der Ostariophysen. *Zeitschrift für Vergleichende Physiologie* 47: 111-147.

Pitcher, T.J. 1983. Heuristic definitions of shoaling behavior. *Animal Behaviour* 31: 611-613.

Plath, M. and I. Schlupp. 2008. Parallel evolution leads to reduced shoaling behavior in two cave dwelling populations of Atlantic mollies (*Poecilia mexicana*, Poeciliidae, Teleostei). *Environmental Biology of Fishes* (online first).

Plath, M., J. Parzefall and I. Schlupp. 2003. The role of sexual harassment in cave-and-surface dwelling populations of the Atlantic molly, Poecilia mexicana (Poeciliidae, Teleostei). Behavior Ecology and Sociobiology 54: 303-309.

Plath, M., J. Parzefall, K.E. Körner and I. Schlupp. 2004. Sexual selection in darkness? Female mating preferences in surface- and cave-dwelling Atlantic mollies, *Poecilia mexicana* (Poeciliidae, Teleostei). *Behavior Ecology and Sociobiology* 55: 596-601.

Plath, M., K.U. Heubel, F.J. Garcia de Leon and I. Schlupp. 2005. Cave molly females (*Poecilia mexicana*, Poeciliidae, Teleostei) like well-fed males. *Behavior Ecology and Sociobiology* 58: 144-151.

Plath, M., U. Seggel, H. Burmeister, K.U. Heubel and I. Schlupp. 2006a. Choosy males from underground: male mating preferences in surface- and cave-dwelling Atlantic mollies (*Poecilia mexicana*). *Naturwissenschaften* 93: 103-109.

Plath, M., M. Rhode, T. Schröder, A. Taebel-Hellwig and I. Schlupp I. 2006b. Female mating preferences in blind cave tetras *Astyanax fasciatus* (Characidae, Teleostei). *Behaviour* 143: 15-32.

Plath, M., I. Schlupp, J. Parzefall and R. Riesch. 2007. Female choice for large body size in the cave molly, *Poecilia mexicana* (Poeciliidae, Teleostei): influence of species- and sex-specific cues. *Behaviour* 144: 1147-1160.

Plath, M. and M. Tobler. Subterranean fishes of Mexico (*Poecilia mexicana, Poeciliidae*), this volume.

Poulson, T.L. 1963. Cave adaptation in amplyopsid fishes. *American Midland Naturalist* 70: 257-290.

Poulson, T.L. 1985. Evolutionary reduction by neutral mutations: plausibility arguments and data from amblyopsid fishes and lyniphiid spiders. *NSS Bulletin* 47(2): 109-117.

Poulson, T.L. and W.B. White. 1969. The cave environment. *Science* 165: 971-981. Proudlove G. Biodiversity and distribution of the world.

Pradhan, R.K., A.K. Pati and S.M. Agrawal. 1989. Meal scheduling modulation of circadian rhythm of phototactic behaviour in cave dwelling fish. *Chronobiology International* 6: 245-249.

Reis, R.E., E. Trajano and E. Hingst-Zaher. 2006. Shape variation in surface and cave populations of the armoured catfish *Ancistrus* (Siluriformes: Loricariidae) from the São Domingos karst area, Upper Tocantins River, Brazil. *Journal of Fish Biology* 68: 414-429.

Secutti, S. and E. Trajano. 2009. Reproductive behavior, development and eye regression in the cave armored catfish, *Ancistrus cryptophthalmus* Reis, 1987 (Siluriformes: Loricariidae), breed in laboratory *Neotropical Ichthyology* 7(3): 479-490.

Schemmel, C.H. 1980. Studies on the genetics of feeding behaviour in the cave fish *Astyanax fasciatus* f. anoptichthys. *Zeitschrift für Tierpsychologie* 53: 9-22.

Thinés, G. and J.M. Legrain. 1973. Effects de la substance d'alarme sur le comportement de poissons cavernicoles *Anoptichthys jordani* (Characidae) et *Caecobarbus geertsi* (Cyprinidae). *Annales de Spéléologie* 28: 271-282.

Timmermann, M., I. Schlupp and M. Plath. 2004. Shoaling behaviour in a surface- and a cave-dwelling population of a barb, *Garra barreimiae* (Cyprinidae, Teleostei). *Acta Ethologica* 7: 59-64.

Trajano, E. 1989. Estudo do comportamento espontâneo e alimentar e da dieta do bagre cavernícola, *Pimelodella kronei*, e seu provável ancestral epígeo, *Pimelodella transitoria* (Siluriformes, Pimelodidae). *Revista Brasileira de Biologia* 49: 757-769.

Trajano, E. 1991. The agonistic behaviour of *Pimelodella kronei*, a troglobitic catfish from Southeastern Brazil (Siluriformes, Pimelodidae). *Behavioural Processes* 23: 113-124.

Trajano, E. 2003. Ecology and ethology of subterranean fishes. In: *Catfishes*, V. 2. G. Arratia, B.G. Kapoor, M. Chardon and R. Diogo (eds.). Science Publs., Enfield, 812 p.

Trajano, E. 2005. Evolution of lineages. In: *The Encyclopedia of Caves*. D.C. Culver and W.B. White (eds.). Elsevier Academic Press, San Diego, pp. 230-234.

Trajano, E. and L. Menna-Barreto. 1995. Locomotor activity pattern of Brazilian cave catfishes under constant darkness (Siluriformes, Pimelodidae). *Biological Rhythm Research* 26(3): 341-354.

Trajano, E. and L. Menna-Barreto. 1996. Free-running locomotor activity rhythms in cave-dwelling catfishes, *Trichomycetus* sp., from Brazil (Teleostei, Siluriformes). *Biological Rhythm Research* 27(3): 329-335.

Trajano, E. and F.A. Bockmann. 1999. Evolution of ecology and behaviour in Brazilian cave Heptapterinae catfishes, based on cladistic analysis (Teleostei: Siluriformes). *Mémoires de Biospéologie* 26: 123-129.

Trajano, E. and F.A. Bockmann. 2000a. Ecology and behaviour of a new cave catfish, genus *Taunayia*, from northeastern Brazil (Siluriformes, Heptapterinae). *Ichthyological Explorations of Freshwaters* 11: 207-216.

Trajano, E. and L. Menna-Barreto. 2000b. Locomotor activity rhythms in cave catfishes, genus *Taunayia*, from eastern Brazil (Teleostei: Siluriformes: Heptapterinae). *Biological Rhythm Research* 31(4): 469-480.

Trajano, E. and R. Borowsky. 2003. Behavior of cave fishes from Thailand (Cypriniformes: Balitoridae, Cyprinidae). In: *2003 Joint Meeting of Ichthyologists and Herpetologists*, ASIH.

Trajano, E., L. Duarte and L. Menna-Barreto. 2005. Locomotor activity rhythms in cave fishes from Chapada Diamantina, northeastern Brazil (Teleostei: Siluriformes). *Biological Rhythm Research* 36(3): 229-236.

Trajano, E., N. Mugue, J. Krejca, C. Vidthayanon, D. Smart and R. Borowsky. 2002. Habitat, distribution, ecology and behavior of cave balitorids from Thailand (Teleostei: Cypriniformes). *Ichthyological Explorations of Freshwaters* 13(2): 169-184.

Volpato, G.L. and E. Trajano. 2006. Biological rhythms. In: *Fish Physiology*, Volume 21. *The Physiology of Tropical Fishes*, A.L. Val, V.M. Val and D.J. Randall (eds.). Elsevier, San Diego, pp. 101-153.

Weber, A., G.S. Proudlove and T.T. Nalbant. 1998. Morphology, systematic diversity, distribution, and ecology of stygobitic fishes. In: *Encyclopaedia Biospeologica*, v. II, C. Juberthie and V. Decu (eds.). Société de Biospéologie, Moulis, pp. 1179-1190.

Wilkens, H. 1972. Über Präadaptationen für das Höhlenleben, untersucht am Laichverhalten ober- und unterirdischer Populationen des *Astyanax mexicanus* (Pisces). *Zoologischer Anzeiger* 188: 1-11.

Wilkens, H. 1988. Evolution and genetics of epigean and cave *Astyanax fasciatus* (Characidae, Pisces). Support for the neutral theory. *Evolutionary Biology* 23: 271-367.

The Evolutionary Genetics of Cave Fishes: Convergence, Adaptation and Pleiotropy

Richard Borowsky

Department of Biology, New York University, NY 10003, USA
E-mail: rborowsky@nyu.edu

ABSTRACT

Cave animals have converged repeatedly on a troglomorphic suite of characters including regression of eyes and pigmentation, augmentation of other senses, and changes in morphology, metabolism and behavior. Convergences are evolution's "experimental replicates" and their study permits the testing of hypotheses about the evolutionary process. The genetic bases for these changes are the focus of this review. We cover the Mendelian, quantitative and molecular genetics of troglomorphic evolution, and the population genetics of cave forms, with an emphasis on the Mexican tetra, *Astyanax mexicanus*. Molecular data is useful for reconstructing the relationships of cave populations, but gene flow into and out of cave populations can obscure relationships. We examine the roles of selection and drift in troglomorphic evolution and conclude that both processes are important. Direct selection against eyes in the cave probably reflects energy conservation and the high metabolic cost of maintaining the retina. Indirect selection through pleiotropy is probably a major factor in all evolutionary adaptation to new environments, including the cave.

INTRODUCTION

The greatest tool of modern science is the replicated experiment, which permits the testing and informed revision of hypotheses. Evolutionary biology, however, is generally an observational science and most often lacks this tool to test hypotheses about the evolutionary process. As detailed by other authors in this volume and elsewhere, the typical evolutionary responses to cave life, regardless of taxonomic group, are the reductions of eyes and pigmentation, the augmentation of other compensatory senses, and behavioral and metabolic changes relating to reduced food supply. Thus, cave animals converge on a troglomorphic suite of characters, and it is this convergence that makes them uncommonly valuable for the study of evolutionary processes.

Figure 1 illustrates an example of this convergence, showing three individual *Astyanax mexicanus* (Mexican tetra; Characidae), one from a surface population and two from independently derived cave populations. The two examples of cave fish are clearly convergent in reduction of eyes and melanin pigmentation. Mutations in how many genes are involved in these changes? Are they the same genes in the two different populations? Are the mutations in coding or regulatory sequences? Can these examples of evolution in response to a defined ecological change tell us something general about the evolutionary process?

The cave fish system is a powerful one to approach such questions because of convergence, which brings to evolutionary biology much of the power of the replicated experiment. In effect, each cave species is a replicate of an experiment that asks what happens when you turn out the lights. Given replication, hypotheses formed by the study of one cave species can be tested through their predictions by using other cave species. Unlike parallel evolution, where several populations evolve towards the same phenotype having started with similar or identical genetics (Schluter *et al.* 2004), the ancestral population genetic states of convergent populations may be completely uncorrelated. Thus, converging replicates can be completely independent and permit us to learn about the different genetic ways that evolution can bring about phenotypic change. The power of this approach in revealing the genetic changes that underlie multitrait evolution is one of the main points of this chapter.

This chapter will touch on each of the three distinct but related areas of evolutionary genetics as they relate to cave animals: the genetic changes underlying phenotypic macroevolution, how ecology influences population genetics, and the use of genetic information to infer the phylogenetic relationships among populations or species. Most examples will come from the well studied *Astyanax mexicanus* system, but information from other cave animals will be added, where appropriate. We begin with a discussion of what is known of the genetic bases of troglomorphic traits, move on to discuss the population genetic and phylogeographic implications, and finally what these findings reveal about troglomorphic evolution, and evolution in general.

Fig. 1 Individuals of *Astyanax mexicanus* from a surface population and from the Molino and Pachón cave populations. The cave populations independently converged on eye reduction, albinism, and decreased numbers of melanophores. *Color image of this figure appears in the color plate section at the end of the book.*

THE GENETIC BASIS OF CAVE FISH EVOLUTION AND ADAPTATION

Sadoglu was the first to study the genetics of troglomorphy in *A. mexicanus* by crossing cave (in this paper, "cave" forms are troglobites) and surface forms to obtain F_1 and F_2 generations (Sadoglu 1956, 1957b). For most characters that differ between cave and surface *A. mexicanus*, the F_1 is intermediate between the two parents and the F_2 exhibits a range of variation. Some traits, however, exhibit unifactorial inheritance. The F_2 progeny (n = 1065) derived from a cross between an albino Pachón cave fish and a surface fish exhibited a 3:1 ratio of melanic ($ab^+/\_$) to albino individuals (ab/ab), demonstrating that albinism in this population was inherited as a simple recessive (Sadoglu 1957a). All melanic individuals at seven days post-fertilization had black eyes while all albinos had pink eyes. Thus, the same gene affecting melanin formation in the melanophores also controlled melanin production in the retinal pigmented layer of eye (RPE). The melanic fish exhibited a full range of pigmentation from almost no melanophores in the skin to as many as in surface fish. Both classes had individuals exhibiting a broad range of eye sizes from greatly reduced to ones as large as in surface fish.

A second locus ("brown") that affected the amount of melanin produced was also documented by classic genetics (Sadoglu and Mckee 1969). Progeny of cave/surface crosses that were able to make melanin ($ab^+/\_$) either had black or brown eye color as seven day larvae, according to their genotype at this second locus ($bw^+/\_$ or bw/bw, respectively). Thus, the presence or absence of melanin as well as its quantity or state were both shown to be under control of single unlinked loci, while control of melanophore numbers and eye size were demonstrably multifactorial.

The locus responsible for albinism in the Pachón and Molino populations was later shown to be *oca2* (Protas *et al.* 2006). The loss-of-function mutations in both cave populations are deletions of whole or major portions of exons in *oca2*, but interestingly, the mutational sites in the two populations are different. Thus, albinism in the two populations was achieved independently, although the same gene is involved. Albinism is also known from the Japonés, Yerbaniz, Sabinos, and Curva populations (Sadoglu 1957a, Wilkens and Strecker 2003, Protas *et al.* 2006, unpublished observations, R. Borowsky). The Japonés albino gene does not complement the Pachón albino mutation, suggesting its basis is also loss of *oca2* function, but the *oca2* coding sequence in the Japonés population is intact and the basis for albinism in the population remains unclear (Protas *et al.* 2006). The molecular bases of albinism in Sabinos, Yerbaniz and Curva populations are unknown. Interestingly, *oca2* is the locus responsible for ocular and cutaneous albinism, the most common form of albinism in humans (King *et al.* 2001).

The frequency of the albino allele may have changed in the Pachón population in recent times. Sadoglu's wild stock from Pachón was obtained prior to 1956. She reports the genotypes of six fish used in three crosses as five heterozygotes (ab/ab^+) and one homozygote (ab/ab) and in the discussion it is implied that the six were a random sample from her stock (Sadoglu and Mckee 1969). The frequency of the wild type allele in this sample was 0.417. In contrast, in a recent collection (2008) of cave fish from Pachón only four out of 69 individuals were melanic, giving an estimate of the frequency of the wild type allele of 0.029 (unpublished observations, R. Borowsky).

Breeding and complementation studies have shown that the brown mutation is present in the Pachón, Yerbaniz, Piedras, Sabinos, and Curva caves, but not the Molino cave population (Sadoglu and Mckee 1969, Wilkens and Strecker 2003). In Pachón, it is at high frequency (Sadoglu and Mckee 1969). The gene responsible for the phenotype is as yet unidentified.

Studies of the polarities of quantitative trait loci (QTL) suggest that eyes are selected against in caves and that this selection is direct (Protas *et al.* 2007). Selection may also operate against the production of melanin in cave animals, but it might be indirect, as a consequence of the loss or alteration of melanin in the eye. Melanin is an important constituent of the retinal pigmented epithelium (RPE), which is in intimate contact with the photoreceptor layer of the retina. Changing or eliminating melanin may alter the ability of the RPE to interact properly with the photoreceptors and alter eye structure. Thus, *oca2* or *bw* alleles affecting melanin production or structure may be selected for in the cave environment through negative pleiotropic affects on eye development. This potential pleiotropic pathway would be independent of genes affecting the number or distribution of melanophores.

In contrast to the genes that affect melanin production, other troglomorphic traits studied in this species are under multigenic control; the F_2 of cave/surface crosses exhibit broad variation in trait values because of independent assortment of cave and surface alleles during meiosis in the F_1. Wilkens estimated the minimum numbers of genes affecting four troglomorphic traits in several cave populations using Lande's modification of the Castle/Wright quantitative genetic method (Lande 1981, Wilkens 1988). He determined that six to eight genes controlled eye size differences between cave and surface populations, two to three controlled differences in melanophore numbers, two controlled differences in feeding behavior, and 11 to 12 controlled differences in numbers of taste buds. These estimates are based on the assumptions of full additivity and equal effects of the genes and, thus, are underestimates of the true number of genes involved. Wilkens' work clearly established that many troglomorphic traits in *A. mexicanus* have a multigenic basis and evolved because of changes at numerous loci. Presumably, this is generally true of troglomorphic vertebrates.

Quantitative trait locus analyses of both the Molino and Pachón populations confirmed that multiple unlinked genetic loci controlled phenotypic differences between cave and surface fish for many traits (Borowsky and Wilkens 2002, Protas *et al.* 2006, 2007, 2008). These include the classic regressive traits of eye size and number of melanophores, other traits that are presumably adaptive, like condition factor, olfactory sensitivity to dissolved amino acids, and numbers of maxillary teeth, and those of unknown significance such as depth of the caudal peduncle, skull bone structure and number of anal fin rays. Differences between cave and surface populations for these traits are controlled by between two and thirteen identifiable QTL, and they are more numerous than the earlier estimates from quantitative genetics. Thus, different components of the suite of troglomorphic traits range from unifactorial to greatly multifactorial.

Hybrids between cave populations of *A. mexicanus*, especially geographically distant ones, exhibit some recovery of wild type phenotype compared to their parents. Eye rudiments, for example, are larger in F_1 hybrids between Sabinos and Pachón caves than they are in the parental stocks (Wilkens 1971). This recovery of phenotype is exhibited in numbers of melanophores as well (Wilkens and Strecker 2003), and derives from complementation, reflecting the fact that the genes involved in character change differ among populations. Hybridization between individuals from Molino and Tinaja cave produced F_1 with recovery of the eye so complete that up to 40% of them had visual function as fry (Borowsky 2008). Complementation analyses revealed that the loci involved in eye loss differ nearly completely among Molino, Pachón and Tinaja caves (Borowsky 2008). Thus, evolution's palette is enormously varied and numerous genes are available to effect convergent trait changes.

GENETIC DIVERSITY IN CAVE FISH POPULATIONS

Populations of troglobites generally have lower genetic variability than those of related surface species, but exceptions to the generalization exist, especially among cave invertebrates (Culver 1982). For fishes, however, the pattern of lower diversity in cave populations appears consistent.

Three studies of allozyme variation in amblyopsids, characids, and trichomycterids directly compared related troglobitic, troglophilic and epigean forms and showed that the troglobitic populations were consistently less variable, averaging heterozygosity levels of about 0.01, one sixth the average in surface populations (Avise and Selander 1972, Swofford *et al.* 1980, Perez and Moodie 1993). Nucleotide diversity (π) was estimated using RAPD technology (Borowsky 2001) in four cave and eight surface species of balitorid fishes from Thailand (Borowsky and Vidthayanon 2001). The cave species had significantly lower π values, averaging about 20% of those found in surface species. Thus, in four different families, estimates of nuclear DNA diversity using a variety of methods revealed a consistent pattern of reduced variability in cave fishes.

Astyanax mexicanus is particularly interesting in this regard because it has multiple cave populations, some of independent origin. Thus, it allows for replicate measures of the effects of cave occupation on population genetic diversity. Comparisons of mitochondrial gene diversity between multiple surface and eight cave populations of *A. mexicanus* showed that the majority of the cave populations had fewer haplotypes and lower nucleotide diversities than surface populations (Dowling *et al.* 2002, Strecker *et al.* 2003, 2004). There were a few exceptions: some cave populations had diversities as high as in surface populations and the two studied genes, cytochrome b and ND2, did not reveal consistent patterns for the same cave populations. In general, however, the cave populations with higher diversity were those known to receive eyed surface conspecifics, such as Rio Subterráneo and Cueva Chica, or receive waters from arroyos supporting surface populations, such as Tinaja and Sabinos. Isolated, perched populations, such as Molino, Pachón and Curva had no detectable mtDNA diversity. A study of nuclear DNA diversity in seven cave populations of *A. mexicanus* and eight surface populations using both RAPDs and microsatellite typing again showed lower variability in cave than in surface populations (Panaram and Borowsky 2005).

There are three main determinants of genetic variability in isolated populations: effective population size, fluctuations in size (bottleneck effects), and mutation rate (Kimura and Ohta 1971). In general, effective population size in cave species is probably significantly smaller than in surface species because of limitations in extent of available habitat and food supply, and the species' restricted geographic range. Population sizes of balitorid cave species in Thailand were estimated by visual census and estimates of suitable habitat, and were generally on the order of 10^2 to 10^3 (Trajano *et al.* 2002). Mark-release-recapture of *A. mexicanus* in Pachón cave led to estimates of a total population size on the order of 10^3 to 10^4 (Mitchell *et al.* 1977); although this estimate appears large, it is dwarfed by the numbers of the same species in the surface rivers (personal observation). Bottlenecks may also play a significant role in reducing variability in cave populations, but more likely these would be episodic in nature, as opposed to an initial founder effect supposed by some to be significant in the origin of cave forms (Barr 1968). Differences in generation time between surface and cave populations or other factors might lead to differences in mutation rates, but it remains to be tested whether such factors are significant in reducing cave fish genetic diversity.

In *A. mexicanus*, the issue of genetic diversity is complicated by gene flow between cave and surface populations. Populations of *A. mexicanus* in caves known to receive migrants from the surface populations were more genetically variable than those in caves isolated from the surface, suggesting that they had received measurable gene flow from epigean populations (Panaram and Borowsky 2005). The same study showed through analyses of microsatellite variation that gene flow between cave and surface populations was bidirectional, with a greater flow of cave alleles to surface populations than of surface to cave. These results are significant for two reasons: First, gene flow from surface to subterranean populations greatly increases their evolutionary potential by seeding them with the genetic variation needed to evolve. Even a modest flow of genes into a cave population from a much larger pool of surface fish would introduce variants at a rate far exceeding that from mutation alone. Second, the opportunity for genes to leave cave populations and return to the surface means that evolutionary novelties evolved in cave populations might have futures independent of their source populations.

As a single, although perhaps atypical, example of influx of genetic material into cave populations from the surface, the Pachón cave mitochondrion was derived from recent surface populations and is distinct from those of more closely related cave populations (Dowling *et al.* 2002, Strecker *et al.* 2003, 2004). No example of the ancestral cave haplotypes has been detected in the Pachón population, although they may exist at low frequency. Thus, it appears that the surface haplotype was fixed, or nearly so, by a selective sweep. In contrast, the nuclear markers from Pachón show clear affinities to other cave populations. One possible explanation for this state is that early evolutionary changes in Pachón ancestors in nuclear genes related to metabolic efficiency created a genetic background within which the surface mitochondrial haplotype was more efficient than the cave haplotype. Surface migrants into the cave population could have then introduced nuclear and mitochondrial genes. Selection against surface nuclear genes and for the surface mitochondrion would have led to the present day situation. The phylogeography of cave populations and the placement of Pachón are discussed below.

How can cave genes enter into surface populations? Cave fish have little chance of survival in surface waters and less chance of successful reproduction. The example of Cueva Chica suggests the answer. Chica's entrance is an insurgence, but one that receives runoff from an intermittent arroyo, which contains no fish. The first pools encountered upon entering the cave typically have only troglomorphic fish and it is only deeper in the cave that hybrids between troglomorphic and surface fish are found. Eyed fish are typically found in the last pool, which sumps (Breder 1943). There is general agreement that eyed fish enter the cave at its deepest pool, confluent with the aquifer and at the same level as the nearby Rio Tampaon (Mitchell *et al.* 1977). Thus, the aquifer is at least sparsely populated with eyed fish, which could serve as a conduit for genes moving from one location to the next, within the aquifer. Surface fish occupying and alternating between the two habitats at the edges of rivers or the shallow pools (= *tinajas*) that are windows into the aquifers, could bring genes out of cave populations.

In summary, cave fish populations generally have lower genetic diversity than surface populations. It is probable that this reflects their relatively small effective population sizes, resulting either from reduced actual population sizes or from periodic fluctuations in numbers. Effective population sizes of cave populations may be somewhat increased by gene flow among cave populations. The balitorid, *Schistura oedipus*, is known from five caves in NW Thailand. Mitochondrial DNA sequence data reveals isolation by distance among the five populations and suggests interpopulation gene flow (Borowsky and Mertz 2001).

Genetic diversity of cave populations may be significantly increased by migration from surface populations of conspecifics, if they exist. If so, this would enhance the evolvability of cave populations. Genes may leak out of cave populations into populations of surface conspecifics, if they exist, potentially allowing cave derived novelties to outlive the cave populations.

MOLECULAR PHYLOGENETICS OF *ASTYANAX* *MEXICANUS* POPULATIONS

In order to interpret the evolutionary history of cave adaptation in this species and the genetic bases of the striking morphological convergences exhibited by the present day cave populations, we must understand their origins and relationships. How closely are they related to the surface populations and when did they diverge?

The 29 known cave populations of *A. mexicanus* have a geographic range of over 130 km in NE Mexico (Fig. 2). Although most of the populations exhibit similar degrees of troglomorphy, including extreme rudimentation of eyes and pigmentation, there are subtle morphological differences among them that probably make each population unique and discriminable. For example, multivariate statistical analysis of morphometric data from eight of these populations, some geographically proximate, established that every individual studied grouped unambiguously with other individuals from its own population (Mitchell *et al.* 1977). The observation of reliable differentiation among distinct populations that currently experience barriers to gene flow raises legitimate questions of their evolutionary relationships. Was each cave population derived from a separate invasion of subterranean waters or were there fewer invasions, perhaps even one, with subsequent expansion of the range though underlying aquifers? If

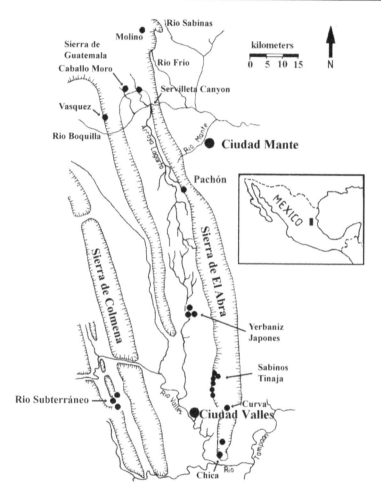

Fig. 2 The Sierra de El Abra region of NE Mexico illustrating the locations of caves and other geographic features mentioned in the text.

there were range expansions, did they occur before the troglomorphic phenotype was attained or was adaptation to subterranean life a necessary step before the species could spread?

Two factors complicate unraveling these phylogenetic relationships: the first is that karst terrain is continuously remodeling, and the present geography and hydrology may differ greatly from that of the past when the cave populations were evolving. As a single example of this, the caves of the Sierra de Guatemala currently have no obvious underground connection with those of the Sierra de El Abra; the ridges are separated by Servilleta Canyon, cut by the Rio Boquilla (Fig. 2). A walk through the canyon, however, affords views of numerous fossil channels of phreatic origin on opposing walls which clearly provided past subterranean connections between what are now the ridges (personal observation). The second complicating factor is that at least eleven of the populations are known to receive immigration from surface populations (Mitchell *et al.* 1977, Langecker *et al.* 1991, Panaram and Borowsky 2005), and there is evidence of bidirectional gene flow between cave and surface populations (Panaram and Borowsky 2005). Thus, information from geography is certainly incomplete and even genetics may be a source of misinformation about phylogeny because of possible lateral transmission of alleles.

Four studies have addressed the phylogenetic relationships of *A. mexicanus* populations using a variety of molecular data including RAPDs, mtDNA sequences and SSCPs, and microsatellite variants. Although the results differed in detail from one to the next and sometimes different cave populations were employed, an overall picture of relationships has emerged. The studies show that there were at least four separate invasions of the underground and that at least two different ancestral stocks of surface *A. mexicanus* gave rise to the present day populations. Subsequent discussion is best followed with reference to Fig. 2, which shows the geographic locations of the Mexican populations.

An early study used RAPDs to examine the relationships of various *A. mexicanus* cave and nearby surface populations (Borowsky and Espinasa 1997). It showed that cave populations of the Sierra de Guatemala (Molino, Vasquez and Caballo Moro) were distinct from those of the El Abra (Pachón and Tinaja), that populations of both of these cave groups were distinct from surface fish, and that the cave fish of Rio Subterráneo clustered with the surface populations and not with the other cave groups. The data did not resolve whether the observed relationships were due to shared ancestry or ongoing gene flow. Jeffery and his coworkers used SSCP and sequence data from the mitochondrial ND2 gene to look at six cave and eleven nearby surface populations (Dowling *et al.* 2002). They found that three of the cave populations, Rio Subterráneo, Cueva Chica and Pachón, clustered strongly with the surface populations while three others, Sabinos, Tinaja and Curva made a separate cluster. The first surprise in their results is that Subterráneo, Chica and Pachón are widely separated geographically and surround the other cave cluster, including Sabinos, Tinaja and Curva (Fig. 2). A second surprising result was the placement of Pachón with surface fish, rather than Tinaja, as in the earlier study.

Strecker *et al.* studied some of the same populations using sequence data from the mitochondrial gene cytochrome b and microsatellite markers and were able to shed light on the apparently discrepant placement of the Pachón population. Cytochrome b data clustered Pachón with Rio Subterráneo, Chica and surface populations and put Sabinos and Tinaja in a separate cluster, as in Dowling *et al.* (2002). In contrast, the microsatellite data clearly placed Pachón with Sabinos and Tinaja, as in the earlier RAPD study. Thus, the difference seems to reflect different histories of mtDNA and nuclear DNA. The authors suggest that Pachón's true affinities are to the other populations of the central El Abra (including Tinaja, Sabinos and Curva) and that Pachón's surface mitochondrial genotype reflects subsequent hybridization and fixation in the population via a selective sweep. They also conclude that the central populations of the El Abra are distinct from the flanking populations (Subterráneo, Chica and Molino) and were derived from an ancestral surface stock inhabiting the region earlier than the present-day stocks, which more recently gave rise to the peripheral populations. Their second, and more extensive study of cytochrome b sequence variation in numerous cave and surface populations from the area and from southern Mexico and Central America (Strecker *et al.* 2004) bears out these conclusions and shows that the present day surface populations of Guatemala and Belize are more closely related to the ancestral stock of the central El Abra cave populations than the surface populations presently surrounding the El Abra.

In sum, these studies suggest that the present day cave populations of *A. mexicanus* come from at least four separate subterranean invasions. The present day populations of both central and northern El Abra originated from the more ancient surface stock. It is not known whether this stock made a single subterranean invasion and then subsequently expanded underground to its present range, from Pachón in the north to Curva in the south, or whether there were multiple invasions. The original surface stock is believed to have gone extinct locally through climate change (Strecker *et al.* 2004). Subsequently, a later invasion of a distinct stock provided ancestors to the present day surface fish and the flanking cave populations to the north (Molino), the south (Chica), and the west (Rio Subterráneo). The geographic separation of the flanking cave populations indicates that they were derived from at least three separate invasions of the underground.

REGRESSIVE EVOLUTION: SELECTION OR DRIFT?

Darwin famously doubted that natural selection played a role in eye loss in cave fishes; "As it is difficult to imagine that eyes, although useless, could be in any way injurious to animals living in darkness, I attribute their loss wholly to disuse" (Darwin 1859). Since then, there has been much discussion and speculation about the roles of selection and drift in the losses of eyes and pigmentation in cave animals [see papers in (Culver 1985) and review in (Culver *et al.* 1995)]. Recent work shows that both drift and selection play important roles in trait regression. Selection may be direct, or indirect, through pleiotropy.

If we focus on any cave associated trait such as eye reduction, we can ask if the derived state came about through selection or genetic drift. We can distinguish between the two possibilities, at least in theory, by determining the polarities of the effects of allelic substitutions at the gene loci involved in determining the trait state. In practice, this is done through QTL analysis, which reveals the approximate genomic positions of the responsible loci. We can define as a positive QTL one in which the homozygote for the cave allele is more troglomorphic than the homozygote for the surface allele, while a negative QTL is the opposite. Orr (1998) observed that QTL analysis of a troglomorphic trait should reveal a consistent positive pattern if the trait evolution was driven by selection but more mixed polarities if the trait evolved through drift.

We compared polarities for eye size (E) and melanophore number (M) QTL in a hybrid cross between Pachón cave and surface fish (Protas *et al.* 2007). QTL affecting eye size were completely consistent in their effects: in every case detected in the Pachón cross, and in others detected subsequently in different crosses (unpublished observations, R. Borowsky), the set of individuals homozygous for the cave allele had smaller eyes on average then the set homozygous for the surface allele (Fig. 3). In contrast, the polarities for M QTL were mixed. Most of the time the cave alleles were associated with decreased numbers of melanophores, but sometimes they were associated with increased numbers of melanophores. The difference between the patterns for E and M was statistically significant and we concluded that eye regression was driven by selection, but that melanophore numbers evolved through drift (Protas *et al.* 2007). By which we mean that melanophore number itself was not subject to selection. It could be that some (or even all) of the allelic substitutions affecting M were driven by selection on pleiotropic characters. While the hypothesis of selection can be tested by experiment or observation, neutrality (drift) is the null hypothesis and cannot be directly tested.

PLEIOTROPY IN MULTI-TRAIT EVOLUTIONARY CHANGE

The term "pleiotropy" has two related but distinct meanings: in the context of functional biology it describes a locus with a gene product participating in two or more developmental pathways (Carroll *et al.* 2005). In the context of quantitative genetics, it refers to the phenomenon in which a single allelic substitution affecting two or more distinct traits (Lynch and Walsh 1998). The second meaning is used here.

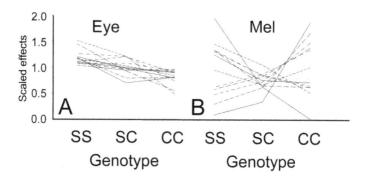

Fig. 3 Scaled phenotypic values for genotypes at QTL affecting eye or lens size (A) or melanophore numbers (B). SS = Surface homozygote; CC = cave homozygote; SC = heterozygote. (From Protas *et al.* 2007.)

Pleiotropy has long been considered a potential mechanism to drive regressive evolution in cave animals through indirect selection (Allee *et al.* 1949, Krekeler 1958, Wright 1964, Barr 1968, Culver 1982, Poulson 1985, Sket 1985, Borowsky and Wilkens 2002, Yamamoto *et al.* 2004) and evidence of its more general importance in multitrait evolution is accumulating (Protas *et al.* 2008). Jeffery (this volume) offers a clear example of a potential source of pleiotropy in *A. mexicanus* regressive evolution: Hedgehog (Hh) gene expression is upregulated during Pachón cave fish development resulting in an expanded midline zone of expression. Increased Hh expression enhances the development of midline structures involved with feeding and gustation, increasing fitness, but at the expense of eye development. Without stabilizing selection to maintain eye structure, the inevitable result of this antagonistic pleiotropy is evolution towards smaller eyes. While the elements of this model have only been tested for the Pachón population and the genetic lesions responsible for eye regression are different in the Tinaja and Molino populations (Borowsky 2008), the proposed mechanism may prove to be an important contributing factor in eye regression in other populations also because numerous genes can affect the Hedgehog pathway.

Other ways in which pleiotropy might come about include signaling systems with multiple signaling molecules and receptors having similar structures and promiscuous binding (Ducrest *et al.* 2008), or through changes in transcription factors with multiple roles (Carroll *et al.* 2005). These are discussed in greater detail, below.

Pleiotropic effects are commonly observed in new mutations (Fitzpatrick 2004) and we should look beyond the narrow question of the degree to which pleiotropy drives regressive evolution and inquire into its more general roles in multi-trait evolution. The *Astyanax* model is a powerful tool to investigate this question because we can identify numerous traits that co-evolved during a defined period of time and also study their genetics.

A key observation is that single trait QTL mapped for numerous traits are significantly clustered in the genome (Protas *et al.* 2008). In addition, more complex analyses cited in the same study detect numerous multitrait QTL. A multitrait QTL is essentially a multivariate phenotype, based on two or more traits, derived from transformation of the initial trait space into a lower dimension in a manner analogous to principal component analysis (Jiang and Zeng 1995, Korol *et al.* 1995, 1998, 2001, Ronin *et al.* 1998, 1999). For example, a model incorporating information on variation in eye size (E), melanophore number (M) and maxillary tooth count (T), in this case named EMT, would find genomic regions that had significant effects on each of the component traits. Five significant EMT QTL were detected in the Pachón cross (Protas *et al.* 2008). Five others were found that control condition factor, number of anal fin rays, and numbers of melanophores. Numerous other multitrait QTL were detected and in some cases different multitrait QTL cluster revealing that as many as eight traits can be influenced by genetic variation in the same region in the genome (Protas *et al.* 2008).

The existence of these clusters and their varied phenotypic associations were unanticipated results of our QTL mapping studies. While multitrait QTL represent areas in the genome where allelic variation influences more than one phenotypic character, they do not necessarily represent clusters of different genes. Any cluster might have a multigenic basis or, alternatively, result from a single genetic locus with pleiotropic effects. In the absence of fine scale molecular characterization of the QTL we do not know how much each mechanism contributes to the overall picture. Nevertheless, we argued that the large numbers of such trait clusters, their rich phenotypic composition, and the relatively short time to create gene clusters through genomic rearrangement, suggest that a good proportion of them must be due to single loci with pleiotropic effects (Protas *et al.* 2008).

An example of how such broad pleiotropic substitution effects could come about is provided by the melanocortin system, in which the product of one gene, pro-opiomelanocortin (POMC), is processed post-translationally to yield a variety of biologically active peptides collectively called melanocortins. These include the melanocyte-stimulating-hormones, α-ACTH, and β-endorphin, and collectively have wide ranging effects on a variety of systems, including skin pigmentation, stress response, energy budgeting, feeding behavior, and the immune response (Takahashi and Kawauchi 2006). The melanocortins stimulate cellular activity through interaction with membrane bound melanocortin receptor molecules (MC1R through MC5R), which have tissue specific expression patterns. Because the MCRs exhibit minor expression in the "wrong" tissues and because of a general promiscuity in binding of melanocortins to receptors, alterations of POMC function may have widespread pleiotropic phenotypic effects (Ducrest *et al.* 2008). As a second example, transcription factors, typically play roles in multiple pathways at different stages of development or in different tissues (Carroll *et al.* 2005). Allelic substitutions in such genes are likely to have pleiotropic effects in diverse pathways.

Wright (1964), building on the observations of Emerson (Allee *et al.* 1949), predicted the existence of multitrait clusters and their importance for regressive evolution, but his argument was more general. Wright suggested a "practically universal pleiotropy," using "pleiotropy" to refer to the multiple affects of a single allelic substitution. He argued that the selective coefficient of any new mutation would depend, therefore, upon the aggregate fitness effects of its component traits.

Figure 4 illustrates the effects of a single mutation at a locus as vectors showing the polarity of multiple phenotypic changes resulting from substitution of the new allele for the old. Increased or decreased phenotypic expression can be either beneficial or deleterious. In the hypothetical example illustrated, beneficial changes include increased sensitivity to dissolved amino acids in the water (A) and decreased eye size (E), while deleterious changes include increased rate of weight loss on low rations (W) and fewer maxillary teeth (T). The selective effect of the deceased numbers of melanophores (M) is unknown. Depending upon the relative importance of the beneficial and deleterious effects, this new mutation could be selected for or against.

Figure 5 illustrates similar clusters of trait effects detected in the F_2 of a cross between a Pachón cave individual and a surface fish (Protas *et al.* 2008). In each of seven different linkage groups we identified regions with QTL affecting at least four of the following traits, E, A, T and M (defined above). In six of the seven clusters there were effects on eye size. Consistent with our previous observations (Protas *et al.* 2007), in all six cases the substitution affect of the cave allele was to decrease eye size, strengthening our earlier conclusion that eyes are directly selected against in the cave environment. In three of the four clusters that had affects on olfactory sensitivity, the cave allele conveyed greater sensitivity (presumably advantageous for finding food in the dark). In one case however, LgP14, sensitivity decreased, but this deleterious affect could be offset by the other affects of decreased eye size and increased numbers of maxillary teeth. In each of the seven clusters the cave genotype always was associated with phenotypic changes conferring fitness benefits. Reading

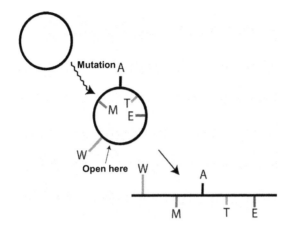

Fig. 4 A hypothetical example of the pleiotropic effects of mutation. An allele at ground state (unmodified circle) is mutated to an allele with pleiotropic effects (negative or positive) on five phenotypic traits: Amino acid sensitivity (A), number of maxillary teeth (T), relative eye size (E), rate of weight loss on reduced rations (W), and melanophore count (M). At the lower right, the circle is opened and stretched into the form presented in Figure 5.

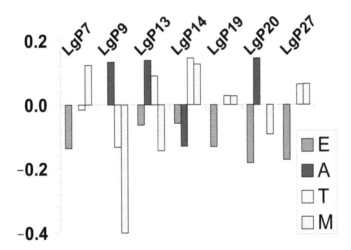

Fig. 5 Scaled values of cave allele substitution effects for four traits in QTL clusters on seven linkage groups in *Astyanax mexicanus*: Amino acid sensitivity (A), number of maxillary teeth (T), relative eye size (E), and melanophore count (M). Substitution effects were scaled against phenotypic range in the mapping F_2 (Pachon × Surface cross). Positive values are phenotypic increases, negative values are decreases. (From Protas *et al.* 2008.)

Color image of this figure appears in the color plate section at the end of the book.

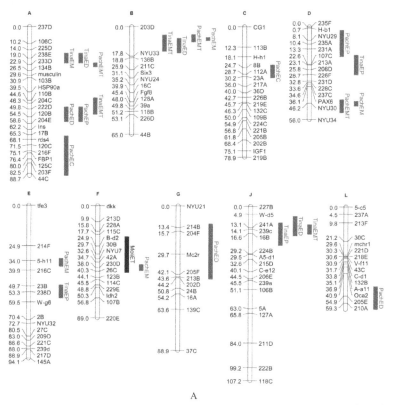

Fig. 6 Contd..

the linkage groups from left to right in Fig. 5, there were: 2+ 1–; 1+ 1–; 3+; 2+ 1–; 2+; 2+; 2+; 2+.

While these clusters could represent regions with tightly linked genes affecting each trait separately, the weight of the evidence is that pleiotropy is making a strong contribution to the phenomenon (Protas *et al.* 2008). Of course, one of the strongest arguments for the role of pleiotropy in these trait clusters is the presence of deleterious effects (LgP7, LgP9, and LgP14). It is hard to explain how alleles causing deleterious effects could come to predominate in the cave environment without indirect selection through pleiotropic beneficial phenotypic correlates.

These observations on multitrait QTL clusters in the Pachón cave fish genome are consistent with Wright's predictions. This hypothesis, which has broad implications for the study of morphological evolution, multitrait adaptation and adaptive radiation, makes testable predictions about the presence and distribution of similar multi-trait clusters in other populations of cave fishes. The predictions are that multitrait QTL of

Fig. 6 Contd..

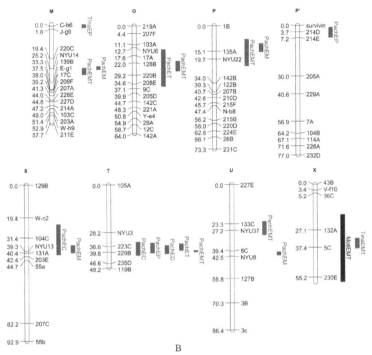

Fig. 6 Genetic map of *Astyanax mexicanus* showing alignment of multi-trait QTL detected in the mapping progenies of crosses between surface fish and cave fish from Pachón, Tinaja, or Molino. QTL were detected as previously described (Protas *et al.* 2008) and hand placed on an integrated map, based on positions determined on cross specific maps. The bars cover ± one LOD intervals. Multitrait QTL were based on combinations of two or three traits: E = eye size; M = melanophore count; T = number of maxillary teeth; D = depth of the caudal peduncle; P = placement of the dorsal fin. QTL are coded, PACH, MOLI and TINA to identify the cross in which they were detected. Traits are fully defined and methods are detailed in (Protas *et al.* 2008).

Color image of this figure appears in the color plate section at the end of the book.

similar trait composition will be observed in crosses between surface fish and other cave populations, but that they will map to different areas of the genome. The similarity in trait composition is expected because the same environmental factors will be driving evolution of the separate cave populations. The expectation is that they will map to different regions reflects the independent origins of the relevant mutations.

These predictions were tested by mapping multiple-trait QTL in separate crosses between surface fish and cave fish from three different populations. Two of the mapping progenies were previously reported on:

Pachón X Surface F_2 and Molino X Surface BC (Protas *et al.* 2006, Protas *et al.* 2008). The third cross was a Tinaja X Surface F_2, with 285 young with 207 markers, not previously reported on. In all of these crosses, multi-trait QTL were detected and mapped using the methods described in Protas *et al.* (2007, 2008). The QTL mapped were EM (eye size and melanophore number), ET (T = number of maxillary teeth), EC (C = condition factor), ED (D = depth of the caudal peduncle), EP (P = placement of the dorsal fin), and EMT. The measurement of these traits and the methods of QTL detection have been described previously (Protas *et al.* 2008). The present analysis is based on an increased number of markers compared to the earlier analyses. Because the mappable markers differed somewhat among the crosses, the QTL were hand placed on a composite map to make their distributions comparable among crosses. Thus, the positions are approximate, but accurate enough to test for positional overlaps between populations.

A total of 44 QTL were detected in the three crosses (Fig. 6). Most (30) were in Pachón, which was the largest cross, and fewest were in Molino (2), which was the smallest cross. Considering all three crosses, eleven sites were found in the genome with EMT QTL. One pair of EMT QTL for Pachón and Tinaja co-mapped to linkage group B (LgB) and another pair for Molino and Tinaja co-mapped to LgX. The remaining three Tinaja and seven Pachón EMT QTL did not co-map. One pair of EP QTL from Pachón and Tinaja overlap in LgD. The remaining three from each of the two populations map to different regions. In addition, one Pachón EMT QTL barely overlaps with Tinaja ED and EM QTL in LgA. Aside from these, there is little evidence that multi-trait QTL from the different crosses that contain the same constituent traits co-map. This supports the pleiotropy hypothesis that predicts that multi-trait QTL can arise in many different places throughout the genome.

SUMMARY AND CONCLUSIONS

The shift from surface life to cave life poses two challenges for animals: they lose the ability to see, and they must survive on relatively restricted rations. Adaptations to cave life include regression of the eyes, augmentation of other senses, and changes in metabolism. Correlated changes include reductions in pigmentation and changes in body proportions and meristic characters. These may or may not be adaptations. In *Astyanax mexicanus*, some of these evolutionary changes, such as albinism or the production of phaeomelanin are due to mutations at single loci. Most of the other changes, such as eye and melanophore regression, changes in body proportions and number of teeth, increased olfactory acuity, and changes in both anabolic and catabolic metabolism, have a multigenic basis.

Populations of cave fishes are generally smaller than populations of related surface fishes, presumably because of reduced habitat and food availability. In consequence, they have lower genetic variability. Bottlenecks caused by periodic reductions in population size may contribute to reduced effective population sizes and lower genetic variability further. In *A. mexicanus*, there is good evidence of bi-directional gene flow between surface and cave populations. Gene flow into cave populations helps maintain genetic diversity and their potential for further evolutionary change. Gene flow from cave to surface populations potentially allows cave derived alleles an evolutionary longevity greater than that of the cave population.

In the case of *A. mexicanus*, phylogeographic analyses show that current cave populations were derived from at least two separate stocks of surface fish and at least four separate subterranean invasions. Thus, fish from the 29 existing surface populations resemble one another not only because of common ancestry, but because of strong convergences.

Evidence from QTL analysis shows that eye regression occurred because of direct selection against eyes in the cave environment. In contrast, the pattern of reduction in numbers of melanophores was more haphazard and does not suggest direct selection against the cell type. Reduction in numbers of melanophores might have come about through drift or indirect selection through pleiotropy.

Pleiotropy plays a significant role in both regressive and constructive evolution, as evidenced by the existence in the cave fish genome of numerous areas having concentrations of QTL for both types of cave related traits. While some of these trait clusters undoubtedly reflect the effects of tightly linked genes, there are too many to explain through linkage. The pleiotropy hypothesis predicts the existence of multitrait QTL with similar trait contents in independently derived populations and that these independently derived multitrait QTL will map to different genomic regions. Data comparing the locations of multitrait QTL in three different cave populations are consistent with this prediction.

While pleiotropy is often cited in discussions of the evolution of troglomorphy as a mechanism to drive trait regression through indirect selection, this is a narrow view of the importance of the phenomenon. Pleiotropy is probably a key feature of multitrait evolution in all species.

One of the principal features of cave species is their evolutionary convergence on troglomorphy. This gives their study some of the power of experimental replication to test hypotheses about evolutionary processes. In *Astyanax mexicanus* there is the extra advantage that the multiple convergent cave populations are interfertile with one another and with eyed surface fish. These features make the *A. mexicanus* system an especially powerful tool for illuminating the genetic bases of cave adaptation and for testing evolutionary hypotheses and, in general.

Acknowledgements

This work was supported by grants from the NIH (R03EY016783-01) and the NSF (IOS-0821939).

References

Allee, W.C., A.E. Emerson, O. Park, T. Park and K.P. Schmidt. 1949. *Principles of Animal Ecology*. W.B. Saunders Co., Philadelphia.

Avise, J.C. and R.K. Selander. 1972. Evolutionary genetics of cave-dwelling fishes of genus *Astyanax*. *Evolution* 26: 1-19.

Barr, T.C. JR. 1968. Cave ecology and the evolution of troglobites. In: Dobzhansky, T., Hecht, M.K. and Steere, W.C. (eds). *Evolutionary Biology*, Plenum Press, New York, pp. 35-102.

Borowsky, R. 2008. Restoring sight in blind cavefish. *Current Biology* 18: R23-24.

Borowsky, R. and L. Espinasa. 1997. Antiquity and Origins of Troglobitic Mexican Tetras, *Astyanax fasciatus*. In: 12th International Congress of Speleology. International Union of Speleology/Swiss Speleological Society, La Chaux-de-Fonds, Switzerland. pp. 359-361.

Borowsky, R. and H. Wilkens. 2002. Mapping a cave fish genome: Polygenic systems and regressive evolution. *The Journal of Heredity* 93: 19-21.

Borowsky, R.B. and L. Mertz. 2001. Genetic differentiation among populations of the cave fish *Schistura oedipus* (Cypriniformes: Balitoridae). *Environmental Biology of Fishes* 62: 225-231.

Borowsky, R.L. 2001. Estimating nucleotide diversity from random amplified polymorphic DNA and amplified fragment length polymorphism data. *Molecular Phylogenetics and Evolution* 18: 143-148.

Borowsky, R.L. and C. Vidthayanon. 2001. Nucleotide diversity in populations of balitorid cave fishes from Thailand. *Molecular Ecology* 10: 2799-2805.

Breder, C.M. 1943. Apparent changes in phenotypic ratios of the characins at the type locality of *Anoptichthys jordani* Hubbs and Innes. *Copeia* 1943: 26-30.

Carroll, S.B., J.K. Grenier and S.D. Weatherbee. 2005. *From DNA to Diversity: Molecular Genetics and the Evolution of Animal Design*. Blackwell Science, Oxford. 2nd edition.

Culver, D.C. 1982. *Cave Life: Evolution and Ecology*. Harvard University Press, Cambridge, Mass.

Culver, D.C. (ed.). 1985. Regressive Evolution. National Speleological Society Bulletin 47, No. 2.

Culver, D.C., T.C. Kane and D.W. Fong. 1995. *Adaptation and Natural Selection in Caves*. Harvard University Press, Cambridge, Mass.

Darwin, C. 1859. *On the Origin of Species by Means of Natural Selection, or the Preservation of Favoured Races in the Struggle for Life.* John Murray, London.

Dowling, T.E., D.P. Martasian and W.R. Jeffery. 2002. Evidence for multiple genetic forms with similar eyeless phenotypes in the blind cavefish, *Astyanax mexicanus. Molecular Biology and Evolution* 19: 446-455.

Ducrest, A.L., L. Keller and A. Roulin. 2008. Pleiotropy in the melanocortin system, coloration and behavioural syndromes. *Trends in Ecology & Evolution* (Personal edition) 23: 502-510.

Fitzpatrick, M.J. 2004. Pleiotropy and the genomic location of sexually selected genes. *American Naturalist* 163: 800-808.

Jiang, C.J. and Z.B. Zeng. 1995. Multiple-trait analysis of genetic-mapping for quantitative trait loci. *Genetics* 140: 1111-1127.

Kimura, M. and T. Ohta. 1971. *Theoretical Aspects of Population Genetics.* Princeton University Press, Princeton.

King, R.A., V.J. Hearing, D.J. Creel and W.S. Oetting. 2001. Albinism. In: *The Metabolic and Molecular Bases of Inherited Disease.* C.R. Scriver, A.L. Beaudey, W.S. Sly and D. Valle (eds.). McGraw Hill, New York. pp. 5587-5627.

Korol, A.B., Y.I. Ronin and V.M. Kirzhner. 1995. Interval mapping of quantitative trait loci employing correlated trait complexes. *Genetics* 140: 1137-1147.

Korol, A.B., Y.I. Ronin, E. Nevo and P.M. Hayes. 1998. Multi-interval mapping of correlated trait complexes. *Heredity* 80: 273-284.

Korol, A.B., Y.I. Ronin, A.M. Itskovich, J. Peng and E. Nevo. 2001. Enhanced efficiency of quantitative trait loci mapping analysis based on multivariate complexes of quantitative traits. *Genetics* 157: 1789-1803.

Krekeler, C.H. 1958. Speciation in cave beetles of the genus *Pseudanophthalmus* (Coleoptera: Carabidae). *American Midland Naturalist* 59: 167-189.

Lande, R. 1981. The minimum number of genes contributing to quantitative variation between and within populations. *Genetics* 99: 541-553.

Langecker, T.G., H. Wilkens and P. Junge. 1991. Introgressive hybridization in the Pachon cave population of *Astyanax fasciatus* (Teleostei: Characidae). *Ichthyological Explorations of Freshwaters* 2: 209-212.

Lynch, M. and B. Walsh. 1998. *Genetics and Analysis of Quantitative Traits.* Sinauer, Sunderland, M.A.

Mitchell, R.W., W.H. Russell and W.R. Elliott. 1977. *Mexican Eyeless Characin Fishes, genus* Astyanax: *Environment, Distribution, and Evolution.* Texas Tech Press, Lubbock.

Orr, H.A. 1998. Testing natural selection vs. genetic drift in phenotypic evolution using quantitative trait locus data. *Genetics* 149: 2099-2104.

Panaram, K. and R. Borowsky. 2005. Gene flow and genetic variability in cave and surface populations of the Mexican Tetra, *Astyanax mexicanus* (Teleostei: Characidae). *Copeia* 2005: 409-416.

Perez, J.E. and G.E. Moodie. 1993. Genetic variation in a cave-dwelling Venezuela catfish. *Zoologia* 44: 28-31.

Poulson, T.L. 1985. Evolutionary reduction by neutral mutations: Plausibility arguments and data from amblyopsid fishes and linyphiid spiders. *The NSS Bulletin* 47: 109-117.

Protas, M., M. Conrad, J.B. Gross, C. Tabin and R. Borowsky. 2007. Regressive evolution in the Mexican cave tetra, *Astyanax mexicanus*. *Current Biology* 17: 452-454.

Protas, M., I. Tabansky, M. Conrad, J.B. Gross, O. Vidal, C.J. Tabin and R. Borowsky. 2008. Multi-trait evolution in a cave fish, *Astyanax mexicanus*. *Evolution & Development* 10: 196-209.

Protas, M.E., C. Hersey, D. Kochanek, Y. Zhou, H. Wilkens, W.R. Jeffery, L.I. Zon, R. Borowsky and C.J. Tabin. 2006. Genetic analysis of cavefish reveals molecular convergence in the evolution of albinism. *Nature Genetics* 38: 107-111.

Ronin, Y.I., A.B. Korol and E. Nevo. 1999. Single- and multiple-trait mapping analysis of linked quantitative trait loci: Some asymptotic analytical approximations. *Genetics* 151: 387-396.

Ronin, Y.I., A.B. Korol and J.I. Weller. 1998. Selective genotyping to detect quantitative trait loci affecting multiple traits: Interval mapping analysis. *Theoretical and Applied Genetics* 97: 1169-1178.

Sadoglu, P. 1956. A preliminary report on the genetics of the Mexican cave characins. *Copeia* 1956: 113-114.

Sadoglu, P. 1957a. A Mendelian gene for albinism in natural cave fish. *Experientia* 13: 394.

Sadoglu, P. 1957b. Mendelian inheritance in the hybrids between the Mexican blind cave fishes and their overground ancestors. *Verhandlungen Deutsche Zoologische Gesellschaft, Graz*, 1957. *Zool. Anz.*, Supplement: 21: 432-439.

Sadoglu, P. and A. Mckee. 1969. A second gene that affects eye and body color in Mexican blind cave fish. *Journal of Heredity* 60: 10-14.

Schluter, D., E.A. Clifford, M. Nemethy and J.S. McKinnon. 2004. Parallel evolution and inheritance of quantitative traits. *American Naturalist* 163: 809-822.

Sket, B. 1985. Why all cave animals do not look alike a discussion on adaptive value of reduction processes. *The NSS Bulletin* 47: 78-85.

Strecker, U., L. Bernatchez and H. Wilkens. 2003. Genetic divergence between cave and surface populations of *Astyanax* in Mexico (Characidae: Teleostei). *Molecular Ecology* 12: 699-710.

Strecker, U., V.H. Faundez and H. Wilkens. 2004. Phylogeography of surface and cave *Astyanax* (Teleostei) from Central and North America based on cytochrome b sequence data. *Molecular Phylogenetics and Evolution* 33: 469-481.

Swofford, D.L., B.A. Branson and G.A. Sievert. 1980. Genetic differentiation of cavefish populations. *Isozyme Bulletin* 13: 109-110.

Takahashi, A. and H. Kawauchi. 2006. Evolution of melanocortin systems in fish. *General and Comparative Endocrinology* 148: 85-94.

Trajano, E., N. Mugue, J.K. Krejca, C. Vidthayanon, D. Smart and R. Borowsky. 2002. Habitat, distribution, ecology and behavior of cave balitorids from Thailand (Teleostei: Cypriniformes). *Ichthyological Explorations of Freshwaters* 13: 169-184.

Wilkens, H. 1971. Genetic interpretation of regressive evolutionary processes – Studies on hybrid eyes of 2 *Astyanax* cave populations (Characidae: Pisces). *Evolution* 25: 530-544.

Wilkens, H. 1988. Evolution and genetics of epigean and cave *Astyanax fasciatus* (Characidae, Pisces) – support for the neutral mutation theory. *Evolutionary Biology* 23: 271-367.

Wilkens, H. and U. Strecker. 2003. Convergent evolution of the cavefish *Astyanax* (Characidae: Teleostei): Genetic evidence from reduced eye size and pigmentation. *Biological Journal of the Linnean Society* 80: 545-554.

Wright, S. 1964. Pleiotropy in the evolution of structural reduction and of dominance. *American Naturalist* 98: 65-69.

Yamamoto, Y., D.W. Stock and W.R. Jeffery. 2004. Hedgehog signalling controls eye degeneration in blind cavefish. *Nature* 431: 844-847.

CHAPTER 6

Development as an Evolutionary Process in *Astyanax* Cavefish

William R. Jeffery* and **Allen G. Strickler**
Department of Biology, University of Maryland, College Park,
MD 20742 USA
E-mail: Jeffery@umd.edu

ABSTRACT

Astyanax mexicanus cavefish and their surface dwelling counterparts are a model system for studying the role of development in the evolution of cave related phenotypes. In contrast to many other cave animals, *Astyanax* is an excellent laboratory organism with frequent and copious spawning and large robust embryos. Here we review molecular, cellular and developmental mechanisms underlying the loss of eyes and pigmentation in *Astyanax* cavefish. Eye degeneration involves enhanced activity of the Hedgehog signaling system along the embryonic midline, which induces lens apoptosis and ultimately causes the arrest of optic growth and changes in craniofacial morphology. The loss of melanophore pigmentation is caused by a specific block in the ability of undifferentiated melanoblasts to convert L-tyrosine to L-DOPA, a precursor of melanin. We discuss the importance of a developmental perspective in understanding the evolutionary mechanisms underlying eye and pigment regression.

INTRODUCTION

Evolution creates novel phenotypes by changing development. Therefore, knowledge of development is a prerequisite for understanding the mechanisms of evolutionary change. Most studies of cave animals have focused on problems in evolution, behavior, ecology, taxonomy, and conservation. There are several reasons for the virtual absence of developmental perspectives in cave biology. First, embryonic and juvenile stages are required for the study of development, and they are difficult to collect in caves. Second, many cave animals are known to produce a reduced complement of embryos that develop very slowly relative to their surface dwelling counterparts. Developmental biologists prefer to focus on species that have many progeny and develop rapidly. Third, some cave animals tend to brood embryos rather than freely spawn their eggs, compounding the difficulty in studying early developmental events. Finally, unlike many other areas of cave biology, developmental biology is a laboratory science that requires experimental analysis and an organism that thrives and reproduces in the laboratory. There are few such model organisms available in cave biology. One of these is the Mexican cavefish *Astyanax mexicanus* (sometimes called *A. fasciatus* and referred to here as *Astyanax*), which has many favorable attributes for developmental analysis (Jeffery 2001, 2008).

Astyanax has a surface dwelling form (surface fish) and 30 known populations of cave dwelling forms (cavefish). Surface fish adults have large eyes and three different types of pigment cells, whereas cavefish have reduced or lost these traits, a phenotype shared with many diverse cave animals (Fig. 1). Cavefish have also gained some less obvious constructive features, such as large jaws, more taste buds, and possibly a more highly developed olfactory system. Background information on the morphology,

Fig. 1 *Astyanax* surface fish (top) and Pachón cavefish (bottom) adults. Note the absence of eyes and black melanin pigment in cavefish.

biogeography, evolution, and classical genetics of *Astyanax* cavefish can be found in Wilkens (1988).

Astyanax cavefish are named after their cave of origin (Mitchell *et al.* 1977). For example, Pachón, Chica, Los Sabinos, and Subterráneo, and Molino cavefish are found in La Cueva de El Pachón, La Cueva Chica, La Cueva de los Sabinos, La Cueva de la Subterráneo, and Sótano de El Molino, respectively. In this article, unless another cavefish population is named specifically, reference to cavefish will imply the Pachón population, which has been the subject of considerable developmental analysis. Evidence from a variety of sources suggests that some *Astyanax* cavefish populations have evolved cave related phenotypes independently (Wilkens 1971, Dowling *et al.* 2002, Strecker *et al.* 2003, 2004, Borowsky 2008). Thus, *Astyanax* cavefish are also an excellent model system to study convergent evolution of developmental mechanisms.

Astyanax can be raised on a simple diet, spawns frequently and abundantly, and has a relatively short generation time (about 4-6 months). It has large, clear robust embryos that develop similarly to those of other teleosts and are suitable for a variety of experimental manipulations, such as microsurgery (Yamamoto and Jeffery 2000, Yoshizawa and Jeffery, 2008) and *in vitro* culture of developing tissues and organs (McCauley *et al.* 2004). Cavefish embryos can be contrasted to surface fish embryos in the same way that mutant and wild-type phenotypes are compared in genetic analysis. The fairly close phylogenetic relationship between *Astyanax* and zebrafish permits rapid cloning of genes identified in the latter, providing a host of potential molecular markers for developmental studies. Finally, the procedures for functional gene analysis pioneered in zebrafish can also be used in *Astyanax*.

Cavefish and other cave animals present two major challenges to evolutionary developmental biologists: (1) to understand the developmental mechanisms underlying the phenotypic changes related to cave life and (2) to determine the evolutionary forces responsible for driving these developmental and phenotypic changes. This chapter reviews *Astyanax* embryonic and larval development in the context of these challenges. We focus on the molecular, cellular, and developmental mechanisms responsible for loss of eyes and pigmentation. This information is then used to address the mechanisms underlying the evolution of these regressive traits.

ASTYANAX DEVELOPMENT

Astyanax development involves (1) fertilization and cleavage, (2) epiboly and gastrulation, (3) morphogenesis, segmentation, and formation of the body axis, and (4) organogenesis. During these stages, the fertilized egg becomes an embryo, the embryo hatches and develops into a larva (or fry), and the fry becomes an adult. Cahn (1958) carried out the first detailed analysis of *Astyanax* development comparing surface fish to Chica and Los Sabinos cavefish. She showed that embryonic and larval development resembles that of other teleosts, such as zebrafish. Early development of surface fish and cavefish is superficially similar, except for slight differences in timing events and the reduced size of eye primordia. As the embryo hatches and the fry grow, differences between the cave and surface forms become more pronounced. Below we describe embryonic and larval development in Pachón cavefish.

Figure 2 shows Pachón cavefish development from fertilization to hatching. Eggs are deposited in a clutch of several hundred and fertilized externally. It has been reported that Subterráneo cavefish eggs are slightly larger than those of surface fish (Hüppop and Wilkens 1991). However, egg size does not seem to differ markedly between Pachón cavefish and surface fish. The egg cytoplasm and yolk platelets are distributed homogenously in the unfertilized egg. Fertilization begins a series of contractile movements in which the yolk is concentrated into a large vegetal region, the yolk mass, and the yolk-free cytoplasm is squeezed into a smaller animal region, the blastodisc (Fig. 2, 1-2). Subsequently, incomplete cleavages occur only in the blastodisc, which becomes the embryo proper. The uncleaved yolk mass becomes the yolk sac.

Teleosts exhibit discoidal cleavage, which is characterized by incomplete cleavage furrows that do not extend through the yolk mass. Discoidal cleavage is also meroblastic, meaning that the early blastomeres retain cytoplasmic continuity with each other through the yolk mass. This continuity, which persists until about the 16-cell stage, is important for functional experiments because molecules microinjected into a restricted region of the embryo can disperse into every blastomere. The early cleavages are synchronous, and the cleaving embryo is marked by a rapid increase in cell number with no increase in the size of the blastodisc (Fig. 2, 3-14). The first cleavage furrow divides the blastodisc into two equal sized blastomeres (Fig. 2, 3). The second, third, and fourth cleavage furrows are positioned perpendicular to the first cleavage plane, producing a 4 × 4 array of 16 cells in the blastodisc (Fig. 2, 4-6). After the 16-cell stage, the four inner cells (non-marginal cells) are separated completely from the underlying yolk mass, whereas the 12 outer cells (marginal cells) retain cytoplasmic links to the yolk mass. Every cleavage up to the sixth is in a vertical plane through the blastodisc (Fig. 2, 3-8). The sixth cleavage is the first horizontal cell division (Fig. 2, 8). Beyond the sixth cleavage, there are no stereotypical arrangements of blastomeres (Fig. 2, 9-14). Blastomeres

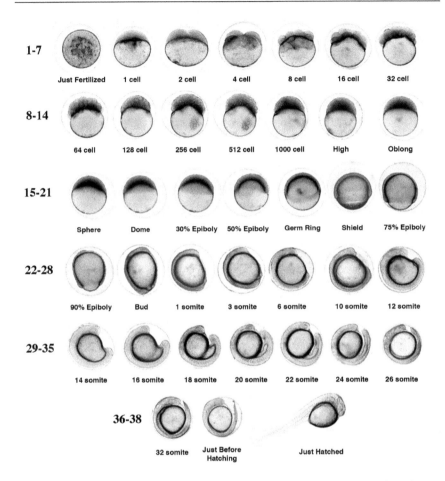

Fig. 2 Early development of *Astyanax* cavefish (Pachón) eggs and embryos. Cavefish embryos are shown from the unfertilized egg to the recently hatched larva. Stages are ordered from left to right in each row. Numbers at the left indicate successive stages to the right in each row. Stages are named below the corresponding photograph.

(marginal and non-marginal) facing the outside of the embryo become the enveloping layer (EVL), whereas cells that have no external surface are known as the deep cells. As cleavage continues, the deep cells greatly outnumber those of the EVL.

The blastula is distinguished from previous stages by the loss of division synchrony, the beginning of cell motility, and other changes associated with the so-called mid-blastula transition (Fig. 2, 12). During the blastula stages, the blastodisc forms a crescent shaped structure on top of the yolk mass (Fig. 2, 13). Blastulae pass through the oblong, sphere, and dome stages. During these stages, the blastodisc elongates and spreads out over the yolk mass (Fig. 2, 14-16). Blastodisc expansion involves increased adherence of the marginal cells to the underlying yolk mass, forming a border zone known as the yolk syncytial layer (YSL). In zebrafish, other teleosts, and presumably *Astyanax*, blastomeres are still uncommitted to particular developmental fates at the blastula stage (Long 1983, Langeland and Kimmel 1997).

During epiboly the EVL moves over the yolk mass to create an embryo whose outer surface is almost completely covered with a thin epithelium. The extent of epiboly is measured as the percentage of yolk mass covered by extending epithelial cells. Epiboly is driven by the YSL, which slowly contracts and pulls the attached EVL over the yolk mass (Fig. 2, 17-22).

Gastrulation is complex in teleosts, and will be only briefly considered here. Gastrulation begins at about 50% epiboly, when cells of the surface layer move under the rim of the extending blastodisc, then converge and extend toward the animal pole, thus forming the embryonic axis. The movements of gastrulation are most extensive in a region called the embryonic shield, which is the teleost counterpart of the amphibian organizer. At the end of gastrulation (and neurulation), the embryonic body plan is clearly visible, with future head and tail on opposite sides of the anterior-posterior axis. This is called the (tail)bud stage (Fig. 2, 22). At the tailbud stage, the embryonic midline divides the embryo into left and right halves. The midline (prechordal plate and the notochord) is an important signaling center throughout embryogenesis. The commitment of cells to distinct fates occurs during gastrulation, and the fate map is now restricted.

During the next stage of embryogenesis, the embryonic body plan is elaborated. This period includes the enlargement and elongation of the embryo, the gradual disappearance of the yolk sac, the formation of organ primordia, and the appearance of somites, the progenitors of body wall muscle and the vertebral column (Fig. 2, 23-37). Somites appear in sequence from the base of the head toward the tail and can be used to classify embryos into different stages (e.g. 20 somite stage, Fig. 2, 32). Many morphogenetic events occur during this period of development, including brain formation from the anterior neural tube and the migration of neural crest cells. Organogenesis begins after the embryo hatches by breaching the wall of the chorion, and thus becomes a free-swimming larva (Fig. 2, 38). During the larval stage, fry develop the phenotypic characteristics of *Astyanax* adult surface fish and cavefish (Fig. 3).

As in other teleosts (Fernald, 1991), the eyes of surface fish grow continuously to match the growth of the body. In contrast, cavefish eyes

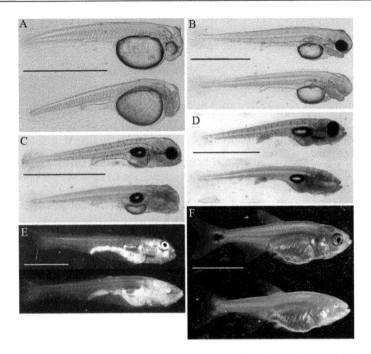

Fig. 3   Post-hatching development of *Astyanax* surface fish and cavefish (Pachón) larvae. Surface fish (above) and cavefish (below) are shown in each frame. A. One day post-fertilization (dpf). B. Three dpf. C. One week post-fertilization (wpf). D. Two wpf. E. Three wpf. F. One month post-fertilization (mpf). Note developmental arrest and progressive loss of eyes in cavefish, development and rapid growth of eyes in surface fish, body pigment cell development in surface fish, and absence of eye and body pigmentation in cavefish. Scale bars are 62.5 mm (A), 125 mm (B), 250 mm (C-D), and 500 mm (F).

reach a certain size and then growth is arrested (Figs. 3, 4A). Because the body continues to grow, the degenerating eyes gradually disappear into the orbits (Fig. 3). Surface fish eyes become pigmented, due to melanin deposition in the retinal pigment epithelium (RPE), the cell layer surrounding the retina, whereas the cavefish RPE either lacks pigmentation or the number of pigment cells is greatly reduced (Fig. 3C-F). In surface fish, melanin-containing body pigment cells also differentiate and form distinctive patterns in the dermal layers of the skin, whereas these cells are absent or decreased in cavefish (see Figs. 1A and 3C-F). Many of the constructive features of cavefish, including changes in jaw structure and taste buds, also appear during the larval stages.

EYE REGRESSION: FROM OPTIC PRIMORDIUM TO CRANIOFACIAL SKELETON

Eye Development and Degeneration

Teleost eyes develop similarly to the eyes of other vertebrates. Three different parts of the ectoderm contribute to the eye. The ocular lens is formed from the lens placode, a thickening in the surface ectoderm. The retina and RPE are formed from the optic vesicles, which are bilateral protrusions of the forebrain neuroectoderm. The optic vesicles arise from bilaterally symmetric optic fields in the anterior neural plate. Each optic vesicle rotates about 90° and then buckles inward to form the optic cup, with the future retina on the concave side and the RPE on the convex side. The isthmus connecting the optic cup to the forebrain becomes the optic stalk, which will form the casing of the optic nerve fibers. Simultaneously, the lens placode reorganizes into a vesicle, which detaches from the surface epithelium and enters the concave opening of the optic cup. The third part of the ectoderm responsible for eye development is the neural crest. Cranial neural crest cells migrate from the developing neural tube region into spaces between the lens and surface epithelium to form the inner parts of the cornea, between the lens and the distal edges of the retina to contribute to the iris and ciliary body, and into the areas surrounding the RPE to form the choroid and sclera. Neural crest cells probably also contribute to the ocular dermal bones that develop much later around the orbit, becoming part of the adult craniofacial skeleton.

The eye primordium consists of three parts, lens, retina, and RPE, which differentiate in concert. The lens vesicle produces fiber cells, which synthesize crystallin proteins, and becomes transparent, leaving behind a layer of undifferentiated stem cells. The retina differentiates into several layers. From distal to proximal, they consist of (1) the ganglion cell layer, which transmits neural signals to the brain via axons extending through the optic stalk into the optic tectum, (2) the intermediate layers, which consist of inter-neurons and glial cells, and (3) the photoreceptor layer, where rod and cone cells translate photons into neural signals. The RPE forms tight connections with the photoreceptor layer and produces melanin pigment. Body pigment cells containing melanin have a different origin from those of the RPE, and will be discussed later in this chapter.

The sequence of events during surface fish and cavefish eye development are compared in Fig. 4A. The cavefish eye primordium is slightly smaller than its surface fish counterpart (Fig. 4B, C; Fig. 6C, D). This difference in size is due to a smaller lens and optic cup, which appears to be missing its ventral sector (Fig. 6G, H). In contrast to the surface fish eye, cavefish

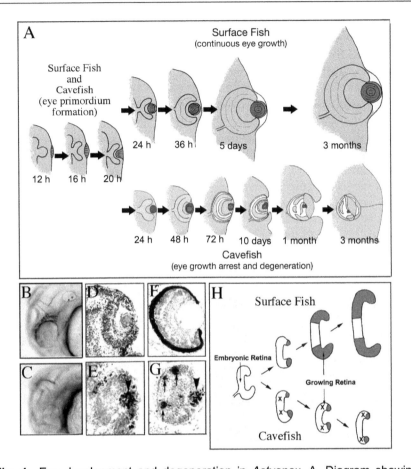

Fig. 4 Eye development and degeneration in *Astyanax*. A. Diagram showing similar development of the eye primordium from up to about 20 hrs post fertilization (hpf) in cavefish and surface fish (left), followed by their subsequent divergence (right). The surface fish eye differentiates and rapidly increases in size (top) from 1 dpf to 1 mpf, whereas the cavefish eye arrests in growth, degenerates, and gradually sinks into the orbit (bottom). B, C. Size differences in the 24 hpf surface fish (B) and cavefish (C) eye primordia. D-G. Sections of 2 (D, E) and 3 (F, G) dpf surface fish (D. F) and cave fish (E, G) eye primordia showing apoptosis (dark stained spots) in various eye tissues. In cavefish, apoptosis begins in the lens (arrowheads) and spreads to the retina (arrows). There is no apoptosis in these tissues in surface fish. H. Diagram showing the roles of cell proliferation and apoptosis during retina/RPE growth in surface fish and cavefish. Clear retinal areas: embryonic retina and central part of growing retina derived from embryonic retina. Shaded area: part of retina derived from cell proliferation at the CMZ after the embryonic stages. X: Apoptotic areas. The surface fish retina grows continuously due to cell proliferation at the CMZ, whereas the cavefish retina is arrested in growth because the products of cell proliferation at the CMZ die before they contribute to the differentiated retina.

optic tissues either fail to be induced (cornea, iris, ciliary body) or begin to differentiate and then partially or completely degenerate (lens, retina, and probably RPE). However, perhaps the most important flaw in the cavefish eye primordium is the absence of optic growth from the larval to the adult stages (Fig. 4A). Eventually, the arrested cavefish eye primordium is overgrown by head epidermis and connective tissue, and disappears into the orbit, making adult cavefish appear superficially eyeless.

Because cavefish eye development involves growth arrest, it is important to consider the origin of new cells. Stem cells in the epithelial layer are the source of new lens fiber cells. The source of most new retinal and all new RPE cells is stem cell niche at the edge of the optic cup, a region known as the ciliary marginal zone (CMZ). As the eye enlarges during larval development, it is surrounded by orbital bones, which form a part of the craniofacial skeleton. The number, size, and organization of the orbital bones is distinct between surface fish and cavefish, and even among different cavefish populations (Alvarez 1947, Yamamoto *et al.* 2003). As described below, the presence or absence of a functional eye is critical in the morphogenesis of the orbital bones and organization of the craniofacial skeleton.

Cellular Mechanisms of Eye Degeneration

A block in cell proliferation, an increase in programmed cell death, or a combination of these processes could cause the arrest of cavefish eye development. In many developing systems, programmed cell death or apoptosis is an alternative to cell differentiation. Investigations of apoptosis during *Astyanax* eye development (Jeffery and Martasian 1998, Yamamoto and Jeffery 2000) have led to a breakthrough in understanding the regulation of eye degeneration.

If cell death is restricted to a single eye tissue, or begins in one tissue and later spreads to others, then the tissue that dies first is a candidate to initiate the degeneration process. The early cavefish eye primordium is largely free of cell death, except for one tissue: the lens (Fig. 4E). No apoptosis is detected in the surface fish lens (Fig. 4D). Although cell proliferation does not cease in the cavefish lens, the rate of apoptosis is very high, eventually obliterating the lens, or reducing it to a tiny vestige in the adult (Soares *et al.* 2004). A few days after the initiation of lens apoptosis, the cavefish retina also begins to undergo apoptosis (Fig. 4G) (Alunni *et al.* 2007, Strickler *et al.* 2007a). Cell death is restricted to the intermediate retinal layers and regions adjacent to the CMZ. Later in development, dying cells are also seen in the cavefish RPE (Strickler *et al.* 2007a). Like in the lens, cell death is not observed in the surface fish retina (Fig. 4F) or RPE. Thus, the lens is the first tissue to undergo cell death in the cavefish eye primordium, suggesting that its absence may be responsible for eye degeneration.

The cessation of retinal growth in cavefish could be caused by failure of the dying lens to produce a growth-promoting factor. Surface fish have an active CMZ, in which proliferating cells can be detected by labeling replicating DNA or the presence of DNA replication enzymes, such as proliferating cell nuclear antigen (PCNA). As newly born cells are displaced from the CMZ into the retinal layers and RPE, they differentiate and increase the general mass of the retina/RPE. Surprisingly, cell proliferation markers are expressed normally in the cavefish CMZ during the period in which the retina does not markedly increase in size (Strickler *et al.* 2002, 2007a, Alunni *et al.* 2007). However, the new cells are quickly removed from the retina by apoptosis, which persists during cavefish larval development and into adult life (Strickler *et al.* 2007a). Thus, the cavefish eye is arrested in growth because newly born cells die before they are able to differentiate and join the retinal layers (Fig. 4H). The possibility that lens dysfunction is responsible for retinal cell death and has a general role in eye degeneration was tested by transplantation of embryonic lenses between surface fish and cavefish.

The lens transplantation method is illustrated in Fig. 5A. The embryonic lens is removed from a donor embryo shortly after its formation (about a day before the first detection of apoptosis in the cavefish lens) and transplanted into the optic cup of a host embryo. Lens transplantation is done unilaterally, with the unoperated eye of the host serving as a control, and reciprocally: a surface fish lens is transplanted into a cavefish optic cup and *vice versa* (Yamamoto and Jeffery 2000). When a cavefish lens was transplanted into a surface fish optic cup it died on schedule, just as if it had not been removed from the donor embryo. In contrast, when a surface fish lens was transplanted into a cavefish optic cup it continued to grow and differentiate, just as it would have done in the surface fish host. These results indicate that the cavefish lens is autonomously fated for apoptosis.

After obtaining a surface fish lens, the eyes of Pachón or Los Sabinos cavefish began to grow (Fig. 5C, H) (Yamamoto and Jeffery 2000, Jeffery *et al.* 2003). Eventually, the cornea and iris appeared, and the enlarged retina became more highly organized. Further growth resulted in the presence of a highly developed eye containing a cornea, iris, and photoreceptor cells. In contrast to the eye with a transplanted lens, the unoperated eye of the cavefish host degenerated and disappeared into the orbit (Fig. 5B, G). Likewise, after obtaining a cavefish lens, development of the surface fish eye was retarded, the cornea and iris did not differentiate, and the size and organization of the retina was reduced. The degenerate surface fish eye

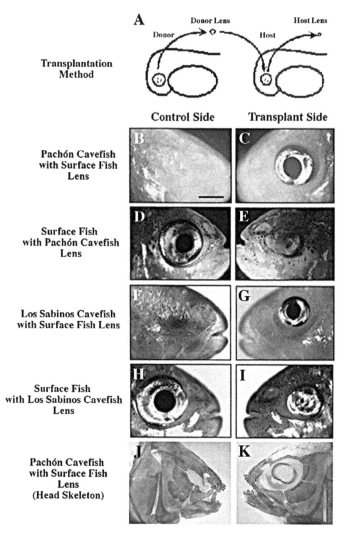

Fig. 5 Lens transplantation experiments. A. Diagram showing the transplantation method in which a donor lens is removed from the optic cup of one form of *Astyanax* embryo and transplanted unilaterally into the optic cup of another form after the host lens is removed. This operation is carried out at about 1 dpf. B-J. Changes in eye development after lens transplantation during embryogenesis. B, C, F, G. Surface fish lens was transplanted into a Pachón (B, C) or Los Sabinos (F, G), cavefish host. D, E, I, H. Changes in eye development after a Pachón (D, E) or Los Sabinos (H, I) cavefish lens was transplanted into a surface fish host. J, K. Changes in orbital bone structure after a surface fish lens was transplanted into a cavefish optic cup unilaterally. B, D, F, H, J: Control (unoperated) side. C, E, G, I, K. Transplant side.

eventually disappeared into the orbit, mimicking the cavefish eye (Fig. 5E, I), whereas the unoperated eye developed normally, producing a one eyed surface fish (Fig. 5D, H).

The following conclusions can be made from the lens transplantation experiments. First, the lens is required for normal development of the retina, cornea, and iris. Second, as a result of apoptosis the cavefish lens has lost the ability to organize eye development. Third, the cavefish optic cup (RPE/retina) has retained the ability to respond to signals generated by a normal surface fish lens. Lens transplantation also allowed two further insights. First, the lens has a role in promoting the survival of retinal cells: a transplanted surface fish lens can protect the cavefish retina from apoptosis (Strickler *et al.* 2007a). Second, the lens has an indirect role in determining craniofacial morphology. When a surface fish lens is transplanted into a cavefish optic cup, an orbital bone phenotype is obtained resembling surface fish rather than cavefish (Fig. 5J, K) (Yamamoto *et al.* 2003). Thus, the cavefish host develops with hybrid craniofacial morphology, one side (the lens transplant side) resembling surface fish and the other (the control side) cavefish.

Clearly, the lens has a major role in regulating cavefish eye degeneration. Whether the death of the lens is the only cause of eye degeneration, or other optic alterations, such as independent changes in the retina or RPE, are also involved (Strickler *et al.* 2007a), remains to be determined.

Molecular Mechanisms of Cavefish Eye Degeneration

Understanding the molecular mechanisms of eye degeneration requires identification of the genes involved in this process and how they function during development. Many eye development genes are known in vertebrates, allowing a candidate gene approach to be used for gene identification. Genes encoding transcription factors that function near the top of eye gene hierarchies, as well as structural genes encoding proteins that function at the bottom of these cascades have been isolated in *Astyanax*. *In situ* hybridization was used to determine their expression patterns.

Most of the surveyed genes and/or their encoded proteins do not show expression changes in surface fish and cavefish embryos. For example, the Prox1 transcription factor is expressed normally in the developing lens and retina of cavefish until after the eye begins to degenerate (Jeffery *et al.* 2000). Likewise, prior to lens degeneration, genes encoding ßB- and γM-crystallins and the membrane proteins MIP and MP19 are expressed normally (Jeffery *et al.* 2000, Strickler *et al.* 2007b). Many genes also show the same or similar expression patterns in the developing surface and cave fish retinas (Jeffery *et al.* 2000, Strickler *et al.* 2002, Menuet *et al.* 2007). For example, Langecker *et al.* (1993) noted that an opsin gene is expressed in the degenerating photoreceptor layer of the cavefish retina. The most important changes in optic gene expression patterns during cavefish development are described below.

The *pax6* gene encodes a transcription factor that is expressed in the lens, retina, RPE, and their precursors early in teleost eye development (Krauss *et al.* 1991, Püschel *et al.* 1998). Later, *pax6* expression becomes restricted to the lens epithelial cells, some of the retinal layers, and the corneal epithelium. In surface fish embryos, the *pax6* expression domains in the bilateral optic fields connect across the midline at their anterior margins (Fig. 6 A, B). In cavefish embryos, however, the corresponding *pax6* domains are diminished in size and show a large gap across the midline (Strickler *et al.* 2001) (Fig. 6C, D). The division of the optic vesicle into the optic cup and stalk is controlled by reciprocal antagonistic intereactions between the Pax6, Pax2, and Vax1 transcription factors (Schwarz *et al.* 2000). Pax6 directs optic cup development, whereas Pax2 and Vax1 control optic stalk development. Accordingly, a reduction of *pax6* levels (or an increase in *pax2* and *vax1* levels) increases the optic stalk at the expense of the optic cup. The reduction of *pax6* expression coupled with the overexpression of Pax2 and Vax1 accounts for the ventrally reduced optic cup in cavefish embryos (Fig. 6).

The wider gap between optic fields in the cavefish neural plate provides further insight into how eye degeneration is controlled. During vertebrate development, the presumptive optic cup is initially determined as a single medial optic field, which is subsequently split into two bilateral eye domains by Hedgehog (Hh) signals emanating from the underlying prechordal plate (Macdonald *et al.* 1995, Ekker *et al.* 1995). Hh signaling inhibits *pax6* expression along the midline to divide the original eye domain into bilateral eyes. Teleosts have at least two-*hh* midline signaling genes, *sonic hedgehog* (*shh*) and *tiggy winkle hedgehog* (*twhh*), which show overlapping expression patterns (Ekker *et al.* 1995). Yamamoto *et al.* (2004) compared *shh* and *twhh* expression patterns during surface fish and cavefish development, showing that the midline expression domains of both genes are expanded in cavefish relative to surface fish (Fig. 7A, B). Later in cavefish development, *shh* expression is also expanded anteriorly, curling around the rostrum in the presumptive oral area (Fig. 7C, D). The expression patterns of genes acting downstream of *shh* and *twhh* in the Hh midline signaling pathway, such as *patched*, encoding a Shh receptor, and *nkx2.1*, encoding an Shh dependent transcription factor, are also expanded (Yamamoto *et al.* 2004), suggesting that a general increase in midline Hh signaling has evolved in cavefish.

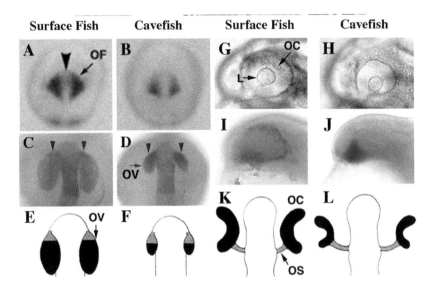

Fig. 6 Optic vesicle (A-F) and optic cup (G-L) development in surface fish and cavefish. A, B. Neural plate stage embryo showing differences in *pax6* expression in the surface fish and cavefish optic fields (OF). Arrowhead shows the midline *pax6* expression gap, which is wider in cavefish. C, D. Optic vesicles (OV) showing size and *pax2a* expression (arrowheads) differences in surface fish and cavefish. In A-D. embryos are viewed dorsally with anterior at the top. E, F. Diagram showing size differences in the surface fish and cavefish optic vesicles. Territories fated to form optic stalk are lightly shaded and those fated to form retina/RPE are darkly shaded. G, H. Surface fish and cavefish optic cups (OC) showing ventral size reduction in the latter. L: lens. I, J. The *vax1* gene is overexpressed ventrally in the cavefish optic cup relative to surface fish. In G-J, embryos are viewed laterally with dorsal at the top. K, L. Diagram showing size and relative optic cup territorial differences between cavefish and surface fish. The optic stalk (OS) is lightly shaded and the optic cup is darkly shaded.

Two additional genes appear to be associated with cavefish lens apoptosis. First, the *heat shock protein90α* (*hsp90α*) gene is specifically activated in the cavefish lens vesicle just prior to apoptosis (Hooven *et al.* 2004). Inhibition of *hsp90α* expression suppresses lens apoptosis and rescues lens differentiation. Second, the α*A-crystallin* gene, which encodes a potent anti-apoptotic factor, is strongly downregulated in the lens vesicles of Piedras (Behrens *et al.* 1998) and Pachón cavefish (Strickler *et al.* 2007b). αA-crystallin may normally protect the lens from apoptosis. Further studies will be necessary to determine if αA-crystallin and Hsp90 α interact in a cascade leading to lens apoptosis.

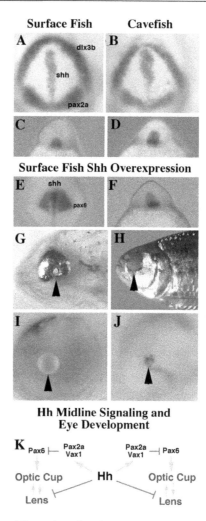

Fig. 7 Role of Hh midline signaling in cavefish eye degeneration. A-D. The cavefish embryonic midline shows a wider *shh* expression domain than its surface fish counterpart. The expression of *dlx3* and *pax2a* marker genes does not change. A, B. Tailbud stage. C, D. Ten somite stage. E-J. Effects of *shh* overexpression in surface fish. E, F. Increased *shh* expression (compare F with C) and reduced *pax6* expression (E) on one side of the midline of an embryo injected with *shh* mRNA. G, H. As a result of *shh* overexpression, the optic cup (retina/RPE) is missing its ventral sector (G) and the adult eye has degenerated (H). Arrowhead in G: missing ventral sector of the retina. Arrowhead in H: missing eye. I, J. Lens apoptosis (J) after injection of an embryo with *shh* mRNA. Arrowheads: lens. H. Diagram showing antagonistic relationship between Pax6, Pax2, and Vax1 transcription factors, ventralization of the optic cup, and lens apoptosis in cavefish. Arrows: activations. Blocked lines: inhibitions.

The Key Role of Hedgehog Signaling

The role of enhanced Hh signaling in cavefish eye degeneration was further investigated by Yamamoto *et al.* (2004). Shh signaling was increased by injection of *shh* mRNA into surface fish embryos. When *shh* mRNA was injected into one side of a cleaving embryo, *shh* expression was expanded along that side of the prechordal plate (anterior embryonic midline), and *pax6* expression was down regulated unilaterally in the corresponding optic field (Fig. 7E, F). When surface fish embryos overexpressing Shh developed into larvae and adults, the eye was missing on one side of the head (Fig. 6G, H). Thus, blind cavefish were phenocopied by increasing the levels of *shh* gene expression in surface fish, demonstrating a key role for Shh signals in eye degeneration. Importantly, lens apoptosis is also induced by *shh* overexpression in surface fish embryos (Yamamoto *et al.* 2004) (Fig. 7I, J). A diagram of the gene network leading to eye degeneration via hyperactive Shh signaling, reduction of the optic vesicle, and lens cell death is shown in Fig. 7K.

It can be concluded that a sequence of regulatory events beginning with expanded midline signaling, proceeding through reduction in size of the eye primordia, lens apoptosis, and retinal apoptosis, and resulting in arrested eye growth and alteration of craniofacial morphology, is responsible for cavefish optic degeneration (Fig. 8). Alterations in the activity of many different genes and their upstream regulators control these changes. Recent genetic analysis has shown that at least 12 quantitative trait loci (QTL) are involved in the loss of eyes in Pachón cavefish (Protas *et al.* 2007).

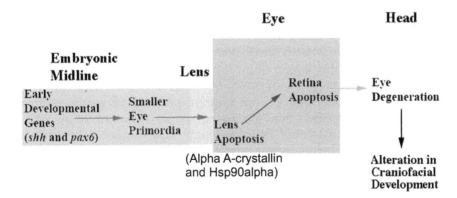

Fig. 8 Summary of early and late events in cavefish eye degeneration and consequences on craniofacial development.

PIGMENT CELL REGRESSION: FROM NEURAL CREST TO MELANOGENESIS

Astyanax surface fish has three types of body pigment cells: light reflecting iridophores, yellow-orange xanthophores, and black melanophores. Pigmentation normally functions in protection from the damaging effects of sunlight, in camouflage, and in species and sex recognition. Selective pressure for retaining these functions is relaxed in the absence of light. What are the consequences in *Astyanax* cavefish?

The early studies of Rasquin (1947) showed that melanophores are decreased but xanthophores seem to be present at the same levels in cavefish. Very little is known about changes in iridophores. Of the three types of pigment cells, most is known about melanophores (Wilkens 1988, McCauley *et al.* 2004). Subterráneo cavefish show a modest reduction in melanophore pigmentation, Curva, Los Sabinos, and Tinaja cavefish show substantial decreases in melanophore pigmentation, and Molino and Pachón cavefish show little if any melanophores. In addition to changes in the number of melanophores, cavefish also show defects in the ability to produce melanin, the pigment found in melanophores.

Migratory Neural Crest Cells in Cavefish

All types of body pigment cells are derived from the neural crest, migratory cells that arise from the border of the neural tube and surface ectoderm (Erickson 1993, LeDouarin and Kalcheim 1999). Vertebrate neural crest cells have a myriad of different derivatives, including sensory ganglia, the peripheral nervous system, cranial cartilage and bone, as well as body pigment cells. Cell tracing, immunological, and tissue culture methods were used to follow neural crest development in cavefish (McCauley *et al.* 2004, Jeffery 2006). In both cavefish and surface fish, similar numbers of labeled neural crest cells migrate into the epidermis (Fig. 9A-D), suggesting that there is no defect in neural crest cells during cavefish development.

Neural crest cells that do not migrate properly or receive normal differentiation signals often die by apoptosis (Morales *et al.* 2005). Therefore, apoptosis could remove neural crest derived precursors in cavefish embryos before they differentiate into pigment cells. When this possibility was tested only a few dying neural crest cells were observed in cavefish embryos, and their number was about the same as in surface fish embryos (Jeffery 2006). Therefore, melanophores or their progenitor cells do not undergo massive apoptosis during cavefish embryogenesis. Cavefish pigmentation defects must arise downstream of the generation, migration, and divergence of pigment cell types.

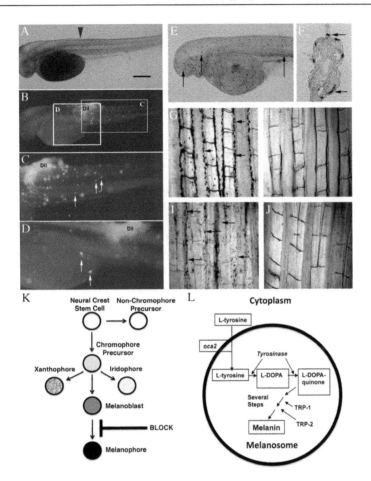

Fig. 9 Neural crest and melanophore development in cavefish. A-C. Detection of migratory neural crest cells in cavefish embryos by DiI injection and subsequent tracing of labeled cells. A. A 1.5 dpf cavefish embryo showing the site of DiI injection (arrowhead). B. Fluorescence image of the embryo in A showing migration of DiI injected cells. DiI: original injection site. C, D. Higher magnification images of insets in B showing morphology of injected cells (arrows). E, F. A cavefish embryo after L-DOPA assay showing tyrosinase positive melanoblasts (arrows). E. Whole mount viewed laterally. F. Section through the trunk. G-J. Whole mounts of tail fins of adult surface fish (G) and cavefish (H-J). G. The surface fish fin has melanophores (arrows). H. The cavefish fin lacks melanophores. I. The cavefish fin has melanoblasts (arrows) that can convert exogenously supplied L-DOPA to melanin. J. Cavefish melanoblasts lack the ability to convert L-tyrosine to L-DOPA and melanin. K. Diagram of pigment cell development from the neural crest derived precursor cells showing the location of the pigmentation block in cavefish. L. The melanin biosynthesis pathway showing the relationships between L-tyrosine, tyrosinase, OCA2, L-DOPA, downstream substrates and enzymes, and melanin.

Tyrosinase-Positive Melanoblasts in Cavefish

The early events of pigment cell formation and diversification are not completely understood in vertebrates. However, it seems likely that iridophores, xanthophores, and melanophores are derived from the same neural crest cell lineage (Fig. 9K). Since cavefish lacking melanophores contain appreciable numbers of other pigment cell types (Rasquin 1947, McCauley *et al.* 2004), the lesion in melanophore development probably lies downstream of the split between the pigment cell progenitors (Fig. 9K).

Melanophore differentiation involves the initial formation of colorless melanoblasts, which subsequently synthesize black melanin pigment and become functional melanophores. The biochemical steps involved in melanin synthesis during the transition from melanoblast to melanophore are well known (Fig. 9L). First, the essential amino acid L-tyrosine is transported from the cytoplasm into the melanosome, where it is converted to L-DOPA by the multifunctional enzyme tyrosinase. Next, L-DOPA is converted into melanin within the melanosome by a series of enzymatic reactions, the first of which is also catalyzed by tyrosinase. Most subsequent reactions in this biosynthetic pathway are spontaneous.

Tyrosinase is the key enzyme in melanogenesis. Do cavefish pigment progenitor cells have functional tyrosinase? Tyrosinase activity was determined by the L-DOPA assay, in which melanin production is determined after exogenous L-DOPA is provided to fixed specimens. The L-DOPA assay showed that Pachón, Chica, Los Sabinos, Tinaja, and Curva cavefish all exhibit active tyrosinase in cells resembling melanoblasts in their morphology and location within the embryo (Fig. 9E, F). Tyrosinase positive melanoblasts were also observed in the scales and fins (Fig. 9G-I) of adult cavefish. The inability of these melanoblasts to synthesize melanin is due to a block in the melanogenic pathway immediately upstream of the tyrosinase dependent steps.

Cavefish Melanoblasts are Unable to Convert L-tyrosine into L-DOPA

The first step in melanin synthesis is the conversion of L-tyrosine to L-DOPA, which is also catalyzed by tyrosinase (Fig. 9L). Cavefish must have L-tyrosine itself because it is required for protein synthesis. Because cavefish seem to lack endogenous melanosomal L-DOPA there may be a deficiency in the ability of L-tyrosine to be converted to L-DOPA. This possibility was investigated by a modified L-DOPA assay in which exogenous L-tyrosine was provided to fixed specimens instead of L-DOPA

(McCauley *et al.* 2004). If cavefish are able to convert L-tyrosine to L-DOPA then black pigment would be deposited in the same cells that have active tyrosinase. However, melanin deposition could not be detected in cavefish exposed to excess L-tyrosine (Fig. 9H-J). The results show that cavefish melanoblasts are unable to convert L-tyrosine to melanin, implying that melanogenesis is blocked because cytoplasmic L-tyrosine cannot be transported into cavefish melanosomes (Fig. 9L).

Developmental and Genetic Studies Converge to Define the Defect in Cavefish Melanogenesis

In lacking the ability to convert L-tyrosine to L-DOPA, cavefish resemble the most common type of human tyrosinase-positive albinism: type II oculocutaneous albinism (OCA2) (Oeting and King 1999). OCA2 albinism is caused by mutations in the *oca2* gene, the human homologue of the mouse *pink-eyed dilution* (*p*) gene, which encodes a 100-kDa integral membrane protein (Rosenblatt *et al.* 1994). Although its precise function is unclear, the OCA2 protein has been proposed either to facilitate L-tyrosine transport into the melanosome or to generate a proton flux regulating melanosome pH, which is important in melanin synthesis (Brilliant 2001, Toyofuku *et al.* 2002). A single recessive gene controls cavefish tyrosinase-positive albinism (Sadoglu 1957). QTL analysis and a candidate gene survey show that different mutations in the *oca2* gene are responsible for albinism in Molino and Pachón cavefish (Protas *et al.* 2006). Thus, genetic and developmental studies have pinpointed the defect in cavefish melanophore pigmentation: the conversion of L-tyrosine to L-DOPA within the melanosome mediated by OCA2.

DEVELOPMENTAL INFERENCES INTO EVOLUTION

Comparative studies of *Astyanax* surface fish and cavefish development provide important insights into the processes and mechanisms of regressive evolution. In the final section of this chapter, we discuss these insights.

Developmental Constraints

Developmental constraints restrict the amplitude of evolutionary changes, or make them unlikely or impossible, by limiting developmental flexibility. This lack of flexibility appears to have a very important role in cavefish regressive evolution. If the eye is ultimately lost, why is it necessary to construct an eye in the first place? The answer may be that early steps in eye development are required to induce other parts of the embryo, and thus the modification of these steps would be fatal. To our knowledge, eyes are initially formed and then degraded during larval or adult development in all cave dwelling vertebrates (Eigenmann 1909, Durand 1976, Berti *et al.* 2001). Indeed, we feel that cave vertebrates lacking embryonic eye primordia will not be discovered because of this strong developmental constraint.

Exactly how the eyes are lost, which we have seen to be under the control of midline signaling in *Astyanax*, reflects another developmental constraint. Because all vertebrates have bilateral eyes arising from a single medial optic field, the subsequent separation of optic fields is likely to be an ancient vertebrate trait that evolved in concert with other head features. Thus, if the Shh midline-signaling pathway is altered, as we have seen in cavefish, there may be automatic consequences on eye development, in this case leading to degeneration (Jeffery 2005).

The way in which retinal development is arrested in cavefish may be a further example of developmental constraint. We have shown that the arrest of retinal development is not caused by inhibition of cell division at the CMZ, which would seem to be the simplest way to stop growth. Instead, retinal growth is curtailed by apoptosis of newly born cells (Strickler *et al.* 2007a). This must be a very costly process in terms of energy expenditure, so why has inhibition of cell proliferation, the most simple and least expensive route to preventing retina development, not been taken? The probable answer lies in the fact that the retina is actually a part of the brain. In both retina and brain, stem cells replenish the laminated areas through the same course of action, which may not be easy to modify. Accordingly, killing new cells after they proliferate in the retina may be more allowable than blocking stem cell division in the CMZ because of an ancient constraint on how different parts of the brain grow in concert during development.

A final example of a developmental constraint occurs in the melanophore development pathway. In cavefish, a block in the pigment cell-generating pathway occurs relatively late in the pigment cell development pathway, during the conversion of melanoblasts into melanophores. Earlier steps in this pathway, such as the determination and migration of neural crest cells, the restriction to pigment cell fate, and the diversification of different pigment cell lineages is apparently not changed, even though the usefulness of any pigment cell type is questionable in cavefish. The reason neural crest cells must be formed is clear: they have many critical derivatives, the loss of which would be lethal. Why the other pigment cell types are formed in cavefish, albeit possibly at reduced levels, is a mystery. They may have some unknown and indispensable function even in the dark cave environment.

It is clear that cavefish regressive evolution is channeled to a large extent by developmental constraints. Only those pathways that are not linked to other essential pathways can be degraded or truncated.

Developmental Amplification

A major question in contemporary evolutionary developmental biology is how large-scale changes in the phenotype can occur rapidly during evolution. Cavefish show that slight changes in early development can have major impacts at later stages and in the adult. An example is the craniofacial skeleton.

The differences in craniofacial skeletons between cavefish and surface fish, in particular the ocular bones surrounding the eye, are so extreme that they were formerly used to support their designation as separate genera (Alvarez 1947). However, the majority of these changes are related to whether or not a large eye punctuates the craniofacial skeleton. When the eye is absent from the surface of the head, as in cavefish, the craniofacial skeleton is patterned differently from when an eye is present. Major changes in the craniofacial skeleton can be elicited by transplanting a surface fish lens into a cavefish optic cup during early development (Yamamoto *et al.* 2003). The sequence of events is as follows: a normal lens induces anterior eye parts and promotes the growth of a normal retina, producing a large growing eye, which in turn dictates the morphology of surrounding bones in the adult (Fig. 8). This scenario illustrates that amplification of small changes in early development can produce large differences in adult morphology.

Pleiotropy

Pleiotropy, the condition in which a single gene may control multiple, seemingly unrelated phenotypes, has long been considered as a possible mechanism for the evolution of regressive traits in cave animals (Barr 1968). Accordingly, if downregulation of genes controlling eye development simultaneously increases the development of a beneficial trait, such as olfaction or another sensory system, the latter might be adaptive and subject to natural selection. The discovery of enhanced midline signaling mediated by highly pleiotropic *hh* genes (Yamamoto *et al.* 2004) opens many possibilities that may be able to explain eye degeneration in cavefish. As we have seen, Hh overexpression has a negative effect on eye development, and it is known from studies on other vertebrates that Hh signaling has positive effects on many other developmental traits. Thus, selection for the positive traits would automatically affect the negative ones. In the future, it will be important to determine the identity of positive traits influenced by Hh signaling. Pleiotropic events could also explain the loss of melanophores, if the precursors of these cells are used for another purpose that benefits survival in the cave environment.

Evolutionary Forces

Two hypotheses have been championed to explain the regressive features of cave animals: (1) the accumulation of selectively neutral (loss of function) mutations and (2) indirect selection based on energy conservation and/or antagonistic pleiotropy (Culver 1982, Jeffery 2005). Although neither hypothesis has been proved in the case of *Astyanax* cavefish, developmental studies generally support one or the other for loss of pigmentation and eyes respectively.

In the case of eye loss, the developmental information seems to support selection over neutral mutation. First, the genes involved in eye development that have been studied thus far do not appear to have mutated to a degree in which they have lost function. In addition, the restoration of eyes by lens transplantation suggests that all genes that act downstream of lens function are present and potentially active in cavefish. Also supportive of selection is that most genes with modified expression patterns, such as those in the Shh signaling pathway and *hsp90α*, increase rather than decrease their activity in cavefish.

In contrast to eye regression, developmental studies on loss of pigmentation could support either selection or neutral mutation. On one hand, the accumulation of neutral mutations resulting in loss of melanophores might be possible, especially if the *oca2* gene is not pleiotropic and its disruption does not affect other important developmental pathways. Recent genetic analysis, in which individual QTL governing the extent of melanophore development has been shown to either increase or decrease melanophore abundance, support the role of neutral mutation and genetic drift (Protas *et al.* 2007). On the other hand, as we have discussed above, melanogenesis may be disrupted because it is adaptive, allowing pigment cell precursors to be shunted into other, more beneficial differentiation pathways. Some of these possibilities are testable and predict a bright future for developmental biology in addressing both how and why regressive features have evolved in *Astynanx* cavefish.

Acknowledgements

The research from the Jeffery laboratory described in this article was supported by grants from NIH (R01-EY014619) and NSF (IBN-0542384).

References

Alunni, A., A. Menuet, E. Candal, J-B. Pénigault,, W.R. Jeffery and S. Rétaux. 2007. Developmental mechanisms for retinal degeneration in the blind cavefish *Astyanax mexicanus*. *Journal of Comparative Neurology* 505: 221-233.

Alvarez, J. 1947. Descripción de *Anoptichthys hubbsi* caracinindo ceigo de La Cueva de Los Sabinos. *San Luis Potosi Sociedad Mexicana de Historia Natural* 8: 215-219.

Barr, T. 1968. Cave ecology and the evolution of troglobites. *Evolutionary Biology* 2: 35-102.

Behrens, M., H. Wilkens and H. Schmale. 1998. Cloning of the αA-crystallin genes of the blind cave form and the epigean form of *Astyanax fasciatus*: A comparative analysis of structure, expression and evolutionary conservation. *Gene* 216: 319-326.

Berti, R., J.P. Durand, S. Becchi, R. Brizzi, N. Keller and G. Ruffat. 2001. Eye degeneration in the blind cave-dwelling fish *Phreatichys andruzzi*. *Canadian Journal of Zoology.* 79: 1278-1285.

Borowsky, R. 2008. Restoring sight in blind cavefish. *Current Biology* 18: R23-R24.

Brilliant, M.H. 2001. The mouse p (pink-eyed dilution) and human P genes, oculocutaneous albinism type 2 (OCA2), and melanosomal pH. *Pigment Cell Research* 14: 86-93.

Cahn, P.H. 1958. Comparative optic development in *Astyanax mexicanus* and in two of its blind cave derivatives. *Bulletin of the American Museum of Natural History* 115: 73-112.

Culver, D. 1982. *Cave Life. Evolution and Ecology*. Harvard, Cambridge, MA.

Dowling, T.E., D.P. Martasian, and W.R. Jeffery 2002. Evidence for multiple genetic lineages with similar eyeless phenotypes in the blind cavefish, *Astyanax mexicanus*. *Molecular Biology and Evolution* 19: 446-455.

Durand, J.P. 1976. Ocular development and involution in the European cave salamander, *Proteus anguinus* Laurenti. *Biological Bulletin of the Marine Biological Laboratory, Woods Hole* 151: 450-466.

Eigenmann, C.H. 1908. The eyes of the blind vertebrates of North America. V. The history of the eye of blind *Amblyopsis* from its appearance to its disintegration in old age. *Contributions of the Zoological Laboratory, Indiana University* Mark Anniversary Volume: 167-204.

Ekker, S.C., A.R. Ungar, P. von Greenstein, J.A. Porter, R.T. Moon and P. Beachy. 1995. Patterning activities of vertebrate hedgehog proteins in the developing eye and brain. *Current Biology* 5: 944-955.

Erickson, C.A. 1993. From the crest to the periphery: Control of pigment cell migration and lineage segregation. *Pigment Cell Research* 6: 336-347.

Fernald, R.D. 1991. Teleost vision: Seeing while growing. *The Journal of Experimental Zoology* 5: 167-180.

Hooven, T.A., Y. Yamamoto and W.R. Jeffery. 2004. Blind cavefish and heat shock . protein chaperones: A novel role for hsp90α in lens apoptosis. *International Journal of Developmental Biology* 48: 731-738.

Hüppop, K. and H. Wilkens. 1991. Bigger eggs in subterranean *Astyanax fasciatus* (Charcidae: Pisces). Their significance and genetics. *Zeitschrift für zoologische Systematik und Evolutionsforschung* 29: 280-288.

Jeffery, W.R. 2001. Cavefish as a model system in evolutionary developmental biology. *Developmental Biology* 231: 1-12.

Jeffery, W.R. 2005. Adaptive evolution of eye degeneration in the Mexican blind cavefish. *Journal of Heredity* 96: 185-196.

Jeffery, W.R. 2006. Regressive evolution of pigmentation in the cavefish *Astyanax*. *Israel Journal of Ecology and Evolution* 52: 405-422.

Jeffery, W.R. 2008. Emerging systems in Evo/Devo: cavefish and mechanisms of microevolution. *Evolution and Development* 10: 265-272.

Jeffery, W.R. and D.P. Martasian. 1998. Evolution of eye regression in the cavefish *Astyanax*: Apoptosis and the *Pax-6* gene. *The American Zoologist* 38: 685-696.

Jeffery, W.R., A.G. Strickler and Y. Yamamoto. 2003. To see or not to see: Evolution of eye degeneration in Mexican blind cavefish. *Comparative and Integrative Biology* 43: 531-541.

Jeffery, W.R., A.G. Strickler, S. Guiney, D. Heyser and S.I. Tomarev. 2000. Prox1 in eye degeneration and sensory organ compensation during development and evolution of the cavefish *Astyanax*. *Genes Development and Evolution* 210: 223-230.

Krauss, S., T. Johannsen, V. Korzh and A. Fijose. 1991. Zebrafish pax[zf-a]: A paired box gene expressed in the neural tube. *European Molecular Biology Organization Journal* 10: 3609-3619.

Langecker, T.G. H. Schmale and H. Wilkens. 1993. Transcription of the opsin gene in degenerate eyes of cave-dwelling *Astyanax fasciatus* (Teleostei: Characidae) and its conspecific ancestor during early ontogeny. *Cell and Tissue Research* 273: 183-192.

Langeland, J.A. and C.B. Kimmel. 1997. Fishes. In: *Embryology: Constructing the Embryo*. S.F. Gilbert and A.M. Raunio (eds). Sinauer, Sunderland, MA.

Le Douarin, N.M. and C. Kalcheim. 1999. *The Neural Crest*. Cambridge University Press, New York. 2nd edition.

Long, W.L. 1983. The role of the yolk syncytial layer in determination of the plane of bilateral symmetry in the rainbow trout, *Salmo gairdneri*, Richardson. *The Journal of Experimental Zoology* 228: 91-97.

McCauley, D.W., E. Hixon and W.R. Jeffery. 2004. Evolution of pigment cell regression in the cavefish *Astyanax*: A late step in melanogenesis. *Evolution and Development* 6: 209-218.

Macdonald, R., K. Anukampa Barth, Q. Xu, N. Holder, I. Mikkola and S. Wilson. 1995. Midline signalling is required for *Pax6* gene regulation and patterning of the eyes. *Development* 121: 3267-3278.

Menuet, A., A. Alunni, J-S. Joly, W.R. Jeffery and S. Rétaux. 2007. Shh overexpression in *Astyanax* cavefish: Multiple consequences on forebrain development and evolution. *Development* 134: 845-855.

Mitchell, R.W., W.H. Russell and W.R. Elliot. 1977. Mexican eyeless characin fishes, genus *Astyanax*: Environment, distribution, and evolution. *Special Publications of the Museum, Texas Tech University* 12: 1-89.

Morales, A.V., J.A. Barbas and M.A Nieto. 2005. How to become neural crest: From segregation to delamination. *Seminars in Cell and Developmental Biology* 16: 655-662.

Oeting, W.S. and R.A. King 1999. Molecular basis of albinism: Mutations and polymorphisims of pigmentation genes associated with albinism. *Human Mutation* 13: 99-113.

Protas, M., M. Conrad, J.B. Gross, C. Tabin and R. Borowsky. 2007. Regressive evolution in the Mexican cave tetra, *Astyanax mexicanus*. *Current Biology* 17: 452-454.

Protas, M.E., C. Hersey, D. Kochanek, Y. Zhou, H. Wilkens, W.R. Jeffery, L.T. Zon, R. Borowsky and C.J. Tabin. 2006. Genetic analysis of cavefish reveals molecular convergence in the evolution of albinism. *Nature Genetics* 38: 107-111.

Püschel, A.W., P. Gruss and M. Westerfield. 1992. Sequence and expression pattern of pax-6 are highly conserved between zebrafish and mice. *Development* 114: 643-651.

Rasquin, P. 1947. Progressive pigmentary regression in fishes associated with cave environments. *Zoologia* 32: 35-44.

Rosenblatt, S., D. Durham-Pierce, J.M. Garner, Y. Nakatsu, M.H. Brilliant and S.J. Orlow. 1994. Identification of a melanosomal membrane protein encoded by the pink-eyed dilution (type II oculocutaneous albinism) gene. *Proceedings of the National Academy of Sciences of the United States of America USA* 91: 12071-12075.

Sadoglu, P. 1957. A Mendelian gene for albinism in natural cave fish. *Experientia* 13: 394.

Schwarz, M., F. Cecconi, G. Berneir, N. Andrejewski, B. Kammandel, M. Wagner and P. Gruss. 2000. Spatial specification of mammalian eye territories by reciprocal transcriptional repression of Pax2 and Pax6. *Development* 127: 4325-4334.

Soares, D., Y. Yamamoto, A.G. Strickler and W.R. Jeffery. 2004. The lens has a specific influence on optic nerve and tectum development in the blind cavefish *Astyanax*. *Developmental Neuroscience* 26: 308-317.

Strecker, U., L. Bernachez and H. Wilkens. 2003. Genetic divergence between cave and surface populations of *Astyanax* in Mexico (Characidae, Teleostei). *Molecular Ecology* 12: 699-710.

Strecker, U., V.H. Faúndez and H. Wilkens. 2004. Phylogeography of surface and cave *Astyanax* (Teleostei) from Central and North America based on cytochrome *b* sequence data. *Molecular Phylogenetics and Evolution* 33: 469-481.

Strickler, A.G., Y. Yamamoto and W.R. Jeffery. 2001. Early and late changes in *Pax6* expression accompany eye degeneration during cavefish development. *Development Genes and Evolution* 211: 138-144.

Strickler, A.G., K. Famuditimi and W.R. Jeffery. 2002. Retinal homeobox genes and the role of cell proliferation in cavefish eye degeneration. *International Journal of Developmental Biology* 46: 285-294.

Strickler, A.G., Y. Yamamoto and W.R. Jeffery. 2007a. The lens controls cell survival in the retina: Evidence from the blind cavefish *Astyanax*. *Developmental Biology* 311: 512-523.

Strickler, A.G., M.S. Byerly and W.R. Jeffery. 2007b. Lens gene expression analysis reveals downregulation of the anti-apoptotic chaperone α A crystallin during cavefish eye degeneration. *Development Genes and Evolution* 217: 771-782.

Take-uchi, M., J.D. Clarke and S.W. Wilson. 2003. Hedgehog signalling maintains the optic stalk-retinal interface through the regulation of Vax gene activity. *Development* 130: 955-968.

Toyofuku, K., J.C. Valencia, T. Kushimoto, G-E. Costin, V.M. Virador, W.D. Viera, V.J. Ferrans and V.J. Hearing. 2002. The etiology of oculocutaneous albinism (OCA) type II: The pink protein modulates the processing and transport of tyrosinase. *Pigment Cell Research* 15: 217-224.

Wilkens, H. 1988. Evolution and genetics of epigean and cave *Astyanax fasciatus* (Characidae, Pisces). *Evolutionary Biology* 23: 271-367.

Yamamoto Y. and W.R. Jeffery 2000. Central role for the lens in cavefish eye degeneration. *Science* 289: 631-633.

Yamamoto, Y., D.W. Stock and W.R. Jeffery. 2004. Hedgehog signalling controls eye degeneration in blind cavefish. *Nature* 431: 844-847.

Yamamoto, Y., L. Espinasa, D.W. Stock and W.R. Jeffery. 2003. Development and evolution of craniofacial patterning is mediated by eye-dependent and-independent processes in the cavefish *Astyanax*. *Evolution and Development* 5: 435-446.

Yoshizawa, M. and W.R. Jeffery. 2008. Shadow response in the blind cavefish *Astynanx* reveals conservation of a functional pineal eye. *The Journal of Experimental Biology* 211: 292-299.

CHAPTER 7

Subterranean Fishes of North America: Amblyopsidae

Matthew L. Niemiller[1] and Thomas L. Poulson[2]

[1]Department of Ecology and Evolutionary Biology, University of Tennessee, Knoxville, Tennessee, 37996, USA
E-mail: mniemill@utk.edu
[2]Emeritus Professor, University of Illinois-Chicago
E-mail: tomandliz@bellsouth.net

INTRODUCTION

The Amblyopsid cavefishes, family Amblyopsidae, have been viewed as a model system for studying the ecological and evolutionary processes of cave adaptation because the four cave-restricted species in the family represent a range of troglomorphy that reflect variable durations of isolation in caves (Poulson 1963, Poulson and White 1969). This group has both intrigued and excited biologists since the discovery and description of *Amblyopsis spelaea*, the first troglobitic fish ever described, in the early 1840s. Other than the Mexican cavefish (*Astyanax fasciatus*), cave Amblyopsids are the most comprehensively studied troglobitic fishes (Poulson, this volume).

The Amblyopsidae (Fig. 1) includes species with some unique features for all cavefish. *Typhlichthys subterraneus* is the most widely distributed of any cavefish species. Its distribution spans more than 5° of latitude and 1 million km^2 (Proudlove 2006). *Amblyopsis spelaea* is the only cavefish known to incubate eggs in its gill chamber. In fact, this species is the only one of the approximately 1100 species in North America with this behavior. The Amblyopsidae is the most specious family of subterranean fishes in the United States containing four of the eight species recognized. Two other

Fig. 1 Members of the Amblyopsidae. The family includes (A) the surface-dwelling swampfish (*Chologaster cornuta*), (B) the troglophile spring cavefish (*Forbesichthys agassizii*), and four troglobites: (C) the southern cavefish (*Typhlichthys subterraneus*), (D) the northern cavefish (*Amblyopsis spelaea*), (E) the Ozark cavefish (*A. rosae*), and (F) the Alabama cavefish (*Speoplatyrhinus poulsoni*). Photos courtesy of Uland Thomas (A), Dante Fenolio (E), and Richard Mayden (F).

Color image of this figure appears in the color plate section at the end of the book.

families have species adapted to subterranean habitats, the Ictaluridae and Cottidae. Two species of ictalurids are endemic to the Edwards Aquifer of Texas, the widemouth blindcat (*Satan eurystomus*) and the toothless blindcat (*Trogloglanis pattersoni*); and troglomorphic populations of sculpins (*Cottus* sp.) are known from Missouri and Pennsylvania.

Our primary aims of this chapter are to provide an up-to-date, comprehensive review of the pertinent literature and unpublished research (Niemiller on phylogenetics and phylogeography and Poulson on pigment, foraging patterns, metabolic efficiencies, and metabolic acclimation to temperature) regarding Amblyopsid cavefishes, including their surface and troglophilic relatives in the family. We begin with a synopsis of the history of Amblyopsid research from the early 1840s to the present day. Here we discuss the significant studies and researchers that have shaped the knowledge base. Next, we review the systematic relationships and taxonomy of the family. The Amblyopsidae have only recently been the subject of phylogenetic examination using molecular markers and we include preliminary data here. We then discuss the biogeography, habitat, and distribution of the family. In the next section, we review morphology of the family in general and morphology of cave adaptation in particular. We include some new data on pigment systems. Subsequently, we summarize aspects of ecology, life history, and behavior including topics such as reproduction, diet, parasites and disease, longevity, metabolism, and demography. Here we also include some new data. Finally, we end with a discussion of the conservation status and threats to members of the family. In each section, we suggest avenues of future research that are needed.

Our review is intended to be as detailed and comprehensive as possible and to include the majority of relevant references. With this in mind, we hope that this chapter will serve as both a stimulus for future research and an exhaustive bibliographic reference regarding Amblyopsid biology.

I. HISTORY OF AMBLYOPSID RESEARCH

In this section, we provide a brief history of Amblyopsid research beginning with the discovery of the first member of the family in the early 1840s. By no means is this review meant to encompass all papers on Amblyopsids. Instead, we attempt to outline the major studies and prominent players in Amblyopsid research, particularly during the early years from 1840 to 1910. For more complete reviews on the history of hypogean fish research including Amblyopsids, readers should peruse the works of Romero (2001) and Proudlove (2006).

The description of the northern cavefish, *Amblyopsis spelaea* (= *A. spelaeus*), from the River Styx in Mammoth Cave, Kentucky, by DeKay (1842) represents the first scientifically acknowledged report of a troglomorphic fish species (Romero 2001, Proudlove 2006). However, probable reports of *A. spelaea* occurred earlier. James Flint in 1820 recounted a conversation with a local in Indiana who stated that a neighbor found 'blind fishes' when digging a well (Flint 1820, Romero 2001, Proudlove 2006). The first published account of *A. spelaea* from the Mammoth Cave area appeared in the book *An Excursion to the Mammoth Cave, and the Barrens of Kentucky* by Robert Davidson (Davidson 1840). His observations represent the first biological observations of the species, and of a troglobitic fish (Proudlove 2006).

The original description of *A. spelaea* by DeKay was not very detailed but incited the interest of other researchers. Wyman (1843ab, 1851, 1854ab, 1872) conducted first detailed examination of the internal anatomy and brain. Tellkampf (1844) provided detailed descriptions of *A. spelaea* from Mammoth Cave. Amblyopsids were of particular interest to Jean Louis Agassiz who published the first insight on the potential importance of troglomorphic fishes to biological research in 1847 (Romero 2001). Agassiz (1847) described a research plan involving *A. spelaea* to investigate the embryology, anatomy, and effects of light on the species. Although he never followed through with his plans, several of his students investigated aspects of Amblyopsid biology (see below). Agassiz (1853) did, however, describe the epigean swampfish, *Chologaster cornuta* (= *C. cornutus*).

Between 1850 and 1900, troglomorphic fishes, and Amblyopsids in particular, were prominent in the debate on evolution. Several prominent biologists including Frederick Ward Putnam (a former student of Agassiz), Alpheus Packard, Edward Lankester, Alpheus Hyatt, Jeffries Wyman, and Carl H. Eigenmann discussed the evolution of troglomorphy in cave fauna, and of cavefish especially. Their influence on and contributions to cave biology are eloquently detailed by Romero (2001). Also during this time period, four Amblyopsid species were described (three are currently recognized). The southern cavefish, *Typhlichthys subterraneus*, was described by Girard (1859) from a well near Bowling Green, Kentucky. Girard thought this new species was a transitional form between *A. spelaea* and the entirely epigean species, *C. cornuta*. In 1872, Frederick Ward Putnam described the spring cavefish from a well near Lebanon, Tennessee (Putnam 1872). Ten years later, Forbes (1882) described *C. papilliferus*.

At the end of nineteenth and into the twentieth century, Carl H. Eigenmann was the prominent figure in cave research. Between 1887 and 1909, much of his work focused on understanding the loss of visual structures in cave vertebrates. Amblyopsids were included in many of his studies. In 1898, Eigenmann described the Ozark cavefish, *Amblyopsis rosae* (= *Typhlichthys rosae*). In 1905, Eigenmann (1905) described both *T. osborni* and *T. wyandotte*. Both species would later be synonymized under *T. subterraneus*. Between 1898 and 1905 alone, he published at least 39 papers and abstracts on cave vertebrates focusing on the morphology and development of loss of vision (Romero 2001). His work culminated in his 1909 book titled *Cave Vertebrates of North America: A Study in Degenerative Evolution* (Eigenmann 1909) in which he detailed the structure, development, and degeneration of the eye in several cave-dwelling vertebrates known to science at the time. Amblyopsids were a major focus of his investigations in this volume. The volume also includes an important chapter by Fernandus Payne on laboratory observations and experiments on the feeding and sensory capabilities of *A. spelaea*. Between the writings of Eigenmann and the early 1960s, few studies centered on Amblyopsids.

However, the family received renewed interest from biologists beginning with the systematic investigation by Woods and Inger (1957). In 1955, Thomas L. Poulson began a comparative study of the Amblyopsidae. His Ph.D. dissertation included aspects of morphology, physiology, life history, ecology, and behavior. Using Amblyopsids as a 'natural evolutionary experiment', Poulson (1960, 1963) inferred evolutionary and ecological patterns of cave adaptation and attempted to outline a step-by-step process of increasing subterranean (troglomorphic) specialization from preadapted surface to obligate cave-dwelling species. Poulson has continued to study evolutionary and ecological aspects of cave adaptation to the present day. His multiple hypotheses and tests of these hypotheses have influenced many other researchers. In fact, the most recent Amblyopsid to be described, the Alabama cavefish (*Speoplatyrhinus poulsoni*), was named in his honor (Cooper and Kuehne 1974), partly because of his predictions of what the next stage of troglomorphy would look like. Much of our knowledge about the biology of Amblyopsid cavefishes has been obtained by the numerous studies he has conducted, many of which are referenced in more detail in this chapter. He continues to study these fishes and several new workups of old unpublished data are included in this chapter.

Amblyopsid cavefishes continue to intrigue and excite biologists attempting to discern the ecological and evolutionary facets of cave adaptation. Although many studies in the last twenty years have focused on the demography, distribution, conservation status, and threats of the cave-dwelling species, other aspects of biology including the phylogenetic relationships, biogeography, and life history continue to evoke interest. We have attempted to summarize what we feel are the most relevant references for each of these areas in subsequent sections of this chapter. We apologize in advance to authors of any extensive or insightful studies we may have missed and hope the authors will work with us to redress any such omissions.

II. TAXONOMY AND SYSTEMATICS

The Amblyopsidae have been known to science since the 1840s. The family-group name (Amblyopsidae-Amblyopsini) was first used by Bonaparte (1846). Previously, the family names Heteropygii and Hypsaeidae were used by Tellkampf (1844, 1845) and Storer (1846) respectively, but these names are considered unavailable because they were not formed from the stem of an available genus-group name (Poly and Proudlove 2004). Aspects of taxonomy and systematics of the family have been examined by Cox (1905), Eigenmann (1909), Woods and Inger (1957), Swofford *et al.* (1980), Swofford (1982), Bergstrom *et al.* (1995), Bergstrom (1997), and Niemiller and Fitzpatrick (2008), Near *et al.* (unpublished data). The latter four studies were molecular in nature. More recently, Poly and Proudlove (2004) reviewed taxonomy and systematic relationships of the Amblyopsidae.

Presently five genera and six species are recognized within the Amblyopsidae (Fig. 1). These include the epigean swampfish (*Chologaster cornuta* Agassiz), a troglophile, the spring cavefish (*Forbesichthys agassizii* Putnam), and four troglobitic species in order of presumed increasing time of isolation in caves: southern cavefish (*Typhlichthys subterraneus* Girard), northern cavefish (*Amblyopsis spelaea* DeKay), Ozark cavefish (*A. rosae* Eigenmann), and the Alabama cavefish (*Speoplatyrhinus poulsoni* Cooper and Kuehne). We will briefly discuss the higher-level relationships of the Amblyopsidae before discussing nomenclature and interspecific relationships within the family.

A. Higher-level Relationships

The Amblyopsidae have been considered the sister group to the pirate perches, family Aphredoderidae (Rosen 1962, Patterson 1981, Patterson and Rosen 1989) but also share a close affinity with the trout perches, family Percopsidae (Rosen 1962). Together, these three families have been included in the order Percopsiformes (Greenwood *et al.* 1966, Nelson 1984, 2006). However, some have questioned the monophyly of this order. Murray and Wilson (1999) suggested Amblyopsids might be more closely related to the Anacanthines and proposed recognition of Amblyopsids as a distinct order, the Amblyopsiformes. Poly in Poly and Proudlove

(2004) suggested that Amblyopsids might be related to Gobioids because of the distribution of Amblyopsids in relation to the former Mississippi Embayment (see below), similarities in patterns of neuromasts on the body, and similarities in morphology, particularly when compared with troglobitic Gobioids. McAllister (1968) also suggested a possible relationship between Gobioids and Amblyopsids. Wiley *et al.* (2000) resolved a sister relationship between trout perches and pirate perches based on a combined analysis of morphological and molecular characters but did not include any Amblyopsid species. More recently, a molecular study using both mitochondrial and nuclear markers and including members of all three families supports the inclusion of the Amblyopsidae in the order Percopsiformes with a sister relationship between cavefishes and pirate perches (Smith and Wheeler 2006). Immunological evidence also supports the inclusion of the Amblyopsidae in the order Percopsiformes (Kalayil and Clay 1976).

B. Taxonomy

Agassiz (1853) described the swampfish (Fig. 1A) from ditches in rice fields in South Carolina. Since the description, taxonomy has remained stable. Only two synonyms exist: *Chologaster cornutus* (Agassiz 1853) and *C. avitus* (Jordan and Jenkins in Jordan 1889). The swampfish generally is considered the most basal Amblyopsid. Like the other Amblyopsids, little molecular work has been conducted on the species. The most substantial is the allozyme study by Swofford (1982) who found considerable differentiation among *C. cornuta* populations both among and within drainages. Preliminary evidence using mitochondrial and nuclear DNA markers also indicate significant genetic differentiation across the species range (Niemiller and Fitzpatrick 2008, Near *et al.*, unpublished data). Poulson (1960, 1963) also found considerable variation in adult body size and meristic characters among drainages.

The spring cavefish (Fig. 1B) was originally described as *C. agassizi* by Putnam (1872) from a well near Lebanon, Wilson Co., Tennessee. Later, Forbes (1882) described *C. papilliferus* from a spring in western Union Co., Illinois, on the basis of coloration differences between the Tennessee and Illinois populations. Jordan and Evermann (1927) erected a new genus, *Forbesella*, citing that the subterranean nature of spring cavefish warrants separate recognition. Jordan (1929) later replaced *Forbesella* with *Forbesichthys*, as the former was preoccupied in tunicates. This genus is still considered a junior synonym of *Chologaster* by some authors, however. Woods and Inger (1957) noted that populations of spring cavefish from southern Illinois, central Kentucky, and central Tennessee all differed slightly but did not warrant specific or subspecific designation. Therefore,

C. papilliferus was synonymized under *C. agassizi* and their revision has been followed by most subsequent authors with the exception of Clay (1975) who maintained that *C. agassizi* and *C. papilliferus* are specifically distinct. Allozyme analyses by Swofford (1982) revealed considerable differentiation between populations that justified resurrection of the genus *Forbesichthys*, which was elevated later by, Page and Burr (1991). Spring cavefish have not been the subject of phylogenetic studies since Swofford (1982); however, recent evidence suggests that populations from Illinois (formerly *papilliferus*) and Tennessee may be phylogenetically distinct at both mitochondrial and nuclear markers (Near *et al.*, unpublished data). Kentucky populations have yet to be examined.

The southern cavefish (Fig. 1C) was described by Girard (1859) from a well near Bowling Green, Warren Co., Kentucky. Later, Eigenmann (1905) described both *T. osborni* and *T. wyandotte* based on differences in head width and eye diameter. *Typhlichthys osborni* was described from Horse Cave, Kentucky. *Typhlichthys wyandotte* was described from a well near Corydon, Indiana, that was later destroyed. Recently, a well-like entrance into a cave on the property of a car dealership in Corydon was discovered and is believed to represent the type locality (Black in Lewis 2002). Regardless, this species is generally considered invalid and was not listed as a locality in Woods and Inger (1957). Recent surveys in the vicinity of Corydon have failed to document *T. subterraneus*, finding only *A. spelaea* (Lewis 1998, Lewis and Sollman 1999). *Typhlichthys eigenmanni* (*nomen nudeum*) was described as a fourth species in the genus from Camden Co., Missouri (likely River Cave). Recently, Parenti (2006) proposed that *T. eigenmanni* Charlton (1933) is a subjective synonym of *T. subterraneus*. Woods and Inger (1957) synonymized all species under *T. subterraneus* on the basis of lack of any clear geographic pattern in morphological variation. A population from Sloans Valley Cave, Pulaski Co., Kentucky, differs in several ways from populations to the southwest in Tennessee and may represent an undescribed species (Cooper and Beiter 1972, Burry and Warren 1986). However, further details regarding this population have not been published.

Because of the extensive distribution of *T. subterraneus* and the results of molecular studies of other troglobites (especially Culver *et al.* 1995), some authors have speculated that *T. subterraneus* actually represents several independent invasions and, therefore, distinct lineages (Swofford 1982, Barr and Holsinger 1985, Holsinger 2000, Poulson, this volume). Indeed, electrophoretic allozyme analyses by Swofford (1982) showed considerable differentiation among morphologically similar populations of *Typhlichthys*, even those that are geographically close, suggestive of multiple, independent invasions and limited gene flow. In fact, six of the thirteen populations Swofford sampled were monomorphic and for different allozyme alleles. However, owing to small sample size, Swofford's study was limited in its ability to distinguish modular or hierarchical subdivision from a continuous relationship between genetic and geographic distance. More recently, Bergstrom *et al.* (1995) and Bergstrom (1997) investigated phylogenetic relationships of populations west of the Mississippi River using mitochondrial DNA. Although limited, these studies revealed considerable variation among populations. Likewise, Niemiller and Fitzpatrick (2008) examined genetic variation among eastern populations of *Typhlichthys* in Alabama, Georgia, and Tennessee. Again, significant genetic divergence was observed and both mitochondrial and nuclear DNA variation was structured among hydrological drainages. Molecular and morphological evidence also indicate that Arkansas populations warrant recognition of a distinct species (Graening *et al.*, unpublished data). These studies support the hypothesis that morphological similarity is the result of parallel evolution rather than significant dispersal and gene flow across major drainage and river divides (Cooper and Iles 1971, Barr and Holsinger 1985).

The northern cavefish (Fig. 1D) was the first troglobitic fish ever described in the scientific literature. The formal description is based on fish from Mammoth Cave, Edmonson Co., Kentucky by DeKay (1842). Originally spelled *Amblyopsis spelaeus*, Woods and Inger (1957) corrected the name to *A. spelaea* and provided the most important definition of the species (Proudlove 2006). Swofford (1982) found that populations north and south of the Ohio River are monomorphic for the same allozyme alleles with no heterozygosity. The data do not contradict Poulson (1960) who observed a distinct split in morphological variation across the Ohio River. Rather than indicating pre-Pleistocene dispersal as Poulson (1960) proposed, Swofford concluded that populations north and south of the river likely are of more recent origin during the Pleistocene and insufficient time has elapsed for significant genetic differentiation. Clearly more work is needed to elucidate the biogeographic history of the species. No studies to date have examined mitochondrial or nuclear DNA differentiation within the species.

The Ozark cavefish (Fig. 1E) was first reported from specimens collected from caves near Sarcoxie, Jasper Co., Missouri, by Garman (1889) and identified as *T. subterraneus*. The name *Typhlichthys rosae* was applied to the species in several of Eigenmann's papers in 1898 (for a review see Poly and Proudlove 2004). The Ozark cavefish was reclassified as *Troglichthys rosae* by Eigenmann (1898, 1899a) and this name stood until Woods and Inger (1957) placed the species in the genus *Amblyopsis*, on the basis of morphological similarity with *A. spelaea*. Phylogenetic studies that included samples of *A. rosae* have revealed considerable divergence across the range. Although Swofford (1982) sampled just two populations, one from the Neosho River drainage in Jasper Co., Missouri, and the other from the Illinois River drainage in Benton Co., Arkansas, these populations were well differentiated and with very low heterozygosities for allozyme loci. The phylogenetic studies of Bergstrom *et al.* (1995) and Bergstrom (1997) also revealed distinct genetic structure associated with each major drainage. Bergstrom (1997) proposed that mitochondrial genetic divergences were so pronounced that at least subspecific designation was warranted. At least four genetically differentiated groups associated with distinct drainage basins were identified: 1) Illinois River drainage in northwestern Arkansas, 2) White River drainage in southwestern Missouri, 3) Neosho River drainage in southwestern Missouri, and 4) Neosho River drainage in northeastern Oklahoma. Genetic variation within drainages also suggested that many localities are genetically isolated and, therefore, constitute distinct endemic populations (Noltie and Wicks 2001).

Cooper and Kuehne (1974) described the Alabama cavefish (Fig. 1F), *Speoplatyrhinus poulsoni*, from Key Cave, Lauderdale Co., Alabama, in honor of Thomas L. Poulson. Little is known about how this species is related to other Amblyopsids. Because *S. poulsoni* is the most troglomorphic of all the subterranean Amblyopsids and thought to represent a relict population, the single cave endemic probably is not closely related to any of the other species (Proudlove 2006). It has at least three features that are completely unlike other Amblyopsids: arrangement and relative sizes of otoliths, size and extent of caudal papillae (Cooper and Kuehne 1974), and absence of any tactile receptors on the head (Poulson, personal observation). In addition, there are two extreme neotenic features: the absence of branched fin rays in adults and the absence of a bilateral supraopercular papilla opening to the head subdermal lateral line system (Cooper and Kuehne 1974).

C.   Interspecific Relationships

Although Amblyopsids have received considerable attention from evolutionary biologists and ichthyologists, comparatively little work has been conducted on the systematics of the family, particularly at the molecular level. Prior to the study of Woods and Inger (1957), nine species in five genera were recognized. As mentioned above, the authors synonymized all species of *Typhlichthys* under *T. subterraneus* and moved *Troglichthys rosae* into the genus *Amblyopsis* and *Forbesichthys* into *Chologaster*. Likewise, *C. papilliferus* was synonymized with *C. agassizi*. Woods and Inger (1957) suggested two phylogenies representing the interspecific relationships of the family (Fig. 2A–B) relying primarily on the presence or absence of two characters: (i) scleral cartilages in the eye and (ii) postcleithrum in the pectoral girdle. However, a reanalysis by Swofford (1982) using the same characters as Woods and Inger (1957) does not support either proposed phylogeny. Instead, Swofford (1982) proposed the phylogenies presented in Fig. 2C–D. Accordingly, the resurrection of the genus *Forbesichthys* was recommended because *C. agassizi* and *C. cornuta* are considerably divergent morphologically according to meristic characters (Woods and Inger 1957) and genetically (Swofford 1982). This treatment was preferred over the more drastic alternative, which involved synonymizing *Typhlichthys* under *Chologaster*.

Uyeno (pers. comm.) examined karyotypes of three Amblyopsid species. Both *F. agassizii* from Rich Pond, Kentucky, and from Wolf Lake, Illinois, and *A. spelaea* had 2n = 24 chromosomes noting that both of these species had very similar karyotypes. Uyeno also examined the karyotype

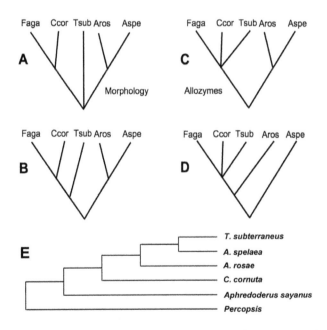

Fig. 2 Phylogenetic hypotheses of the interspecific relationships of the Amblyopidae and according to the (A and B) morphological examinations of Woods and Inger (1957), (C and D) allozyme studies of Swofford (1982), and (E) mitochondrial DNA (Niemiller and Fitzpatrick 2008, Near *et al.*, unpublished data). Faga = *F. agassizii*, Ccor = *C. cornuta*, Tsub = *T. subterraneus*, Aros = *A. rosae*, and Aspe = *A. spelaea*. *Forbesichthys agassizii* and *S. poulsoni* are not represented in the mtDNA phylogeny.

of two populations of *T. subterraneus*, Cave City, Kentucky, and Thomason Cave, Alabama. In *T. subterraneus*, the karyotype was 2n = 46 including one pair of rather large metacentric chromosomes. Karyotypes have not been examined for *C. cornuta*, *A. rosae*, and *S. poulsoni*.

Bergstrom *et al.* (1995) and Bergstrom (1997) were the first to examine interspecific relationships of Amblyopsids using DNA sequence data. Using a portion of the mitochondrial NADH-dehydrogenase subunit-2 (NAD2) gene, their analyses resolved a sister relationship between *A. rosae* and all other Amblyopsids (*S. poulsoni* was not included), but low levels of differentiation were found between *C. cornuta*, *F. agassizii*, *A. spelaea*, and some *T. subterraneus* populations. Their ND2 phylogeny clearly differs from the phylogenetic hypotheses of Woods and Inger (1957) and Swofford (1982). Unfortunately, DNA samples from *C. cornuta*, *F. agassizii*, and *A. spelaea* were extracted from formalin-preserved tissues and were contaminated. Therefore, the relationships resolved likely do not reflect the true history of the ND2 gene. However, the intraspecific relationships reported within *A. rosae* and western populations of *T. subterraneus* are believed to be accurate given that extractions were made from freshly collected tissue.

Recent phylogenetic work utilizing both mtDNA and nuclear DNA markers has revealed a different relationship (Fig. 2E) among the Amblyopsids (Niemiller and Fitzpatrick 2008, Near *et al.*, unpublished data). MtDNA supports the basal position of *C. cornuta* within the Amblyopsidae. Monophyly of *Amblyopsis* is not support with the mtDNA dataset. However, other nuclear markers need to be examined before taxonomic revisions are made.

The Alabama cavefish has not been included in any systematic treatment of the family. Boschung and Mayden (2004) state that *Typhlichthys* and *Speoplatyrhinus* form a monophyletic group that is sister to a clade comprising *Forbesichthys* and *Chologaster*. This statement likely is a typo and *Amblyopsis* should be replaced for *Speoplatyrhinus*. Proudlove (2006) speculated that *S. poulsoni* is probably not closely related to other species in the family. On several counts it is the most neotenic, most troglomorphic, and most distinct species in the family (Cooper and Kuehne 1974, Poulson 1985; and Taxonomy). The exact placement of *Speoplatyrhinus* within the family remains unknown and will remain so until a fresh tissue sample can be obtained.

III. DISTRIBUTION, BIOGEOGRAPHY AND HABITAT

The present-day distributions of the Amblyopsid species have been influenced by climatic, geological, and ecological factors throughout their evolution. All Amblyopsids occur in eastern North America (Fig. 3). All but one species, the epigean *C. cornuta*, occur in the Interior Low Plateau or Ozark Plateau where the ancient former position of the Cretaceous Mississippi Embayment bounds their collective distributions to the south and the Pleistocene glacial advance bands the distribution to the north. In this section we discuss the distribution, phylogeographic history, and habitat preferences of the each species within the Amblyopsidae beginning with epigean swampfish and advancing our discussion toward the troglobitic forms. In the process, we discuss the several hypotheses that have been posited explaining the biogeographic patterns observed.

The swampfish is the only entirely epigean Amblyopsid and the only species of the family found in the Coastal Plain of the southeastern United States (Fig. 4). It occurs from southeastern Virginia to east-central Georgia (Cooper and Rohde 1980, Rohde *et al.* 1994). It typically occurs in heavily vegetated and shaded lowland swamps, swampy creeks, and backwater habitats that are tannin-stained and acidic (Cooper and Rohde 1980, Rohde *et al.* 1994, Ross and Rohde 2003). Water temperatures in these habitats rarely exceed 25°C (Poulson 1963) and pH ranges 5.7–6.8 (M.D. Norman in Jenkins and Burkhead 1994). Swampfish also have been collected from sites with salinities up to 5% in North Carolina but are not found in brackish water. Woods and Inger (1957) hypothesized that the ancestor to *C. cornuta* migrated from the Interior Low Plateau southward and around the southern Appalachians via continuous swamp habitat that fringed the location of the border of the ancient Mississippi Embayment. However, recent molecular evidence suggests an opposite route, as *C. cornuta* is the most basal member of the Amblyopsidae (Niemiller and Fitzpatrick 2008, Near *et al.*, unpublished data). A more plausible scenario would be that a *C. cornuta*-like ancestor inhabited swamps and wetlands adjacent to the karst regions of the Interior Low and Ozark Plateaus and likely the Coastal Plain during the Miocene. As the climate became more arid during this period, some populations became isolated in springs and caves eventually giving rise to the troglobitic forms, whereas other populations remained living in a swamp-like habitat.

However, the absence of *C. cornuta* from the Gulf Coastal Plain has puzzled past researchers. Woods and Inger (1957) felt that *C. cornuta* formerly inhabited this area in the past because several community associates, such as *Aphredoderus sayanus* and *Umbra pygmaea*, have distributions that range into the Gulf Coastal Plain. They speculated that *C. cornuta* might have been extirpated from the Gulf Coastal Plain when their swamp-like habitat dried up because of prolonged drought and was not able to recolonize because of poor dispersal ability. According to the scenario outlined above, we also believe *C. cornuta* or a *cornuta*-like ancestor inhabited portions of the Gulf Coastal Plain and was extirpated. Meristic data (Poulson 1960)

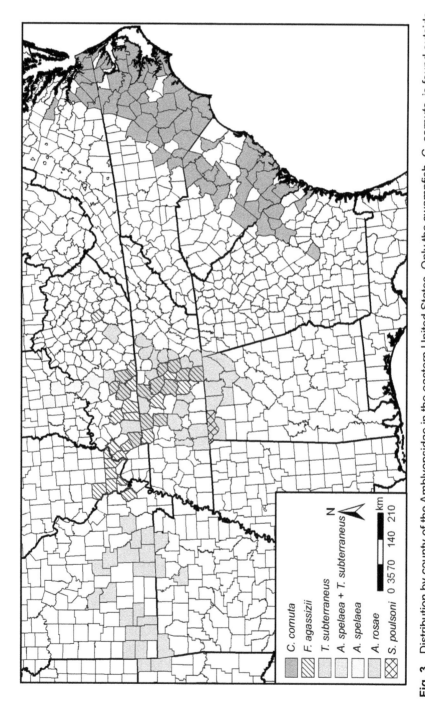

Fig. 3 Distribution by county of the Amblyopsidae in the eastern United States. Only the swampfish, *C. cornuta*, is found outside the Interior Low Plateau or Ozark Plateau.

Color image of this figure appears in the color plate section at the end of the book.

and both allozyme data (Swofford 1982) and DNA sequence data (Near and Niemiller, unpublished data) indicate that dispersal in *C. cornuta* is limited. Swofford (1982) observed significant differentiation even among populations in the same hydrological drainage and concluded that *C. cornuta* is highly specific in its habitat requirements, and, consequently, is unable to disperse over long distances unless corridors of suitable habitat are available. Because *C. cornuta* cannot tolerate brackish water (Woods and Inger 1957), dispersal may only be achieved by headwater exchange (Swofford 1982). However, fine-scale genetic analyses are needed to test this hypothesis.

The spring cavefish occurs from south-central Tennessee northward into central and western Kentucky, then westward following the Shawnee Hills of southern Illinois and the Benton Hills west of the Mississippi River in southeastern Missouri (Fig. 3). Spring cavefish occur at the interface of epigean and subterranean habitats bridging the threshold toward a troglobitic life. Although it occurs in caves throughout most of its range, *F. agassizii* is most abundant in springs, spring runs, and spring seeps (Smith and Welch 1978). Both habitat types are utilized, however, as fish often emerge from subterranean haunts at dusk to feed and then later retreat back underground before dawn. Fish also can be found underneath rocks in springs and spring runs during the day. In Tennessee, *F. agassizii* is common in dense vegetation associated with springs and spring-fed streams (Etnier and Starnes 1993) and have been collected in very low numbers in caves. Illinois populations in the LaRue-Pine Hills Ecological Area have been taken at springs in the swamp during winter (Smith and Welch 1978), but it is unknown whether fish actively use or successfully disperse through swampy, lentic habitats.

In Kentucky and Tennessee, spring cavefish occur in springs and caves, including the Mammoth Cave system, from the Highland Rim physiographic province of the Tennessee River drainage, middle and lower Cumberland River drainage, and the upper Barren-Green River system of Kentucky (Etnier and Starnes 1993). Spring cavefish have not been collected in Alabama, but the close proximity of a population in the Elk River drainage in Franklin Co., Tennessee (Armstrong and Williams 1971), led Boschung and Mayden (2004) to speculate that the species likely occurred in springs associated with the Elk River system in Alabama in the past. *Forbesichthys agassizii* has been collected from springs issuing from the base of limestone bluffs of the Shawnee Hills and Benton Hills at the edge of the Mississippi River alluvial plain in southwestern Illinois (Weise 1957) and southeastern Missouri (McDonald and Pflieger 1979) respectively. Until recently, the Shawnee Hills and Benton Hills populations likely were continuous and were isolated around 2,000 years ago when the Mississippi River was diverted through the narrow gap between these regions (McDonald and Pflieger 1979).

Forbesichthys shows a little less allozyme differentiation than *C. cornuta* with an average heterozygosity of 0.028 compared to 0.040 (Swofford 1982). This suggests that *F. agassizii* populations in Kentucky and portions of the Highland Rim in Tennessee utilize surface corridors for dispersal (Swofford 1982), particularly when compared with more differentiation in troglobitic *T. subterraneus* populations over a similar geographic area. However, molecular evidence indicates that populations of *Forbesichthys* from Illinois and Tennessee are distinct and likely isolated and suggestive that long distance dispersal is rare. Although cave populations of *F. agassizii* exist, subterranean dispersal seems unlikely, given low abundance in caves, low tolerance to starvation, relatively poor food finding ability, and short life span (see below). But surface dispersal along rivers between springs may be possible since Woods and Inger (1957) report several specimens collected from surface streams.

Of the four troglobitic Amblyopsids, three occur in the Interior Low Plateau and the other occurs in the Ozark Plateau (Fig. 3). Collectively, their distributions lie north of the boundary of the ancient Mississippi Embayment and south of the Pleistocene glacial boundary. Woods and Inger (1957) proposed that a surface ancestor entered caves sometime during the Tertiary Period and gradually dispersed into the areas where they occur today.

However, it is unlikely that subterranean dispersal alone can account for the widespread collective distribution of the troglobitic Amblyopsids, and of *T. subterraneus* in particular. Dispersal through surface watercourses is unlikely (Poulson 1960, Woods and Inger 1957) given the separation of some populations by over 1000 surface river miles (Woods and Inger 1957). Fish are very rarely observed in surface streams just downstream of springs after flood events. Although most fish likely perish, some might find their way back underground further downstream. It is much more likely that fish could move through cavernous river bottoms or solution channels (Poulson 1960), however. Although little direct evidence supports this avenue as a significant form of dispersal, indirect evidence indicates that surface rivers do not significantly impede dispersal of subterranean species. First, the thick horizontal Ordovician and Mississippi limestone formations of the Interior Low Plateau and Ozark Plateau and the groundwater systems contained therein pass underneath major surface rivers, including the Cumberland, Tennessee, Ohio, and Mississippi rivers. Populations of the troglobitic species occur on either side of these big rivers (e.g., *A. spelaea* north and south of the Ohio River; *T. subterraneus* north and south of the Tennessee River in Alabama) and, therefore, rivers do not appear to pose significant barriers to dispersal (Woods and Inger 1957). Rotenoned Southern cavefish apparently floated to the river surface from conduits in a limestone riverbed during a fish survey at a prospective dam site (Mohr and Poulson 1966). Second, populations on opposite sides of major rivers often show little genetic variation. Populations of *T. subterraneus* in the same hydrological drainage but on opposite sides of a river are poorly genetically differentiated (Niemiller and Fitzpatrick 2008) indicating that either these populations have been recently isolated or that some migration is occurring between them.

Contemporary surface drainage divides also may not impede subterranean dispersal. Woods and Inger (1957) reported the presence of *T. subterraneus* from a spring well near the crest of the Arkansas/White River watershed divide. The spring well lies at an elevation of 1360 ft at the top of the Springfield Plateau and fish potentially can move from one watershed to the other. This is supported by the observation of very genetically different southern cavefish within the same population in the Salem Plateau (Bergstrom 1997). Although this is consistent with movement between watersheds, there are other explanations for considerable genetic diversity within a population such as lineage sorting. The recharge basins in this area are large and contiguous with few dry or faulted barriers between basins. Therefore, some individuals can potentially move between recharge basins (Noltie and Wicks 2001). However, other molecular evidence indicates that many eastern populations of *T. subterraneus* are confined to distinct hydrological drainage basins (Niemiller and Fitzpatrick 2008), as genetic structure is highly correlated with hydrological structure. Poulson (1960) argued that populations of *T. subterraneus* along the Eastern Highland Rim and western escarpment of the Cumberland Plateau were more similar than elsewhere in the range. He hypothesized that the observed homogeneity was because of occasional dispersal by cave headwater stream capture. However, both the electrophoretic data of Swofford (1982) and genetic analyses of (Niemiller and Fitzpatrick 2008, Niemiller *et al.*, unpublished data) indicate that morphological similarity of these populations is not due to homogenizing gene flow.

The southern cavefish is the most widely distributed troglobitic fish in North America. Its range is discontinuous and divided into two main components east and west of the Mississippi River: an eastern component that extends along the Cumberland Plateau and through Interior Low Plateau from central Kentucky (Mammoth Cave region) southward into central Tennessee, northern Alabama, and northwestern Georgia, and a western component that includes the Ozark Plateau of central and southeastern Missouri and northeastern Arkansas (Fig. 3). The distribution of the species was once thought to include southern Indiana and northeastern Oklahoma. These records are now thought to be erroneous. Mayden and Cross (1983) showed that all specimens of southern cavefish from northeastern Oklahoma are in fact *A. rosae*.

Similarly, records of *T. subterraneus* from Green Co., Missouri, are incorrect and actually represent *A. rosae* (Jones and Taber 1985). Therefore, *A. rosae* and *T. subterraneus* should no longer be considered sympatric (Proudlove 2006). Eigenmann (1905) described *T. wyandotte* from a well near Corydon, Indiana, in the heart of the range of *A. spelaea* and this was almost certainly a misidentification.

There do not appear to be clear differences among the habitats of Amblyopsid cavefish. Most cavefish habitats are at or near the water table (Poulson 1963) and the types of organic and inorganic substrates vary enormously within and among caves for all four troglobitic Amblyopsids (e.g., Fig. 3 in Poulson 1963). Also, water velocity among seasons and among years varies tremendously for most cavefish habitats. Nevertheless, some intriguing regional differences in geology and related size and integration of caves in the Ozarks exist for *T. subterraneus* and *A. rosae*. These differences may relate to food supply and degree of isolation (see below and Poulson, this volume). In addition, the geology of the karstic strata inhabited by *T. subterraneus*, at least in the Salem Plateau, suggests that southern cavefish reside at appreciable depth below the surface (Noltie and Wicks 2001). As mentioned above, evidence suggests that this species can move between recharge basins; however, dispersal between basins appears to be low (Swofford 1982, Niemiller and Fitzpatrick 2008, Niemiller *et al.*, unpublished data). In contrast, *A. rosae* lives in cave systems that are smaller, shallower, and much less likely to be interconnected (Noltie and Wicks 2001).

The extensive but highly disjunct geographic range of *T. subterraneus* may be the result of multiple, independent invasions and subsequent dispersal (Swofford 1982, Barr and Holsinger 1985, Holsinger 2000, Poulson, this volume). Woods and Inger (1957) hypothesized a single invasion of subterranean waters by *Typhlichthys* and a separate invasion for *Amblyopsis*. However, others have proposed a multiple-invasion scenario for *Typhlichthys* as well as the other troglobitic species. Swofford (1982) offered two possible scenarios to explain the distribution of *T. subterraneus*. First, the ancestor of *T. subterraneus* was a widely distributed epigean species preadapted to cave life and possibly similar to *F. agassizii* today. During the Pleistocene, it originally occupied a more northern distribution but was forced southward during glacial periods then underground during the warmer and drier interglacial periods. Alternatively, the surface ancestor may have resembled *F. agassizii*, having occupied a variety of habitats from streams to springs to caves. As climatic conditions changed, surface populations were extirpated leaving only cave populations scattered over a broad area but genetically isolated from one another. In both of these scenarios, some range expansions through subterranean dispersal may occur accounting for some of the current distribution patterns that are difficult to interpret otherwise (Swofford 1982).

Swofford's alternative scenarios also can be applied to both *Amblyopsis* species. Willis and Brown (1985) hypothesized a similar vicariant scenario in which the surface ancestor of cave Amblyopsids inhabited swamps during the wetter periods of the Pleistocene becoming restricted to swamp-like springs and caves during drier periods. Rather than just one or two invasions, possibly each current cavefish population represents a separate invasion event. The actual number of independent subterranean invasions likely falls somewhere in between the two extremes. Recent molecular evidence also indicates that the major lineages within the Amblyopsidae originated well before the Pleistocene (Bergstrom 1997, Near *et al.*, unpublished data), however the glacial cycles likely had a profound effect on the current distributions and intraspecific relationships within each species. These studies also support a multi-invasion scenario and support the hypothesis that morphological similarity is the result of parallel evolution rather than significant dispersal and gene flow across major drainage and river divides.

If current taxonomy is indicative of past evolutionary history, then a plausible biogeographic hypothesis explaining the distributions of both *Amblyopsis* species is difficult to frame. Woods and Inger (1957) proposed two hypotheses to explain the distribution of *Amblyopsis*. First, the disjunct ranges of the two species could have resulted by dispersal from a more northern former distribution above the limit of the glacial ice sheets. The ancestor was subsequently split and diverged as a direct result of Pleistocene glaciations. Since no evidence has ever indicated that any Amblyopsid existed north of the karst regions of the Interior Low Plateau, they favored a second hypothesis that postulated an ancestral *Amblyopsis* that ranged widely through the limestone plateaus and were isolated into two groups by range expansion of the more competitive *Typhlichthys*. Bergstrom (1997) places the timing of the split between *Typhlichthys* and *Amblyopsis* around 8 mya and provides a biogeographic scenario consistent with this and subsequent splits within the respective lineages (Noltie and Wicks 2001).

Recent molecular is inconsistent with the above scenarios since *Amblyopsis*, as currently recognized, does not form a monophyletic group. "with *A. rosae* diverging much earlier." Obviously, more research is needed to elucidate the large-scale biogeographic patterns of the family. At a finer scale, considerable work has been conducted on the distribution and habitat of *A. spelaea* and *A. rosae*.

The northern cavefish follows a narrow arc of karst, from near the city of Bedford in Lawrence Co., south-central Indiana, southward to the Mammoth Cave area in central Kentucky. It has a distribution through portions of ten counties (Fig. 3). Karst landscapes start with the Mitchell

Plain in Indiana and continue as the Pennyroyal Plain in Kentucky. In the most recent assessment of the species distribution, Pearson and Boston (1995) reported 76 localities from five counties in Indiana and 38 localities from five counties in Kentucky; however, some localities no longer exist because of quarry operations and highway construction. Northern cavefish have the largest populations in caves with deep pools and moderately deep shoals. The best habitats have ledges, overhangs, and backwater areas that serve as refugia during floods. Substrate type and particle size are quite variable (Fig. 3 in Poulson 1960). The cave types flood regularly and have high food supplies of plankton and isopods (see below, Food Supplies in Caves).

The distribution of the northern cavefish is bounded by the East Fork White River to the north. Harsh periglacial conditions and alteration and impounding of major rivers north of this area may have limited the northern range of *A. spelaea* (Keith 1988). The southern extent of the species' range is the Mammoth Cave system. It is not clear why its distribution does not extend further to the south or immediately east to the sinkhole plain where the range of *T. subterraneus* starts. In Mammoth Cave itself, *T. subterraneus* and *A. spelaea* occur in different but contiguous habitats.

Mammoth Cave is one of only a few sites with syntopic species of cavefish. Typically only a single species of troglobitic fish is found within a cave system (Weber 2000) and this is the case within the Amblyopsidae. However, two instances of syntopy have been observed in the family. First, *A. spelaea*, *T. subterraneus*, and the troglophile *F. agassizii* co-occur within the Mammoth Cave system in Kentucky. In the Mammoth Cave system, *T. subterraneus* and *A. spelaea* are syntopic only where habitat is sufficiently diverse to allow habitat segregation (Poulson and White 1969, Poulson 1992, Poulson, this volume).

In the Mammoth Cave system, segregation of the three species of Amblyopsids has been hypothesized as follows. Spring cavefish are washed into the cave from sinking streams during flood events but only survive as long as a year when the subterranean food supply is well above average. Although the species can subsist on the increase in food availability, reproduction is unlikely as not enough food can be consumed to support its energetic demands and higher reproductive output compared with the troglobitic species. With regards to the two troglobitic species that inhabit the Mammoth Cave system, their distributions can possibly be explained by competitive interactions (Woods and Inger 1957, Poulson 1992). Southern cavefish are found in the master drains of vertical shafts, whereas *A. spelaea* inhabit deeper water with decreasing food availability further downstream. It remains unclear whether *A. spelaea* is precluded from upstream sections because of its larger size or by aggressive dominance of *T. subterraneus* (Poulson 1992). Alternatively, *T. subterraneus* may be excluded from downstream habitats by decreased food supply or by competitive dominance by *A. spelaea*. The other observed syntopic interaction between two Amblyopsids involves *T. subterraneus* and *S. poulsoni* and is discussed below.

The distribution of *A. rosae* is limited to caves in the Springfield Plateau of the Ozark Highlands Province of northwestern Arkansas, southwestern Missouri, and northeastern Oklahoma. (Fig. 3) The Springfield Plateau encompasses ca. 21,000 km^2 that is drained by the White River to the south and east, the Neosho River to the west, and the Osage River to the north. Historically, *A. rosae* occurred in 24 caves in nine counties with unconfirmed reports increasing occurrence to 52 caves in 14 counties (Brown and Todd 1987). Willis and Brown (1985) confirmed *A. rosae* in just four caves in Arkansas, five in Missouri, and three in Oklahoma. Brown and Todd (1987) reported three additional localities in Benton Co., Arkansas. Currently, *A. rosae* is known from 47 localities (G.O. Graening, unpublished data): 11 caves in one county in Arkansas, 24 in seven counties in Missouri, and 12 in two counties Oklahoma. These localities occur in the main section of the plateau; however, no localities are known from the section of the plateau that extends southeast into Arkansas and from the central area in and around McDonald Co., Missouri (Noltie and Wicks 2001).

Ozark cavefish are confined to the karstic layer at or immediately adjacent to the surface based on the stratigraphy of the Springfield Plateau limestones and underlying shale layer (Noltie and Wicks 2001). The majority of cavefish localities are caves developed in the shallow and thin Boone and Burlington limestones and intersect the water table close to the surface (Noltie and Wicks 2001). Fish are generally observed in two types of subterranean habitats: 1) laterally-oriented cave streams that occupy conduits accessible by humans that intersect the surface near the base of Springfield Plateau escarpment either along the bluff line or of river valley walls, and 2) karst windows and wells that typically intersect conduits that are too small for humans to penetrate (Poulson 1960, USFWS 1989, Noltie and Wicks 2001). Cave streams inhabited by *A. rosae* typically are small. Karst window and well localities often occur toward the interior of the plateau and are not associated with the plateau escarpment. These are windows to human-inaccessible habitat that may be phreatic. These areas may serve as refugia, but also may provide avenues for dispersal and gene flow. Poulson (this volume and below) believes that these areas contain a very small portion of the overall cavefish population. Genetic studies are consistent with both the hypothesis of small numbers and limited dispersal since there is pronounced variation among populations (Swofford 1982, Bergstrom *et al.* 1995, Bergstrom 1997).

Optimum Ozark cavefish habitat occurs in caves with large colonies of gray bats, *Myotis grisescens* (Brown and Todd 1987) or comparatively large sources of allochthonous matter (USFWS 1989). Five of the 14 caves with *A. rosae* populations reported by Willis and Brown (1985) contained bat colonies and another six caves contained guano piles indicating past use. These caves had the highest food supplies based on isopod counts in leaf pack traps and copepods and cladocerans in plankton samples. Ozark cavefish occur in small cave streams where they reside in quiet pools from a few centimeters to 4 m in depth with chert rubble or silt-sand substrates (Poulson 1963).

The Alabama cavefish, *Speoplatyrhinus poulsoni*, is the most troglomorphic Amblyopsid and its single occurrence (Fig. 3) in Key Cave, Lauderdale Co., Alabama, is thought to be a relict population of a once more widely distributed species. It is known from just five pools within Key Cave, Lauderdale Co., Alabama (Kuhajda and Mayden 2001). The maze-like cave system is developed in the Mississippian-aged Tuscumbia Limestone that approximately lies along the ancient Cretaceous shoreline of the Mississippi Embayment (Cooper and Kuehne 1974). The pools inhabited by *S. poulsoni* lie in a zone of seasonal oscillation of the water table where pools that form during high water become isolated during drier conditions (Trajano 2001). Many of the pools are extremely deep with depths up to 5 m (Kuhajda and Mayden 2001) and some are situated adjacent to bat roosts where guano occasionally slides into and likely enriches the aquatic environment (Cooper and Kuehne 1974). Recently, Kuhajda and Mayden (2001) reported the capture in 1995 of a single *T. subterraneus* from Key Cave. Until this observation, the two species were thought to be allopatric. Although the effect of *T. subterraneus* on *S. poulsoni* is unknown, the absence of *S. poulsoni* from adjacent caves where *T. subterraneus* have been observed suggests that competitive interactions have and might currently be influencing the distribution of *S. poulsoni*. This hypothesis has not been examined but Poulson (this volume) argues that it is unlikely. Rather he proposes demographic swamping as an alternative hypothesis.

IV. MORPHOLOGY

A. Family Characteristics

Members of the Amblyopsidae (Fig. 4; Table 1) are characterized by possessing (i) a large, flat head and a tubular, non-streamlined body, (ii) an oblique mouth with the lower jaw protruding beyond the upper jaw, (iii) a segmented premaxilla, (iv) jugular position of the anus and urogenital

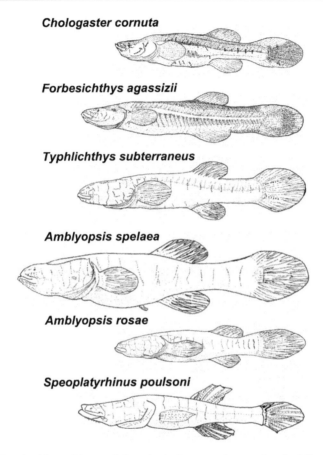

Fig. 4 The Amblyopsid cavefishes include an epigean, troglophilic, and four troglobitic species that exhibit a range of troglomorphy. The fish illustrated are drawn to scale of average-sized adults. Modified from drawing by John Ellis in Romero (2004).

pore, (v) reduced head lateral line canals and absence of the trunk lateral line canal but the presence of superficial papillae (neuromasts) arranged in distinct rows on the head and body, and papillae of unknown function in 2–4 rows on the caudal fin, (vi) small, embedded cycloid scales except on the head, (vii) six branchiostegal rays, (viii) presence of a swim bladder, (ix) tubular anterior nostrils, (x) absence of pelvic fins (except rudimentary in *A. spelaea*). Troglobitic species also can be characterized by (i) lack of externally visible eyes, (ii) reduced pigmentation, (iii) hypertrophy of the superficial lateral line system that includes an extensive system of elevated neuromasts arranged in distinct ridges, (iv) hypertrophy of the semicircular canals and otoliths, and (v) presence of highly-developed

Table 1 Meristic characters for species in the Amblyopsidae.

Species	Dorsal fin rays	Anal fin rays	Pectoral fin rays	Caudal fin rays	Caudal fins	Pelvic fin rays	Vertebrae	Post-cleithrum	Penultimate vertebrae bend
Chologaster cornuta	11 (9–12)	9 (9–10)	10 (10–11)	10 (9–11)	branched	absent	27–29	present	straight
Forbesichthys agassizii	10 (9–11)	10 (9–11)	9–10 (9–11)	14–15 (11–17)	branched	absent	33–35	present	upwards slightly
Typhlichthys subterraneus	8–9 (7–10)	8–9 (7–10)	10–11 (9–12)	12–13 (10–15)	branched	absent	28–29	present	upwards slightly
Amblyopsis spelaea	10 (9–11)	9–10 (8–11)	10 (9–11)	12–13 (11–13)	branched	4 (0–6)[1]	29–30	absent	upwards strongly
Amblyopsis rosae	7–8 (7–9)	8 (8–9)	10 (10–12)	10 (9–11)	branched	absent	28	absent	straight
Speoplatyrhinus poulsoni	9 (9–10)	8 (8–9)	9 (9–11)	22 (21–22)	not branched	absent	27–28	absent	

[1]Frequency distribution of pelvic fin rays in *A. spelaea*:
0 rays – 1, 1 ray – 1, 2 rays – 2, 3 rays – 18, 4 rays – 58, 5 rays – 7, 6 rays – 1

caudal sensory papillae. Amblyopsids are also characterized by having dorsal and anal fins that are similar in shape with the dorsal fin origin anterior to the anal fin origin. All fins lack spines (although the first ray in the dorsal fin of *T. subterraneus* has been reported as spinuous). Dorsal fin rays range 7–12, anal fins 7–11 rays, and pectoral fins 9–12 rays. Pelvic fins are absent except in *A. spelaea* that have 0–6 rays. The caudal fin may be elliptical, lanceolate, or rounded with 9–22 branched rays. No fin rays are branched in *S. poulsoni*. The urogenital pore is positioned just anterior to the anal fin at hatching and migrates anteriad until it occupies a jugular position in adults (Woods and Inger 1957).

B. Sexual Dimorphism

Most Amblyopsids cannot be sexed using external morphology. Spring cavefish cannot be sexed using external morphology, although Weise (1957) noted that in the spring almost all adult individuals could be sexed by observing the gonads through the translucent body wall. This is rarely the case for cave Amblyopsids that seem to have low clutch sizes (see Fecundity).

Male swampfish can be distinguished from females by the presence of a Y-shaped appendage that develops on the snout. This appendage is believed to be a contact-stimulatory organ implicated in courtship behavior (Poulson 1960) and pheromone chemoreception (Jenkins and Burkhead 1994), although its function is unknown.

C. Neoteny

Poulson believes that there is good evidence for neoteny in the Amblyopsidae. He predicted the characteristics of his namesake based on increasing neoteny from *Typhlichthys* to *Amblyopsis*. Absence of bifurcate fin rays in adult *S. poulsoni* is the most extreme neotenic trait for troglobitic Amblyopsids (Cooper and Kuehne 1974, Weber 2000) as this condition is also found in 15-17 mm SL *T. subterraneus*. Moreover, adult *S. poulsoni* and 15-20 mm *T. subterraneus* and *A. spelaea* are similar in head shape and body proportions (Fig. 5 and Poulson in Culver 1982). Some of the specialized morphological and behavioral traits in troglobitic Amblyopsids, such as longer fins, larger heads, and increased exposure of neuromast organs, may be explained by neoteny which is one kind of heterochrony (Cooper and Kuehne 1974, Trajano 2001).

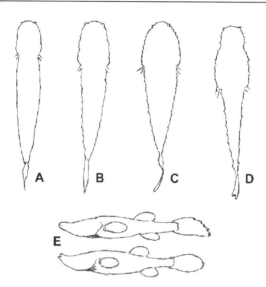

Fig. 5 Scale drawings (Poulson and Cooper) of the body shapes and relative head sizes of Amblyopsid fishes. With increasing cave adaptation heads become relatively larger and bodies smaller. Dorsal views from left to right are *F. agassizii*, *T. subterraneus*, *A. spelaea*, and *S. poulsoni*. That these are neotenic trends is suggested by the side views of a 48 mm SL *Speoplatyrhinus* and a 13 mm SL *Amblyopsis*. Can you tell which is which?

D. Eyes

Carl Eigenmann's classic study (1897) of *The Eyes of the Amblyopsidae* remains the best ever. It is a paradigm for the comparative approach to understanding a process; degeneration of eyes both phyletically, among the five species known at the time, and ontogenetically, across almost the known body lengths of each species. Eigenmann's discussions of mechanisms also are a paradigm for scientific study. He used evidence to falsify some but not reject others of six alternative hypotheses (see Poulson's this volume).

Eigenmann's conclusions (1897, p. 587 and Summary Points 1–23), based on the best-developed eyes for each species, are that phyletically there is increasing degeneration and simplification of eyes from *Chologaster* to *Typhlichthys* to *Amblyopsis spelaea* to *Troglichthys* (=*Amblyopsis*) *rosae*. Of the parts of fish eyes missing or indistinguishable histologically (Table 2) the numbers are *C. cornuta* 0, *F. agassizii* 1, *Typhlichthys* (Kentucky) 7, *A. spelaea* (Indiana) 11, and *A. rosae* (Missouri) 12. The maximum eye sizes in micra are 1100 for *C. cornuta*, 760–930 for *F. agassizii*, 180 for *T. subterraneus*, 200 for *A. spelaea*, and 85 for *A. rosae*. Figure 6 shows these differences in parts and in eye size.

Table 2 Summary of parts of the eye of members of the Amblyopsidae. The eyes of cave Amblyopsids are reduced in different ways but overall the number of parts absent, reduced/vestigial/undefined/merged has increased from *T. subterraneus* to *A. spelaea* to *A. rosae*. This pattern is consistent with the hypothesis that rudimentation of eyes is an index of evolutionary time isolated in caves. From Eigenmann 1897. The eye of *S. poulsoni* has not been histologically examined.

Species	*C. cornuta*	*F. agassizii*	*T. subterraneus*	*A. spelaea*	*A. rosae*
Habitat	Swamp	Spring	Cave	Cave	Cave
Retina					
1. Pigment epithelium (PE)	Defined	Defined	Defined	Defined	Imperfect
Pigment in PE	Present	Present	Absent	Present	Reduced
2. Rods and cones	Defined	Defined	Undefined/Absent	Undefined	Cones absent
3. Outer nuclear layer	1 layer of cells	3 layers of cells	Undefined	Merged	Undefined/Absent
4. Outer plexiform layer	Defined	Defined	Defined	Undefined	Absent
5. Horizontal cells	Defined	Defined	Undefined	Undefined	Undefined
6. Inner nuclear layer	1 layer	3 layers	Defined	Merged	Undefined
8. Inner plexiform layer	Defined	Defined	Defined	Defined	Defined
9. Ganglion layer	Defined	Defined	Defined	Funnel-shaped	Reduced
Optic nerve connection to brain	Yes	Yes	Yes No	No	No
Focusing					
Lens	Present	Present	Vestigial/Absent	Vestigial/Absent	Vestigial/Absent
Ciliary muscles	Present	Absent	Absent	Absent	Absent
Eye muscles	Normal	Normal	Absent	Reduced	Reduced/Absent
Nutritive					
Vitreous body	Present	Present	Vestigial	Vestigial	Vestigial
Hyaloid membrane	Present	Present	Present	Absent	Absent
Scleral cartilages	Absent	Absent	Absent	Present	Present
Pupil	Open	Open	Open	Closed	Closed
Maximum eye diameter (mm)	1.10	0.76 – 0.93	0.18	0.2	0.09
Eye parts absent	0	1	3	3	7
Eye parts vestigial	0	0	4	8	5
Variability within and between individuals	none	none	some	great	great

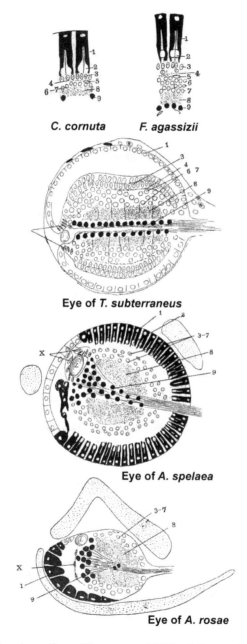

C. cornuta   *F. agassizii*

Eye of *T. subterraneus*

Eye of *A. spelaea*

Eye of *A. rosae*

Fig. 6 Scale drawings (from Eigenmann 1897) of the vestigial eyes of cave Amblyopsids with the retinas of *Chologaster* to the same scale (thus complete eyes would be ~10x as large). The numbers of the retinal layers are identified on Table 2 of eye parts. "x" is the degenerate lens or its parts. Not shown are the vestigial eye muscles of *Amblyopsis spelaea*.

Eigenmann (1897) stressed the fact that eyes of each species "degenerate along different lines". For example, *Typhlichthys* eyes are the least degenerate overall for a cave Amblyopsid but lack eye muscles, scleral cartilages, and pigment epithelium. On the other hand, the overall more degenerate eyes of *A. spelaea* have some eye muscles, clear scleral cartilages, and well developed pigment epithelium. Nonetheless, in general, Eigenmann stated that "the eye of *Amblyopsis spelaea* represents one of the stages through which the eye of *Troglichthys* passed (phyletically)".

Ontogenetically each species' eyes develop to a maximum state of tissue differentiation and then become more simplified and degenerate with advancing age (Eigenmann 1897). In addition, older fish of the same size have more variable eye development and there is more right to left side variation of eyes in the same individual. In *A. spelaea*, for which the largest range of sizes was studied (6 to 130 mm), Eigenmann found that the degeneration started in the earliest stages of eye development and that the eyes had degenerated completely in the largest fish with one eye even being absent in a 130 mm individual. These patterns are consistent with that for cave *Astyanax fasciatus* with increased variability in the largest individuals among individuals and in right to left side asymmetry within individuals.

Poulson (1963, Table 2 in 1985) reported on the sizes and allometric growth constants (*b*) of eyes and brain optic lobes for five species and optic lobe volume in all six species including *S. poulsoni*. The values reported here are for the populations of each species with the greatest available size range since accuracy of allometric constant depends on having a full size range; the size values are for 45 mm SL fish on the fitted line of part size vs. body length. The size ranges in mm SL and localities are as follows: *C. cornuta*, Roquist Creek, North Carolina, 7–50; *F. agassizii* Rich Pond, Kentucky, 15–55; *T. subterraneus*, Shelta Cave, Alabama, 17-65; *A. spelaea*, Letch Cave, Kentucky, 8–90; and *A. rosae*, Cave Springs Cave, Arkansas, 18–55. There are differences among populations related to both locality and growth rate that are discussed elsewhere (Poulson 1960 and this volume).

As for eyes and optic lobe length, there is a decline in optic lobe volume in cubic mm from 2.12 in *C. cornuta* to 1.53 in *F. agassizii* to similar values for the first three troglobites (1.01, 1.37, and 0.96) to the minimum value of 0.38 for *Speoplatyrhinus*. The values for allometric growth constant of optic lobe length are ~0.80 and ~0.60 for the nontroglobitic species to ~0.40 for *Typhlichthys* to ~0.60 for *A. spelaea* to 0.20 for *A. rosae*. There was too small a range in body size of *S. poulsoni* specimens to estimate an allometric constant but it appears to be the lowest of all.

There is a trend of increasing and then decreasing variance of eye and optic lobe size from epigean to troglophilic to the most troglobitic Amblyopsids (Fig. 13 in Poulson 1960, Table XIV in Poulson 1963). Along with this pattern there is a decrease in size and allometric constant for eyes and optic lobe length. In 1960, Poulson interpreted these trends in the context of a balance between natural selection and mutations of rudimentation. Selection for maintenance of vision or against mutations of rudimentation is seen in both non-troglobitic species since size and allometric constants are high but with distinctly greater variability in *F. agassizii*. *Typhlichthys* eyes are about 80 percent smaller with much less positive allometry and still greater variance. Compared to *F. agassizii*, its optic lobe is about 30 percent smaller, variance is the same and allometry is a little less positive. In *A. spelaea*, eyes and optic lobes and their allometric constants are slightly larger than for *Typhlichthys* despite a more degenerate eye and an increased variability of eye size in adults. Poulson's interpretation for *Typhlichthys* and *A. spelaea* was, and still is, of mutations of rudimentation predominately influencing the patterns since eyes had already been reduced so much. In *A. rosae*, eyes and optic lobes and their allometric constants are all lower than for *A. spelaea* but variance for both is as low as for *C. cornuta*! This decreased variability is clearly related to a longer evolutionary time in caves but its mechanistic explanation is elusive. One hypothesis for decrease in visual system variability is that the visual system is already about as reduced as possible (Fig. 2 in Poulson 1985). But *Speoplatyrhinus* has even more reduced optic lobes!

It may be relevant to the trends in visual system variability that the average heterozygosities for 14 allozyme loci also decline with increasing troglomorphy (Swofford 1982). Poulson (1985) hypothesized that "selection for some alleles and chance fixation of the most common alleles gave rise to increasing uniformity." Swofford (pers. comm.) believes that we cannot differentiate between stochastic effects and selection with his data.

E. Pigmentation

Other than the rudimentation of eyes, reduction of pigment cells and pigment are the most commonly observed reductive troglomorphies in vertebrate troglobites (Langecker 2000). Reduction of melanin can occur at many steps with many controlled by single genes (Jeffery, this volume). Size and number of pigmented melanophores is controlled by polygenic systems in an additive manner (Wilkens 1988). In many cases, the genetic bases for reduction are not the same in different subterranean fish species. In the Mexican cavefish, albinism in different populations is caused by independent mutations in a single gene, *Oca2* (Protas *et al.* 2006). Mutations that affect the polygenic system result in continuous variation towards complete depigmentation at slower rates than mutations that result in the loss of the ability to synthesize melanin (Trajano 2007). Accordingly, it has been argued that the loss of pigmentation as a proxy for relative age of subterranean inhabitation should be restricted to related species that have retained melanin where the extent of depigmentation is caused by mutations in the polygenic system (Trajano 2007).

Aspects of the developmental, physiological, and genetic mechanisms of pigment rudimentation have not been investigated in the Amblyopsidae. Only Poulson (Table 15 in 1960, 1963, Table 2 in 1985, and new data herein) has quantified the melanophore sizes and numbers and dispersion of melanin granules within melanophores in the Amblyopsidae. As with eyes, there is a decline for all of these measures in numbers (densities), allometry for number, and size through the epigean, troglophilic, and troglobitic species. All of the data are consistent with increasing durations of evolutionary time and increases in elaborated traits from *Typhlichthys* to *A. spelaea* to *A. rosae* to *S. poulsoni*.

The original observations (Poulson 1960, 1985) were made with a dissecting microscope and the new data are on excised skin wet mounts examined with a compound microscope at 500× and 970× (oil immersion). At 500×, 5–8 melanophores were scored for shape and melanin particle dispersion in each individual. Dispersion scores are as used in zebra fish from 1 punctate (maximum aggregation of melanosomes) to 2 amoeboid to 3 stellate to 4 stellate-reticulate to 5 reticulate (maximum dispersion of melanosomes). At 970×, for each individual 2–3 melanophores were scored for intensity of melanin deposition in melanosomes, the integrity of limiting membranes, presence of a nucleus, and details of the protoplasmic projections (dendrites or filaments).

With the exception of several individuals in one population of *A. rosae* (Roy Pierson Cave, Missouri) and one population of *Typhlichthys* (River Cave, Missouri) with no pigmented melanophores visible with a dissecting microscope (Fig. 7; Table 3), the melanophores in all species appear to have fully developed melanin granules (melanosomes) with dense deposition of melanin (electron microscope examination may reveal differences in the ontogeny of melanophores and melanosomes among the cave species). There is a trend, with increasing troglomorphy among the species, for the troglobites to have an increasing proportion of melanophores in the superficial muscles, body cavity mesenteries, and brain meninges but this has not been quantified. The other trend common to all three troglobites is for expanded melanophores (with dispersed melanin granules) to be less symmetric/stellate and more irregularly elongate. They have more projections having open ends, and they have less distinct limiting membranes overall (Fig. 7).

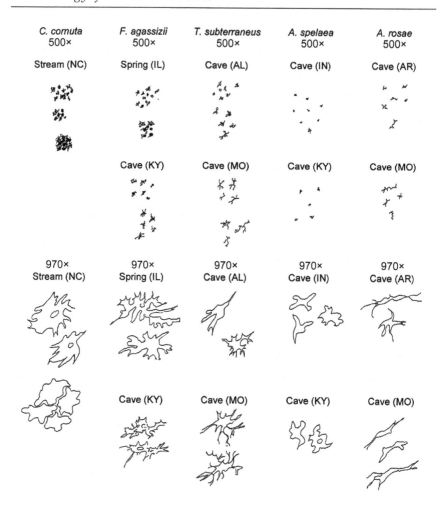

Fig. 7 Drawings of pigmented melanophores. For each of 1–2 localities for each species of Amblyopsid, extremes of melanophore size (microns) and shape are shown for three fish examined at 500×. Also, the range of branching patterns are shown from examination at 970× (oil immersion). See text for methods and interpretations.

Chologaster cornuta possesses the most melanophores that are also the largest. *Chologaster cornuta* is an attractive fish with a striking color pattern due partly to three kinds of chromatophores (Figs. 1A and 7). Despite being nocturnal, the pattern is a combination of disruptive and counter-shading camouflage with the swampfish's habit of resting during the day among dense vegetation with dappled sunlight. In live fish (but not in preserved fish) there appear to be three kinds of chromatophores: xanthophores and,

Table 3 Pigmented melanophores (45mm SL fish). See Fig. 7 for appearance of melanophores at 500× and 970×. See text for methods and detailed results. The decline in density and size of melanophores is consistent with the hypothesis that the amount of rudimentation is an index of evolutionary time isolated in caves.

Trait	C. cornuta	F. agassizii (spring)	F. agassizii (cave)	T. subterraneus	A. spelaea	A. rosae	S. poulsoni
Number on cheek	<=60	35	25	0	0	0	0
b	0.71	0.42	-0.40	-	-	-	-
Number on body	>103	>100	~50	12	6	0.7	0.6
b	+?	+?	+?	1.5	-1.1	-2.4	-?
Micra^2	100	83	50	21	16	15	10?
Micron length							
Mean	32	65	55	42	31	35	20?
Range	15-50	20-150	15-100	33-200	21-45	8-200	na
Dispersion Score (1-5)	5	3-5	4-5	1-5	2-3	1-5	4-5

perhaps, two types of melanophores. Swampfish are chocolate brown dorsally with a sharp border grading to orange-yellowish ventrally and is reddish on the opercles (xanthophores). Two to three stripes are present laterally with one at the brown to orange-yellowish border. Only the melanophores are easily quantified in this species. With full dispersion of melanin (score of 5), the cheek melanophores average 0.10 mm in diameter with up to ≥ 60 in a reference area of 0.175 mm^2. The allometric constant of increase in numbers with standard length is 0.71 for a population that grows up to 50 mm SL. Melanophores on the body are too numerous to count and determine dispersion pattern accurately. Subjectively, however, these melanophores all have fully dispersed melanin with a score of 5.

Swampfish body melanophores are dense, dark, and either packed like paving stones or overlapping, especially along the three black stripes. They range in estimated maximum dimension from 15 to 65 microns with most 40–50. It remains to be seen how dispersion would change under different light and background conditions, but the disruptive color pattern and counter-shading suggests that adapting to light levels and backgrounds may not be necessary.

Forbesichthys agassizii is dull brown dorsally to slightly lighter ventrally (Fig. 1B) with only one kind of chromatophore-melanophores. Fish have poorly defined stripes, one mid-dorsal and two lateral, and slightly more melanophores along the edges of the myomeres than elsewhere. The range in size of melanophores in micra among individuals and populations is 20–150. All melanophores have many projections (10–21 per cell with most blunt-ended and few thinning at the ends) with melanin granule dispersion of $3 < 4 < 5$. The melanophores on the cheek of the Pine Bluffs fish (spring-cave N = 9) increase from 35 to 45 per .175 mm^2 with growth and an allometric constant of 0.42 while for Mammoth Cave spring cavefish (in cave N = 6), cell numbers decline from 25 to 15 per .175 mm^2 with an allometric constant of –0.60.

These data suggest that light intensity will affect number, density and perhaps range of dispersion within a fish ontogenetically. In the spring-cave populations, melanophores have a score of 3–5 for range of dispersion from reticulate to fully dispersed melanin granules. Poulson does not have comparable data for the Mammoth Cave fish but the pale color suggests that the range would include punctate (score of 1). To extend this argument, Poulson expects to see that spring-cave populations will increase melanophore melanin dispersion and density of melanophores if kept in the light.

Live *T. subterraneus* from all localities appear white to pearly opalescent (some guanine?) and in high-resolution photographs outlines of the tiny embedded scales are visible as slightly grayer color (Fig. 1C). Kentucky populations have visible blobs of fat around the vestigial eyes. Compared to *Chologaster*, all *Typhlichthys* populations (three from east of the Mississippi River and two from west of the Mississippi River sample sizes 4–13) have comparable to somewhat higher densities of melanophores but much smaller surface areas of each pigmented melanophore. Unlike in *Chologaster*, it is rare to see melanophores contiguous and there is only one individual where some melanophores overlapped. In preserved specimens, the visible melanophores are concentrated along the dorsal myomere borders and scattered elsewhere on the body. Melanophores are not visible on the cheek and the number in the reference 0.175 mm^2 pectoral area increase from 2 to 12 with growth with an allometric constant of 1.50. Melanin dispersion in these melanophores ranges from mostly stellate (score of 3) to reticulate (score of 4 < 5). Their longest axis (6–130 micra) is much greater than the narrowest (3–20 micra). There are some pigmented melanophores on the surface of muscle under the dermis and relatively more in the internal connective tissues.

For the populations with the greatest size range and sample sizes, Shelta Cave, Alabama (N = 13, 13–60 mm SL) and The Gulf, Missouri (N = 13, 21–30 mm SL), there are interesting differences (Fig. 7), which Poulson hypothesizes are related to the open, partially sunlit, and very deep sinkhole of The Gulf (the same differences are seen with the Shelta fish 21–32 mm SL as a control for size). Of the Gulf fish melanophores, the dispersion scores are one (0), two (0), three (7), four (19) and five (16); whereas for Shelta fish, the scores are one (20), two (9), three (9), four (4), and five (5). The Shelta fish are not only more variable between individuals (e.g., one with no pigmented melanophores, one with all score one and one with all score 5) but also within individuals (e.g. one with all 1, one 1-3, and one 3-5). Poulson suggests that the greater uniformity and darker melanophores in The Gulf are due to the low to medium light levels. This is consistent with substantial increases in pigmentation seen by several workers in *Typhlichthys* kept in the light in the laboratory (e.g., Woods and Inger 1957).

Compared to *Typhlichthys*, *A. spelaea* melanophores are much reduced in numbers, sizes, and ranges of melanin granule dispersion. The data are mean melanophore area in mm^2 × 10^{-3} from *T. subterraneus* to *A. spelaea* (0.21 to 0.06), range of longest dimension in micra of melanophores (8–135 with a median of 75 to 23–65 with a median of 36), number per 0.175 mm^2 reference area (90 to 6), and range of dispersion (1–5 with a mean of 4.4 to 2–3 with a mean of 2.4). In addition, for the same dispersion scores of 3, *A. spelaea* melanophores have about half as many projections (4.0) as *T. subterraneus* (7.0). And unlike for *Typhlichthys*, neither number nor darkness of melanophores increases in the light (Eigenmann 1897,

Poulson, personal observations). In fact, the allometric constant for number of melanophores is negative (–1.10) and some larger fish have no melanophores (vs. 1.50 for *Typhlichthys*). All of these data are consistent with increasing accumulation of rudimentation mutations at the *A. spelaea* grade of evolutionary troglomorphy.

Compared to *A. spelaea*, *A. rosae* has a somewhat different pattern of pigment loss but overall, in contrast to a much greater rudimentation of eyes, shows only somewhat more rudimentation of pigmented melanophores. Mean melanophore area in $mm^2 \times 10^{-3}$ is not as reduced in *A. rosae* compared to *A. spelaea* (0.15 to 0.06) the range of longest dimension in micra of melanophores is greater (12–200 with a median of 42 to 23–65 with a median of 36), and the range of melanin granule dispersion is greater (1–5 with a mean 4.1 to 2–3 with a mean of 3.4). However, melanophore numbers per $0.175\ mm^2$ reference area is less in *A. rosae* (0.7 < 6.0). There was no difference in number of melanophore projections for the same dispersion (5.3 ~= 4.0). As for *A. spelaea*, *A. rosae* does not become darker in the light and the negative allometric constant is still more negative (–2.37) with some large individuals with no melanophores.

F.   Brain Pattern as a Clue to Sensory Systems

In teleost fish, many brain sensory input areas are visible as separate parts so sensory abilities can be inferred from brain pattern (Evans 1940). All the following comparisons are for 45 mm SL Amblyopsid fish from double log plots of size of part versus standard length.

Vision and Reactions to Light

Even the surface-living, nocturnal Amblyopsid with the largest eyes, *C. cornuta*, has very small eyes and optic lobes compared to the diurnal *Fundulus notatus* topminnow (Figure 11 in Poulson 1960). Both estimated eye and optic lobe volumes for *F. notatus* are ~12× that for *C. cornuta*.

Both *C. cornuta* (Poulson, personal observations) and *F. agassizii* (Weise 1957, Poulson, personal observations) hide in vegetation or under objects in even low levels of ambient light. Even blinded *F. agassizii* show no change in swimming patterns or food-finding ability (Payne in Eigenmann 1909). Spring cavefish certainly uses their eyes to detect light intensity since the species exits caves into springs after dark and retreats back underground as dawn approaches (Weise 1957, Hill 1966; Poulson, personal observation). *Amblyopsis spelaea* show very weakly developed negative photokinesis even when vestigial eyes of 15–25 mm TL fish are removed (Payne 1907).

Comparative data on Amblyopsid visual systems can be found in Table 2 and Fig. 6). Maximum eye diameters in mm are greatest in the non-troglobitic species and decrease in the troglobitic species. No data are available for *S. poulsoni* but much smaller eyes are expected based on very small optic lobe length. Optic lobe volumes in mm^3 also are greatest in the nontroglobitic species and smallest in the most troglomorphic species, *S. poulsoni*. If we could measure optic tectum volume, Poulson expects that lamellar organization of visual fish will be absent in all the troglobites even if some optic tracts have been taken over by other sensory systems. Even further optic tectum rudimentation is predicted with increasing evolutionary time in caves. See the details of eye structure rudimentation in the earlier section on eyes.

Response to light has been investigated for several Amblyopsid species. Poulson (1963) noted both *C. cornuta* and *F. agassizii* are negatively phototactic, i.e., directed movement away from light. Among the other Amblyopsid species, *A. spelaea* is reported to be negatively photokinetic, i.e., lower activity in the dark resulting in more time spent in the dark end of a gradient (Payne 1907, Eigenmann 1909, Green and Romero 1997), whereas *T. subterraneus* is indifferent to light (Eigenmann 1909, Verrier 1929, Green and Romero 1997). Photokinetic response may be linked to functioning of the pineal organ in Amblyopsids (Green and Romero 1997). The pineal organ is poorly developed and functionally ineffective in *T. subterraneus* (McNulty 1978b), which is indifferent to light. Although the pineal organ of *A. spelaea* has not been examined, Green and Romero (1997) predict that the organ is photosensitively functional, although role of extraocular, extrapineal photoreception cannot be readily dismissed. Photokinetic responses have not been investigated for *S. poulsoni*. Likewise, variation among populations within species has not been examined.

Poulson believes that the weak and varying photokinetic responses to low light intensities among troglobitic Amblyopsids (e.g., Payne 1907, Poulson, unpublished data) have no ecological or evolutionary significance. Part of his reasoning is that *A. rosae* and *T. subterraneus* change from negatively to positively or neutrally photokinetic with increase in size. This could be because of decrease in transparency of the head tissues above the brain. Nerves and the brain of many organisms respond electrically to bright light. Payne removed eyes of small *A. spelaea* and found no change in their weak negative photokinesis. Furthermore, it is extremely rare that any cave Amblyopsid is exposed to a light gradient where water flows into or out of a cave. In one such case (the caves of Spring Mill State Park where Eigenmann and Payne worked) *A. spelaea* avoids the silty areas in a 150 m stretch before the stream exits a cave but can be seen swimming in the twilight zone of rocky areas at the entrance to the next cave 50 m downstream across a karst window (Payne 1907, Poulson, unpublished data). In another case (The Gulf in Missouri), *T. subterraneus* preferentially forages in the light areas along the sides of a steep-sided and deep karst window, probably because of higher food densities.

Taste

The sense of taste (contact chemoreception) does not seem to be important for Amblyopsids. Anatomically the sense organs on the head do not look like taste buds and neither the skin brain centers (a median 'facial lobe' of the medulla) nor mouth brain center (paired 'vagal lobes' of the medulla) can even be identified visually. Smith and Welch (1978) noted that spring cavefish rely on gustation rather than visual or olfactory cues to determine prey edibility. *Forbesichthys* that ingest and then spit out inedible items that researchers wiggle near them may be using some mouth taste, but there is sometimes inedible organic matter or bat guano as a small component of gut contents in troglobitic Amblyopsids. This suggests that they are not as good as human babies that can selectively sort and spit out peas from a mush of potatoes or oatmeal (Poulson, personal observations).

Olfaction

The sense of smell (distance chemoreception) is inferred to be better developed in all Amblyopsids than in *Fundulus notatus* based both on external olfactory rosette area and internal brain olfactory lobe length (Table 4, Fig. 8, Fig. 11 in Poulson 1960, Fig. 4 in Poulson 1963). All Amblyopsids have a tubular intake and pore output for water that somehow is drawn across the olfactory rosette. Since every rosette surface cell has a cilium (Claude Baker, personal communication) Poulson presumes that the surface area of olfactory rosettes is a good index of olfactory capacity. Among Amblyopsid species, there is no discernable change with increased troglomorphy in olfactory lobe size, no pattern for number of lobes of the olfactory rosette, and only modest increase in estimated olfactory rosette area (Table 4, Fig. 8, Fig. 11 in Poulson 1960, Fig. 4 in Poulson 1963, Table 2 in Poulson 1985). This lack of a trend suggests to Poulson that cave Amblyopsids do not exhibit much enhanced chemoreceptive abilities.

Fernandus Payne (in Eigenmann 1909), Poulson (1960), and Hill (1966) have done simple behavioral feeding experiments and neither *F. agassizii* nor *A. spelaea* react positively to smells of live or dead prey that are the main food items in their guts and in their environments. However, both the troglophile and troglobite react to and even seize moving cotton balls, sticks or wires. This is not surprising with the overwhelming importance of lateral line neuromast and tactile senses (see below).

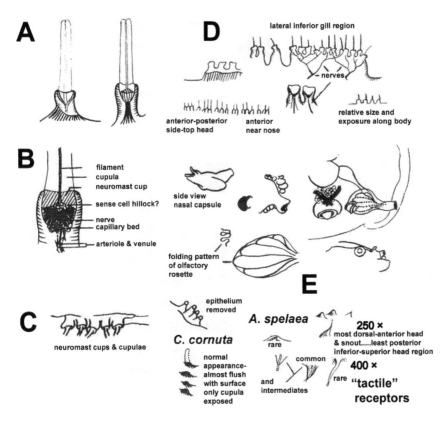

Fig. 8 Drawings by Poulson of sensory system anatomy of Amblyopsid fishes. Shown are "tactile" receptors of *Amblyopsis spelaea*, the olfactory apparatus of *Forbesichthys agassizii*, and the difference in exposure of neuromasts in *Forbesichthys agassizii* and *Chologaster cornuta*.

If not used in feeding, what is the function of olfaction in Amblyopsids? At present this is an unsolved mystery. Experiments on neurophysiological responses to serial dilutions of amino acids have not been conducted but lack of behavioral response to live but immobile prey only millimeters away suggest that olfactory detection of prey is not important. Aumiller and Noltie (2003) provided evidence that *T. subterraneus* detects and is attracted to exudates of both conspecifics and prey (amphipods) in a laboratory setting but the study did not differentiate between olfaction and gustation, nor is there any behavioral responses to an injured fish that suggests the presence of alarm odors (i.e., shreckstoff). Likewise, sex-related variation in responses has not been addressed. However, Aumiller and Noltie suggest that cavefish may rely on chemoreception when locating conspecifics during the reproductive season or detecting

Table 4 Sensory systems of Amblyopsid fishes listed in order of increased (+++) to no (0) evolutionary elaboration with life in caves. Based on Poulson (1960, 1963, 1985, unpublished data) and Cooper (unpublished data). Standardized to 45 mm SL. Where there is a trend, species are given rank from least elaborated (1) to most elaborated (6). Cave species are listed in order of increasing time of isolation in caves (1–4) based on eye and pigment rudimentation.

Species	C. cornuta	F. agassizii	T. subterraneus	A. spelaea	A. rosae	S. poulsoni
Habitat	Swamp	Spring	Cave1	Cave2	Cave3	Cave4
H$_2$O Movement (Lateral Line)						
Lateral line neuromast num. in rows 2–4	21.6 (1)	28.5 (2)	45.6 (6)	37 (4)	32.5 (3)	41.9 (5)
Head surface area (mm^2)	443 (2)	329 (1)	666 (3)	615 (3)	683 (3)	666 (3)
Height of cupula + neuromast (mm)	0.3 (1)	0.66 (2)	0.84 (3)	1.14 (6)	0.8 (3)	0.96 (5)
Forebrain 'computer' volume (mm^3)	0.5 (1)	0.9 (2)	1.15 (3)	1.3 (5)	1.25 (5)	1.15 (3)
Equilibrium (Otoliths & Semicircular Canals)						
Saggitus length (mm)	0.69 (1)	0.96 (2)	2.1 (4)	2.05 (4)	2.07 (4)	1.3 (3)
SS Canal complex (mm)	1.65 (1)	2.15 (2)	4.1 (5)	4 (5)	3.48 (4)	3.15 (3)
Cerebellum 'computer' volume (mm^3)	0.67 (1)	0.7 (1)	1.04 (4)	1.48 (5)	1.49 (5)	0.9 (2)
Caudal Papillae (Unknown Function)						
Total number	24? (1)	36? (2)	89 (3)	96 (5)	111 (6)	88 (3)
Relative size	1	1	1.5	1.5	1.7	3.2
Tactile Receptors						
Number on head (0.175 mm^2)	8 (1)	14 (2)	21 (4)	16 (2)	18 (4)	0 (0)
Relative size	1	1.5	1.5	2.5	2	0
Number of types	1	1	1	2–3	1–2	0
Smell (Distance Chemoreception)						
Number of lobes in olfactory rosette	4	5	6	8	5	5
Olfactory rosette surface area (mm^2)	0.045	0.05	0.062	0.07	0.071	0.066
Olfactory 'computer' lobe length (mm)	0.53	0.55	0.47	0.60	0.62	0.45

potential food sources via exudates dispersed in groundwater flows over long distances. These hypotheses remain to be critically tested.

Poulson (this volume) argues that critical testing of the importance of olfaction must involve realistic scales of volume, mixing of currents, and extreme dilution effects expected in nature. Even in a moderate-sized artificial stream, Bechler (personal communication) saw only occasional agonistic responses of ripe males and females of four Amblyopsid species tested intraspecifically. In addition, Poulson (in Mohr and Poulson 1966) once spent 24 hours watching six presumed pairs of large *A. spelaea* in Upper Twin Cave, Indiana. Even though they were close to each other, he could detect no change in behavior when they closely approached each other, seemingly by chance. Poulson has watched newly hatched *A. spelaea* freeze when approached by an adult in the field. One hatchling that moved was eaten immediately and cannibalism is known to occur when population densities are high (Poulson 1969; Hill 1966). Bechler found that a 'freeze' reaction is a submissive act seen in all Amblyopsid species that show agonistic behavior (Bechler 1983). All of these observations are consistent with the importance of lateral line sensing and not olfaction.

Mechanosensory Lateral Line System

Eigenmann (1909) thought that the large number of lines of pores or ridges on the head and body in all Amblyopsids were tactile organs. In fact, each ridge or line of pores is one part of a lateral line system that is most sensitive to water movement. The lateral line system has sense organs called neuromasts that are part of two systems. The head canal system has only three pores opening to the surface visible with a dissecting microscope. With histological sections there appear to be neuromasts under each of a number of tiny openings all along the subdermal head canals (Claude Baker, personal communication). The much more extensive superficial system has large neuromasts that are arranged in lines on the body surface. In many teleosts, superficial neuromasts are scattered, much smaller, and much more numerous than in the Amblyopsids (Montgomery *et al.* 2001 for *Astyanax fasciatus*; Claude Baker, personal communication, for *Aphredoderus*). In all cases, each neuromast consists of several sensory cells with cilia embedded in a gelatinous cupula (Fig. 8A and B; Fig. 4 in Poulson 1963).

Montgomery *et al.* (2001) provide a short and lucid summary of the mechanosensory lateral line system anatomy and how it works in teleost fish. The authors go on to summarize a variety of neuroanatomical, behavioral, selective pharmacological inhibition, and neurophysiological evidence including some on *Astyanax fasciatus* cavefish that suggest different functions for subdermal canal neuromasts, mostly on the head, and superficial neuromasts, all over the body. The canal system neuromasts filter out low frequency noise and are most sensitive to high frequency water movements generated by localized moving prey. In contrast, the superficial neuromasts are most sensitive to low frequency water movement generated by changes in water flow velocity around the fish ('hydrodynamic imaging') as it approaches an obstacle or swims upstream in a current (rheotaxis). Nonetheless, superficial neuromasts in some species react to water displacement rather than to velocity and this seems to be the case in Amblyopsids.

Poulson's (1960, 1963) data are consistent with the head neuromast ridges detecting live prey. In fact, the cupulae of anesthetized Amblyopsids can be seen visibly to move under dark field of a dissecting microscope to tethered moving prey. For large water fleas (*Daphnia magna*), *F. agassizii* cupulae move at 10 mm distance while for *T. subterraneus* and *A. spelaea* the distance is 20–40 mm. For small amphipods (*Hyalella azteca*), *F. agassizii* cupulae move at 20–30 mm, *A. spelaea* cupulae at 30-45 mm, and *T. subterraneus* cupulae at 40–50 mm. These are a little less than the distances at which live fish orient toward and then capture the same prey items. The longer distance for reaction of the troglobites is related to more elevated neuromast ridges and longer cupulae in *T. subterraneus* and *A. rosae* compared to *F. agassizii*. In *C. cornuta*, the neuromasts are in lines but they are not elevated and the cupulae are very short (Fig. 8 and Fig. 4 in Poulson 1963).

The system of superficial ridges of neuromasts, each with 10–20 neuromasts, is especially hypertrophied in all the troglobites (Figs. 9 and 10) and so Poulson (1960, 1963, 1985) asserted that it was their primary sensory system. This is consistent with relative enlargement of the brain primary sensory nerve projections of lateral line to the cristae cerebelli on the medulla and eminentia granularis on the sides of the cerebellum. Higher sensory processing and integration of the lateral line inputs with other sensory systems is in the forebrain (telencephalon), which is also hypertrophied in all the troglobites (Fig. 9; Fig. 3 in Poulson 1963; Table 2 in Poulson 1985). The many neuromast ridges on the head, with vertical and horizontal orientations of adjacent ridges (Fig. 10) must give rich detail of water movement information for the forebrain to interpret.

In addition to the advantage of a relatively large head having more elaborate superficial neuromasts, there is a side benefit of better 'hydrodynamic imaging'. A larger head allows slower swimming to detect and avoid obstacles. Slower swimming generates less noise for the neuromasts to detect moving prey. Montgomery *et al.* (2001) write, "Although hydrodynamic imaging may not be the sole prerogative of blind cavefish, these fish appear to have evolved rather sophisticated mechanisms for processing and using the images."

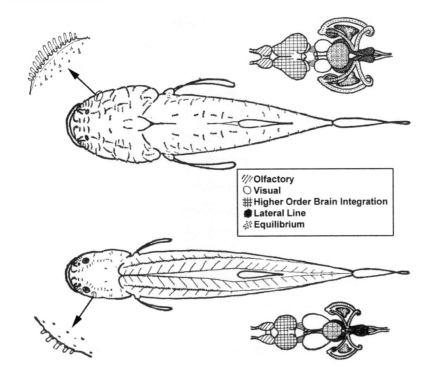

Fig. 9 Scale drawings (from Poulson and White 1969) contrasting the troglobitic *Amblyopsis spelaea* with its troglophilic relative *Forbesichthtys agassizii*. Compared to *Amblyopis*, each of the lateral line rows in *Forbesichthys* has fewer neuromasts, they are less elevated above the skin, and their brain 'computer' centers are less hypertrophied, but its eyes and optic lobes are larger. Also, the presumed 'tactile' receptors (shown by the enlarged lateral line rows), are smaller and fewer in kind than for *Amblyopsis*. Note also the difference in semicircular canals and otoliths. However, there is no difference in olfactory lobes of the brain.

Rheotaxis Depends on the Lateral Line System

Except perhaps for *C. cornuta*, all Amblyopsids show positive rheotaxis (e.g., Tables 13 and 14 in Poulson 1960). *Chologaster cornuta* rarely is found in flowing streams. The difference among species in rheotaxis is of degree. Allochthonous food renewal is almost always associated with late winter/spring increases in water input to caves. *Amblyopsis spelaea* will swim in the strong currents and lives in cave streams that almost always have appreciable current. As floods begin and the current becomes too strong (up to 7 m^3 sec^{-1}) they move to the bottom of deep pools and into backwaters. *Typhlichthys* in L&N Cave, Kentucky, are exquisitely sensitive

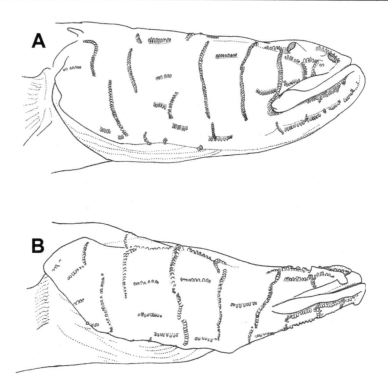

Fig. 10 Scale drawings (courtesy of John E. Cooper) of the heads of *Typhlichthys subterraneus* and *Speoplatyrhinus poulsoni*. Note the relatively longer head and shovel-nosed shape of *S. poulsoni* head, which are neotenic trends. Both species show the convergent similarity in lateral line ridges or "stitches". Each stitch has four to as many as 39 neuromasts cups.

to slight increases in current; they seek shelter hours before we can detect an oncoming flood (Pearson, personal communication). *Forbesichthys* rely on swimming upstream to return to caves every day (Weise 1957) or at the end of the growing season (Hill 1966).

Tactile Receptors and Thigmotaxis

In fish, the brain primary input for tactile information, the somato-sensory lobe, is hard to distinguish from that for lateral line input since they go to the same medulla area and share cranial nerves V and VII. A further complication is that the short neuromasts could function as tactile receptors when Amblyopsids strike at moving prey or are touched by moving prey (think about seeing light when you are hit on the eye in the dark). Also, the small size of what are presumed to be tactile receptors on the head and around the mouth of both *Chologaster* and *Forbesichthys* makes it very difficult to quantify numbers using a dissecting microscope, especially with the heavily pigmented epithelium.

Despite the above problems, Poulson has attempted to quantify putative tactile receptors in all the Amblyopsids. All have what are presumed to be tactile sense organs concentrated on the heads with relative densities on the body << side of head < top of head above the mouth << under the mouth and lips. The incomplete data for number of tactile receptors per 0.175 mm^2 are as follows: for the side of the head, *T. subterraneus* 3–8 = *A. spelaea* 4–16 = *A. rosae* 4–7; for the top of the snout, *A. spelaea* 1–4 << *A. rosae* 11–15; and for under the mouth and lips, *A. spelaea* 16–20 < *A. rosae* 25–35. On the side of the head, both *F. agassizii* and all three troglobites have similar densities of these presumed tactile receptors but they are larger in the troglobites and especially larger and more diverse in shape in *A. spelaea* (Fig. 8, Fig. 4 in Poulson 1963). *S. poulsoni* has none of these receptors anywhere on its head or body.

Despite relatively low numbers of tactile receptors and superficial neuromasts, both *C. cornuta* and *F. agassizii* behaviorally show strong thigmotaxis and *Typhlichthys* shows stronger thigmotaxis than either species of *Amblyopsis*. In spring cavefish, Weise (1957) found that strong thigmotaxis overshadows negative phototaxis. Spring cavefish exhibit strong preferences for cover objects, even if those objects are well illuminated. When resting, *F. agassizii* and especially *C. cornuta* are often in groups touching each other. In aquaria and in the field, *C. cornuta* rests amongst dense vegetation during the day and *F. agassizii* rests under sticks and rocks during the day. In an artificial stream *F. agassizii* spends > 90% of its time under rocks and *Typhlichthys* spends much more of its time under or next to objects resting than either species of *Amblyopsis* (Bechler, personal communication).

During foraging and prey capture, lateral line and touch receptors probably act in a complementary way. Poulson's (1960) and others' observations (Eigenmann 1909, Weise 1957) show that Amblyopsids capture the same live prey by a combination of lunge and grab. They do not appear to use gape and suck feeding. Troglobitic cavefish initiate capture at the same distances that cupulae can be seen to move but *F. agassizii* appears to contact the prey before attack. *Chologaster cornuta* will not capture prey unless it contacts the head or lips; whereas *F. agassizii* will turn and grab prey touching any parts of its body especially if it has previously captured one or more prey. Weise (1957) notes for *F. agassizii* that "an amphipod is taken by a vicious sidewise jerking of the head and is immediately swallowed." Spring cavefish will take three to four 10–15 mm amphipods in a few seconds. *Amblyopsis spelaea* also 'jerks and grabs' as it captures a large prey item.

Static and Dynamic Equilibrium Reception

As with all vertebrates, teleost fish and Amblyopsids have semicircular canals and otoliths. The sensory organs associated with each are modified neuromasts. The semicircular canals detect rates and directions of movement while the otoliths detect body position even at rest.

The three semicircular canals are all filled with a viscous fluid and are oriented at right angles to each other; the neuromasts in the ampullae at the end of each canal detect inertial movement of fluid in the three canals in each of three dimensions as a fish swims or stops. The three calcium carbonate otoliths detect position in three planes even while a fish is not moving. The otoliths rest on a group of neuromasts such that the null or usual body orientation results in no nerve impulses. In the null position the otolith presses evenly on all the sensory hairs but any change in position results in an uneven loading of the otolith on the hairs and increasing frequency of nerve impulses with increasing departure from the null position. The saggitta is the largest otolith and it detects deviation from a horizontal position. The much smaller astericus and lapillus give information when the fish is oriented in other planes.

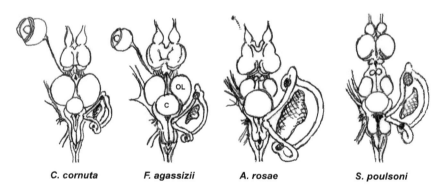

| C. cornuta | F. agassizii | A. rosae | S. poulsoni |

Fig. 11 Scale drawings (Poulson) of eyes and optic nerve (where present), brains, and semicircular canals (SSC) with otoliths of Amblyopsid fishes. *Chologaster cornuta* is the small-eyed swampfish with largest eyes and optic lobes (ol) and smallest forebrain, cerebellum, SSCs and otoliths. *Forbesichthys agassizii* has intermediate brain proportions. *Amblyopsis rosae* shows the convergent proportions also seen in *Typhlichthys subterraneus* and *Amblyopsis spelaea* but has the smallest eyes and optic lobe and the largest forebrain, cerebellum, SSCs and otoliths. *Speoplatyrhinus poulsoni* shows still smaller optic lobes but its cerebellum and SSCs are not as hypertrophied. Note that it has a different relative size of the otoliths than for the other Amblyopsids. Also note that there is no trend in relative development of olfactory lobes at the front of the brain. See Fig. 9 for the relation of brain proportions to body shape and relative development of the lateral line system.

Ordinarily, the static equilibrium system, the otoliths, and dynamic equilibrium system, the semicircular canals, operate in concert with vision. So it is perhaps not surprising among the Amblyopsids that the reduced eye *Forbesichthys*, that cannot see images (Eigenmann 1909), and blind cavefish have hypertrophied semicircular canals, otoliths, and the cerebellum brain integrative area for these two systems (Figs. 8 and 11, Fig. 3 in Poulson 1963). As with the lateral line system and its brain input and integrative centers, there is little further increase in relative size of these systems with increased evolutionary time in caves from *F. agassizii* to *T. subterraneus* to *A. spelaea* to *A. rosae* to *S. poulsoni* (Table 4, Table 2 in Poulson 1985).

It is of interest to discuss why the increase in semicircular canals and otoliths seen in essentially blind and nocturnal *Forbesichthys* is not sufficient to compensate for lack of vision. The answer appears to be that the relatively larger head in all the troglobites has been selected in part to increase the lateral line sensory system (above). The larger head in turn makes it necessary to have further hypertrophy of the semicircular canals and otoliths. Fine *et al.*'s (1987) analysis of the largest known semicircular canals and otoliths in any teleost is instructive. The fish in question is a deep-sea Ophidiid with an even larger head relative to body size than in *S. poulsoni*. It carries to extremes the theoretical need for large canals in fish generally, because fish have no neck and swim in a viscous medium (Jones and Spells 1963). With increased relative head size the radius of curvature and canal diameters must be larger to have the same neuro-physiological gain in sensitivity. In addition, a larger head results in less inertial change when turning and so requires larger canals to detect. The deep sea Ophidiid has a very large head, approximately 10 × the volume of the body. The relative head to body volumes in Amblyopsids range from about equal in *S. poulsoni* to about 10% in *C. cornuta*. Even *C. cornuta* has a relatively large head for teleosts.

A Mystery Sensory System: The Caudal Papillae

All Amblyopsids have a vertical row of papillae at the base of the caudal plus 2–5 horizontal rows extending out along the fin. These are larger and more rounded than the largest presumed tactile sense organs elsewhere on the body. Based on drawings in Woods and Inger (1957), papillae are slightly larger on *T. subterraneus* and both *Amblyopsis* species than on either *C. cornuta* or *F. agassizii*. In addition, total numbers of these papillae are *T. subterraneus* 27 <= *A. spelaea* 30 < *A. rosae* 36. Based on very accurate drawings by John Cooper (Fig. 12), the densities, placement, and sizes of the papillae are strikingly different in *T. subterraneus* (Alabama) and *S. poulsoni*. *Typhlichthys* papillae are much smaller and more crowded than

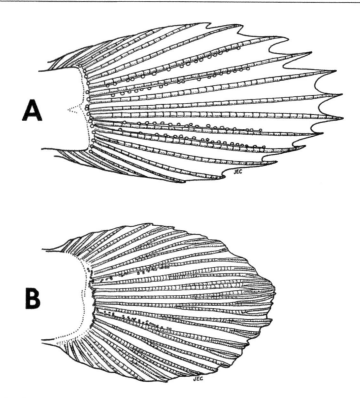

Fig. 12 Scale drawing (courtesy of John E. Cooper) of the caudal fins of *Typhlichthys subterraneus* and *Speoplatyrhinus poulsoni*. The unbranched fin rays of *S. poulsoni* is a neotenic trait. The especially large caudal papillae (function unknown) of *S. poulsoni* are much larger than those of *T. subterraneus* and all other cave Amblyopsids.

in *S. poulsoni*; with shorter total row length (8.5 mm *T. subterraneus* << 24.8 mm *S. poulsoni*) and fewer rows (3 vs. 5) over less fin area. *Typhlichthys* has 70 papillae with 12–17 per mm and *S. poulsoni* has 88 papillae with 3–5 per mm. Obviously, we need equally careful measurements and counts on all the other species to see if there are differences among other species.

V. PREY DETECTION AND AVAILABILITY

All of the available field and gut content evidence is consistent with the laboratory feeding observations (reviewed above) that show that the superficial lateral line system is necessary and probably sufficient to explain the diet of Amblyopsid fish. All items in the gut are certain to have been alive and moving prey, and predominant prey in the gut match the distribution and abundance of prey in the field. This is true in terms of frequency of occurrence, abundance, and volume. If prey are similar in these regards, then the species that is most active, and thus most easily detected with the lateral line system, is more frequent and abundant in fish guts. Thus, the fish are not showing a preference for particular food items because they take whatever they can detect.

A. Data for *Chologaster* and *Forbesichthys* Gut Contents

The data are excellent, in terms of sample sizes (hundreds to thousands of fish) and time of year of sampling for *C. cornuta* (Ross and Rohde 2003) and *F. agassizii* (Weise 1957; Hill 1966). The match of prey in the gut and prey in the field was remarkably similar in *C. cornuta* from Black Swamp Creek, North Carolina, and *F. agassizii* from Pine Hills, Illinois, cave-spring runs. Both sites have dense vegetation and hiding places so visual predators of the Amblyopsids are probably unimportant. In both sites Gammarid amphipods were the most frequent, most numerous, and largest food items in the Amblyopsid guts and in the field. Of the items eaten the amphipods are also the most active and so most easily detectable using the superficial lateral line system. The Rich Pond spring runs did not have amphipods but did have abundant Chironomid midge larvae and small oligochaetes with a high diversity of other potential prey. The *Forbesichthys* guts had more midges than worms. Hill (1966) attributed this to the greater activity levels of the midges, which made them more detectable by the fish.

At Rich Pond, Kentucky, there is a huge difference in food supply and feeding outside the cave and inside (Hill 1966). All surface-collected fish had very high condition factors and full guts. The cave source of the Rich Pond spring run had almost no live prey compared to the super-abundance of prey outside. Guts of fish 35–40 mm SL spending their first winter in the cave were 99.9% empty and fish had low condition factors. Guts of fish spending their second winter in the cave were only 18% empty with the rest having fish (cannibalism) and only slightly poorer condition factors than when they were outside the spring before.

B. Cave Amblyopsid Gut Contents as Samplers of Prey Density

For Amblyopsid troglobites the fish are a better sampler than we are of prey availability in the caves with their very low food supplies (Poulson and Lavoie 2000, Poulson 2005). Poulson (1960, unpublished data) has some gut content data for several localities with sample sizes in the 50–100 range and where we can both easily census fish and crayfish and can occasionally census amphipods and isopods by examining loose rocks. Harpactacoid and Cyclopoid copepods are virtually impossible to census accurately and they are the most frequent and numerous prey in guts of cavefish.

Here we present some representative data for *T. subterraneus* from Shelta Cave, Alabama, *A. spelaea* from Donaldson Cave, Indiana, and *A. rosae* from Cave Springs Cave, Arkansas (a gray bat maternity cave). For four fish each of 37–43 mm SL the gut contents were as follows with total numbers of each prey and % frequency (25, 50, 75, or 100%). All the data were taken at low water times well after the most recent flood. Four *T. subterraneus* had 167 copepods (100% frequency), 13 tiny ostracods (25%), 1 amphipod (25%), and 1 tiny shrimp (25%). Four *A. spelaea* had 65 copepods (100% frequency), 7 isopods (75%), and 1 amphipod (25%). Larger *A. spelaea* had more isopods, some amphipods, and one fish even had a tiny crayfish. Four *A. rosae* had 65 copepods (100% frequency), 8 cladocerans (25%), 4 isopods (50%), and one 10 mm crayfish (25%).

Extensive checks of guts in museum collections for other caves with lower censused numbers of each Amblyopsid species show that troglobites NEVER had an empty rectum or intestine and rarely had an empty stomach. In contrast, six of ten *Forbesichthys* from Mammoth Cave had completely empty guts in the year when Poulson censused fifteen fish that had survived after being washed in from far upstream sinking creeks. For those four individuals with any food in their guts, the maximum was 6 copepods!

C. Is *Forbesichthys* Truly Troglophilic?

Hill's (1966) data on *Forbesichthys* at Rich Pond reviewed above is consistent with our current view that spring cavefish are not ordinarily able to complete their life history in caves because of inadequate food supplies. Rich Pond has many thousands of *Forbesichthys* that eat frequently and grow quickly in the spring run. However, 6–30 mm SL fish in the cave had empty guts and had poor condition factors. Even 40–60 mm individuals had only cannibalized fish in their guts. In caves with the same very low food densities, of copepods >> isopods > amphipods > tiny crayfish or salamanders, all troglobitic Amblyopsids have food in their guts.

D. Gut Contents in Relation to Prey Densities

An early review of relative live food occurrence in relation to gut contents for all the Amblyopsids (Poulson 1960) is consistent with our present generalization that fish eat the most available and fastest moving prey.

The habitats of *C. cornuta* and *F. agassizii* have abundant prey with 2–4 orders of magnitude greater numbers and volume than any cave.

Here are some data for surface habitats for *C. cornuta* and *F. agassizii*. A 0.25 ft^2 Ekman dredge sample, in Rockfish Creek, North Carolina, in April had three odonate nymphs, 15 small amphipods, 121 copepods, 52 ostracods, and many rotifiers. In the guts of the 16 *C. cornuta* collected at the same time there were 17 taxa but they were predominately amphipods (60% frequency, 41% numbers, and 80% volume). In a Pine Bluffs, Illinois, spring run with a cave-spring population of around 150 *Forbesichthys*, large (10–15 mm) *Gammarus troglophilus* have astounding densities year around of 10–25 per 0.25 ft^2. Isopods and flatworms make up the rest of the macroscopic fauna but neither is found in *F. agassizii* stomachs. As Weise (1957) states the fish are monophagous on amphipods.

The best comparative data on food scarcity in caves is from Hill (1966) since he made weekly censuses of *Forbesichthys* abundance, sizes, and gut contents along with Ekman dredge samples in both the cave and surface spring run at Rich Pond, Kentucky. During the spring months as the water table rises and the cave waters flow out into the huge spring run, surface prey became abundant but inside the cave there was almost no food. The extremes of prey in Ekman dredge samples outside in the spring run vs. inside the cave were as follows: chironomid larvae 300–1200 versus 0–6, oligochaetes 600–1200 versus 0–2, nematodes 0–500 versus 0–1, copepods 1–250 versus 0–20, and ostracods 1–120 vs. none. Outside fish guts were completely full with 70–80% chronomids and 10–20% oligochaetes by volume; inside the cave fish guts were empty.

Only some caves with Amblyopsid troglobites studied by Poulson (1960) have more live prey than in the Rich Pond cave. Even the best *A. spelaea* cave (Upper Twin Cave in Indiana) in terms of abundance with 86–130 cavefish (Fig. 4 in Poulson 1960) and 45–80 cave crayfish had 1–5 isopods and 0–1 amphipods in ten Ekman samples. However, plankton samples of 800 liters ranged from 0–300 ml ETOH displacement (Scott 1907). During a spring flood, the 300 ml consisted of 70% silt and 15% leaf fragments by volume. The remaining 45 ml had an incredible number of plankters including 1998 *Bosmina* cladocerans, 410 adult copepods + 319 nauplii, and 717 rotifiers. This cave is fed in part by a sinking stream and both the plankton and the benthos are more than an order of magnitude more than in the most food rich caves Poulson sampled with other species of Amblyopsids. We do not have data on food supply for the two *Amblyopsis spelaea* caves with the highest densities of cavefish known: Under the Road Cave had an estimated 4199 fish per hectare in a 300 m stream segment with virtually no visible organic matter (Table 5, Poulson 1969) and Penitentiary Cave had an estimated 2643 fish per hectare in a

1200 m series of rimstone pools with 10-30 cm of leaf, twig, and acorn fragments (Pearson, personal communication). Seven caves with the highest visually censused populations of *Typhlichthys* had peak numbers of 4, 6, 11, 16, 20, 23, and 64 fish and for *A. rosae*, three caves had 7, 11, and 72 fish. The two most abundant *Typhlichthys* caves had only 2–4 copepods, 0–3 ostracods, and 0–2 rotifers in one 800 liter plankton sample and no fauna in ten Ekman dredge samples. However, 40–50 cave crayfish were sampled. The most abundant *A. rosae* cave had 40 copepods, 6 ostracods, and 1 isopod in an 800-liter plankton sample and the benthos could not be sampled among the large rock rubble; however fish gut samples show that isopods, amphipods, and small crayfish are present.

VI. ECOLOGY, LIFE HISTORY AND BEHAVIOR

In this section, we discuss aspects of Amblyopsid ecology, life history, and behavior, including such topics such as reproduction, metabolism, demography, and agonistic behavior. Much of the information assembled to date is from the works of Poulson, Bechler, and colleagues. However, many questions remain unanswered.

A. Life History Patterns

Hypotheses and Theory

On the basis of scale marks, size-frequency distributions, egg size and number, and metabolic rates Poulson (1960, 1963, 1985) inferred that there is a trend of increasing longevity, decreasing growth rates, and decreasing metabolic rates from the epigean *C. cornuta*, to the troglophile *F. agassizii*, to the three troglobitic cavefish known at the time. For the troglobites, he hypothesized that increasing life span, lowered metabolic rates, and large eggs resulting in large young are adaptations to low food supply in caves. On the basis of gaps in the size-frequency distribution with no small fish and some data on gill surface area, he later inferred that his namesake *S. poulsoni* has an even longer and more energy efficient life history and metabolic pattern.

   In the context of life history theory (Poulson 2001b and this volume) the Amblyopsids represent the extremes of very high r to very low r (instantaneous population growth rate in the logistic equation $dN/dt = rN (K-N/K)$. At the *C. cornuta* extreme of high r, there is a 1–2 year lifespan and one-time reproduction (semelparity) at an age of $<= 1$ year. The tradeoffs are of population booms when times are good and busts when times are bad. Associated with this pattern are spreading of risk of mortality among numerous small eggs that result in small young. *Amblyopsis spelaea* represents the other extreme of low r with a 20–30 year lifespan (see below) and an age at first reproduction no earlier than 6 years. The tradeoffs are of population 'fizz' when times are good and population 'fizzle' when times are bad. Associated with this pattern are spreading of risk of reproductive failure among many attempts (iteroparity) with few large eggs and large resulting young at each reproductive attempt.

Iteroparity likely evolved in cave Amblyopsids, and in many other cave species, because of extreme selective pressures on energy economy at all life history intervals and because adults are top predators and essentially immune from predation (Poulson 2001b). In the surface species, *C. cornuta* and *F. agassizii*, death likely occurs after a single reproductive event shortly after reaching sexual maturity. For both these species, almost all adult females breed in a given year (Poulson 1963, Ross and Rohde 2003, Hill 1966). Older individuals in their second or third years probably represent those that failed to reproduce at an earlier age.

For cave species food limitation is the likely selective force limiting reproduction and reproductive output. In wet years when food availability is higher, a larger proportion of females likely reproduce. Contrastingly, in poor years when food availability is severely limited, perhaps all females fail to reproduce. Iteroparity and increased longevity reduce the risk of population extirpation when food supplies are too low for an extended period of time to allow females to reproduce. However, mark-recapture measures of growth rate (below) show that cave Amblyopsids are much older than Poulson (1963, 1985) originally suggested based on scale marks and so the reproductive potential of each species may be considerably less. This means that conservation of various populations may be more critical than we currently think (D. Bechler, personal communication).

Tests of Hypotheses: Life History Patterns

In cave Amblyopsids, only a fraction of all sexually mature females reproduce in a given year. In the three species for which reproductive data existed, Poulson (1963) estimated that 50% of *T. subterraneus*, 20% of *A. rosae*, and only 10% of *A. spelaea* females reproduce in a given year. Based on better data and inferred longer longevities, his revised estimates (Poulson 1985) were 30% *T. subterraneus*, 5% *A. spelaea*, and 25% *A. rosae*. Based on a thorough review of even more data and the knowledge from growth rates of even greater longevities, he now estimates the percent of mature females reproducing in a year averaged over a decade for all Amblyopsids as 100% *C. cornuta*, 70% *F. agassizii*, 20% *A. rosae*, 10% *Typhlichthys*, and 5% *A. spelaea*. These numbers may be too high for the troglobites if preliminary data hold up showing sex ratios of all Amblyopsids as almost 2 to 1 for adults. The revised percents of females reproducing over a decade then become 13 % *A. rosae* > 7% *T. subterraneus* > 3% *A. spelaea*).

The Ross and Rohde (2003) life history study of *C. cornuta* is a paradigm for rigor and completeness. It redresses the odd situation that this most common and accessible (epigean) species of Amblyopsid was the least studied and least understood. The life history of this Amblyopsid was previously only inferred from size-frequency distributions and ovarian egg counts from small samples from many localities each on usually only one date (Fig. 4 in Poulson 1960; Ross and Rohde 2003). From collections at one site on 25 dates with samples of 24–41 fish over one year, Ross and Rohde could follow the growth rates of young of the year over 65 days in spring from 8–10 mm SL to 13–18 mm SL for males (~4 mm growth per month) and to 23–30 mm for females (~8 mm growth per month). By 14 June, the few surviving adults were gone and presumably dead. The first year fish growth rate slowed in summer and increased in fall. By early January, males were 19–25 mm and females 29–37 mm SL. We now think that troglobitic Amblyopsids of these sizes are as old as 3-6 years even with energetic adaptations to orders of magnitude lower food supplies.

From all of their data, Ross and Rohde (2003) agree with Poulson's earlier inference that *C. cornuta* has a one year life cycle and dies after reproduction (semelparity). Their data show a sharp 2-month decline of mean gonosomatic index from 11% to 1% and from 1% to 0% in the next month (GSI = weight of ovaries as % total body weight less ovary weight). The initiation of this short spawning season coincided with a sharp February to March rise in stream temperature from 5 to 15°C by mid May when 8–10 mm young of the year were sampled. There were only two of this size so the smallest young were undoubtedly missed with relatively coarse meshed seines.

Even Ross and Rohde's extensive data on ovarian egg number and diameter are difficult to interpret and compare to other Amblyopsids. They wisely used only ovaries that were "tight and full of eggs" to estimate fecundity in 158 females 20–47 mm SL. There was huge variation in egg diameter from 1.14–1.78 mm (mean 1.48) and in number of eggs from 6–98 (mean 24.7). Together this indicates great variation in both clutch size and size of smallest free-swimming hatchlings. These data are difficult to compare to other Amblyopsids that grow much larger (females assayed for fecundity are 45–65 mm SL). Females from the population Ross and Rohde studied mostly grew to a maximum of 39 mm with 3 of 156 larger (45 mm = 33 eggs, 41 mm = 71 eggs, and 47 mm = 98 eggs). From a sample size of only three females from two other populations where average maximum SL is in the mid 40s, the fecundities are much higher (42 mm = 339 and 45 mm = 426 eggs in one population and 44 mm = 252 eggs in another population) than the fecundities Poulson used for comparison to other Amblyopsids (Table 6 in Poulson 1960). Also the egg diameters are higher than the 0.9 –1.2 mm Poulson (1960) reported and used to estimate reproductive risk (Poulson 1985).

Loren G. Hill's (1966) Ph.D. dissertation on *F. agassizii* is in many ways the most complete study ever done on one population of Amblyopsid. In his Rich Pond, Kentucky, study site there is a population size of many thousands of fish and he collected samples of up to 100 on each date throughout the year. He followed growth, feeding, gut contents, condition factor, fecundity, and ontogeny of vent migration and scale development weekly for the growing season. He studied fish both in the cave and in the spring-run into which the fish moved in late winter, as water started to flow from the cave. Fish returned to the cave in summer, as the surface waters dried up, and spawned in the cave the following winter.

Rich Pond has similarities and differences to the Pine Bluffs, Illinois, spring-cave populations (Weise 1957). Pine Bluff populations are similar in fecundity, growth rates, and 3–4 year classes (size cohorts and scale annuli). However, it is different in the much smaller habitat area and much smaller populations (Weise 1957, Smith and Welch 1978). In eight Pine Bluffs springs studied by Petersen mark–recapture, the population estimates were 1–302 with a conservative total of 862 (Smith and Welch 1978). The food available was abundant under rocks and in watercress in epigean spring runs with 291–902 individuals per square foot of five taxa. In every monthly sample over a year a large *Gammarus* species (10–15 mm) comprised 50–80% of numbers and 90–95 % of volume in the field and > 98% by volume in adult fish guts (Weise 1957). Both Weise and Smith and Welch agree that most fish exit the underground into the spring a half hour after dark and return about one hour before dawn (a few remain under rocks outside during the day). Thus, they are extremely negatively phototactic and positively thigmotactic. They also agree that spawning occurs underground since gravid fish disappear from the surface in January and February and no gravid fish are found on the surface after that.

Tests of Hypotheses: Growth Rates and Age

Since Poulson's (1985) latest summary of data and hypotheses based on scale marks and presumed cohorts, various workers have done mark-recapture studies to obtain data on growth rates of *F. agassizii* (Smith and Welch 1978), *A. spelaea* (N. Welch, personal communication), *T. subterraneus* and *A. spelaea* (W. Pearson, personal communication), and *A. rosae* (Brown and Johnson 2001). All of these studies show that the ages and growth rates match those inferred from scale marks and distinct size classes in

Forbesichthys (2–3 years) but are underestimates of age by 2–4 × for all the troglobites. Thus, potential lifespans are 16–24 years for *Typhlichthys*, 30–45 years for *A. spelaea*, 8–12 years for *A. rosae*, and 16–24 years for *Speoplatyrhinus*. A large range in age for the same size fish within each species is due in part, Poulson now believes, to the great variation in growth rate in fish eating just copepods, fish also eating a few isopods and amphipods, and fish that show huge growth rate spurts when they are lucky enough to capture and eat a large prey item (Poulson 2001b).

Tests of Hypotheses: Fecundity, Egg Size and Cost of Reproduction

Cave Amblyopsids tend to have not only fewer, but also larger and potentially more nutrient-rich eggs than their surface counterparts. As in other cave organisms, larger eggs also contain more yolk and produce larger larvae. By investing more yolk volume per egg, larvae hatch at larger size and are larger in size when all yolk is absorbed. Therefore, individuals have greater starvation tolerance, greater swimming ability when foraging but also when avoiding predation, and can accommodate a wider range of prey sizes when all yolk reserves are depleted. All of these are adaptations that increase survival in a food-poor environment.

How well does a more careful evaluation fit Poulson's (1985) hypotheses about the advantages of increasing egg size with decreasing clutch size with cave adaptation? Data on size of fry when their yolk sac are absorbed is a better measure of sizes of eggs laid than is the sizes of eggs in ovaries of gravid females. The reason is that *Forbesichthys* ovarian eggs at Rich Pond range from 1.5 to 2.0 mm but the sizes of hatchling fry after yolk sac absorption are all 6.0 mm (Hill 1966) suggesting that laid eggs are 2.0 mm. The only other species where we are certain about size of post yolk absorption fry is in *A. spelaea* where we can occasionally find females brooding 2.3 mm eggs in their branchial cavities and find free-swimming 8.5 mm hatchlings (e.g., Eigenmann 1909, Poulson 1960, Pearson, this volume). Thus, an increase in egg diameter from 2.0 to 2.3 translates to an increase in post yolk absorption fry size from 6.0 to 8.5 mm SL. What does this size difference mean to fry?

From data on fish (Blaxter 1983, Hempel 1965, Hunter 1972) an increase in egg size of 2.5 mm translates to the following advantages for fry. There is an estimated 2.5-fold increase in maximum size of prey and increase in range of food item sizes that can be eaten (mouth size), a 3-fold increase in volume of water searched per time (swimming speed), a better ability to escape predators (swimming speed), and a 3-fold decrease in density of prey needed to maintain weight (resistance to starvation). This is about the advantage of increased egg size in troglobitic Amblyopsids (from

1.8 mm eggs in *C. cornuta* to 2.0 mm eggs in *F. agassizii* to 2.2 mm eggs in *A. rosae* to 2.3 mm eggs in *T. subterraneus* and *A. spelaea*) that Poulson (1985) calculated. It does not seem like a big difference but it still seems evolutionarily significant with the extremely low food supplies in caves.

For an egg-laying female, the evolutionary reduction in estimated clutch volume from *C. cornuta* to *A. rosae* is an important aspect of overall increase in energy economy. Poulson has seen photos or read about live females with egg-swollen bodies and easily visible eggs but only for *C. cornuta* and *F. agassizii* (Ross and Rohde 2003, Weise 1957). Poulson now believes that the estimates of clutch size based on ovarian eggs (i.e., Table 2 in Poulson 1985) are too high for cave species but we have no way of knowing by how much.

Poulson has recalculated aspects of reproductive cost that he presented in 1985. The new data (Table 5) are based on average clutch sizes, maximum ova diameter, and average weight of reproducing females. For a *C. cornuta* of 42 mm SL and a weight of 1.0 g with a clutch of 152 eggs of 1.8 mm diameter, the total egg volume is 466 mm^3 with 466 mm^3 per gram of female. For a *F. agassizii* of 60 mm SL and a weight of 3.0 g with a clutch of 154 eggs of 2.0 mm diameter, the total egg volume is 645 mm^3 with 215 mm^3 per gram of female. For a *Typhlichthys* of 50 mm SL and a weight of 1.6 g with a clutch of 58 eggs of 2.3 mm diameter, the total egg volume is 370 mm^3 with 231 mm^3 per gram of female. For an *A. spelaea* of 65 mm SL and a weight of 5.2 g with a clutch of 65 eggs of 2.3 mm diameter, the total egg volume is 414 mm^3 with 80 mm^3 per gram of female. Finally, for an *A. rosae* of 45 mm SL and a weight of 2.7 g with a clutch of 34 eggs of 2.2 mm diameter, the total egg volume is 190 mm^3 with 70 mm^3 per gram of female. In summary, the cost of eggs measured as mm^3 per gram of female declines with increasing cave adaptation with *C. cornuta* > *F. agassizii* <= *T. subterraneus* >> *A. spelaea* >= *A. rosae*.

We get a different perspective if we compare these volumes converted to calories per day over four months (the time observed for eggs to mature *in vivo* in nature) with routine metabolic rate converted to calories per day (Table 5). The data are calories per day routine metabolic rate + calories per day to grow eggs (% due to egg growth): *C. cornuta* 18.4 + 2.9 (16%) << *F. agassizii* 45.0 + 4.0 (9%) >> *A. spelaea* 18.7 + 2.6 (14%) > *T. subterraneus* 13.4 + 2.3 (17%) > *A. rosae* 9.8 + 1.2 (12%). Thus, from this perspective Poulson's (1969) statement would seem to be wrong: "The high cost of reproduction is a crucial problem to be solved in adapting to the cave environment". The low absolute values for *A. rosae* only helps explain why it has the highest estimated percent of females reproducing each year (20% compared to 10% in *T. subterraneus* and 5% in *A. spelaea*). But comparing this to *C. cornuta* (100%) and *F. agassizii* (70%) it does appear

Table 5 Calculation of reproductive costs and comparison to percent of females reproducing and maximum population density. See text for methods. The advantage of large egg size for fry size is not offset by a high cost of egg production and total mm^3 of eggs per g of female declines. In addition, the restrictions of low food supply in cave (*F. agassizii* spring vs. cave) are offset among the cave species by increasing efficiencies that is reflected by increasing population densities and percent females breeding.

Species	C. cornuta	F. agassizii	F. agassizii	T. subterraneus	A. spelaea	A. rosae	S. poulsoni
Habitat	Surface	Spring	Cave	Cave	Cave	Cave	Cave
Free-swimming fry (mm SL)	4.5[a]	6.0	na	8.5[a]	8.5	8.0[a]	
Maximum egg diameter (mm)	1.8	2.0	na	2.3	2.3	2.2	
Egg volume (mm^3)	3.05	4.19	4.19	6.38	6.38	5.58	
Clutch size	152	154	0	<58?	65	<34?	
Female size (mm SL)	42	60	40	50	65	45	
Egg volume per g female mass	466	254	na	231	80	70	
Cal eggs	2.9	4.0	na	2.3	2.6	1.2	
RMR Cal	18.4	45.0	na	13.4	18.7	9.8	
Estimated % females breeding/yr	100	70	0	10	5	20	
Estimated reproductions/lifetime	1	1	0	2–3	3–5	2–3	
Lifespan	1–2	2–4	2–4	16–24	30–45	16–24	16–24
Max density hectare^{-1}	42,000	80,000	0–56	864	4199	2134	

[a] estimated

that the statement is correct. The reason is that caves have such low food supplies compared to swamp and spring-run habitats.

With the low volume of eggs per gram of female in troglobites, we would not expect to notice large swollen bodies in females ready to spawn. Also, there would not seem to be a high energetic risk given the great energetic economy in all the species. Poulson now predicts a flexibility of clutch size in all the troglobites based on differences among years in food supply (e.g., Pearson this volume for a rare year with ~20% of adult female *A. spelaea* with branchial eggs and hatchlings versus usual maximum of 5%) and his lucky large meal hypothesis (Poulson 2001b).

B. Branchial Brooding and Reproductive Behavior

There is a morphological reason to expect all Amblyopsids to branchially brood eggs but it has only been seen in *A. spelaea*. The jugular position of the genital papilla in adults and the attachment of gill membranes to the isthmus in all Amblyopsids suggest that all species brood eggs in their branchial cavities (Woods and Inger 1957, Poulson 1963). Some have thought *A. rosae* brood eggs in the branchial cavity (Breder and Rosen 1966, Balon 1975). It has been suggested that non-incubated eggs are more susceptible to predation in the cave environment (Noltie and Wicks 2001); however, direct evidence is lacking. Conversely, evidence against branchial incubation in the other Amblyopsids exists. At least two researchers have suggested that total egg volume in ripe Amblyopsids exceeds branchial volume in Amblyopsids other than *A. spelaea*. Here we do some calculations to determine whether we agree.

We start by calculating branchial volume of a 65 mm SL incubating *A. spelaea* with 60 eggs each 2.3 mm diameter: $6.38 \text{ mm}^3 \times 60 = 383 \text{ mm}^3$. This is about 4 × the branchial volume Poulson (2001a) estimated from linear estimates for a 45 mm fish. If we proportionally increase the 45 mm estimate to 65 mm/45 mm the estimate is still only 121 mm^3 and about a third of the observed volume of eggs in the females branchial cavity. We suggest that this continued discrepancy is due to the likelihood that 121 mm^3 is a large underestimate because brooding females have obviously puffed up cheeks with a completely distended branchial cavity. If we apply a proportional correction to Poulson's estimate of *A. rosae* branchial volume for a 45 mm SL female we get 260 mm^3 for a branchial cavity that is completely distended (383 mm^3/121 mm^3 × 82 mm^3 linear estimate for *A. rosae*). If this 45 mm fish had a clutch size of 34 eggs each 2.2 mm in diameter (Table 2 in Poulson 1985) the clutch volume would be 190 mm^3 or about 73 percent of the estimated distended branchial volume. On this basis, we disagree with Ginny Adams (personal communication) that total ovarian egg volume of *A. rosae* exceeds its distended branchial volume. In fact, we believe that Amblyopsis clutch volumes are even less (see reasoning in the previous section). On the other hand, we agree with Jenkins and Burkhead (1994) that 339 and 426 ova would not fit in the branchial cavities of female *C. cornuta* of 42 and 46 mm SL. Our estimate of distended branchial volume of 42–46 mm SL *C. cornuta* is about 110–130 mm^3 and using a 1.8 mm egg of 3.05 mm^3 volume (since we believe that females only lay eggs at or near maximum size) the clutch volumes are 600 mm^3 and 754 mm^3. Unlike the case for either *Amblyopsis* species, these clutch volumes hugely exceed our newly estimated distended branchial volume (~1220 mm^3 >>>> 130 mm^3). Ross and Rohde's (2003) observations are consistent with our conclusion that *C. cornuta* cannot carry its large clutches of eggs in its branchial cavity. Ross and Rohde were very careful to hold many newly captured large gravid females in jars but never saw eggs or fry in or released from a branchial cavity in 13 years of field study. They speculate that swampfish likely deposit eggs somewhere in the stream. Swampfish have been observed building a nest-like depression under a rock in an aquarium (J.S. Smith, personal communication in Ross and Rohde 2003).

Poulson thinks that the especially low reproductive effort of eggs per gram for *A. spelaea* (~80) may be related to the constraint of branchial brooding that has been verified many times for this species (Eigenmann 1898, Poulson 1960, Pearson, this volume). Clearly there could be no more eggs than observed in fish with extremely distended branchial cavities. In this context, it is of interest that Eigenmann reported a few cases of decrease of eggs in the branchial chamber in three cases where he kept careful track ($57 \rightarrow 34$, $61 \rightarrow 43$, and $65 \rightarrow 51$). This could be due to failure of development and/or swallowing some eggs. It seems clear that the + tradeoff of branchial incubation is protection of eggs and new hatched fry but the potential – tradeoff of not being able to feed while incubating may not be important. The metabolic rate of *A. spelaea* is low and fish survive more than a year without feeding. Whatever the negatives, it is clear that females could not hold any more eggs in their fully distended branchial cavity than the reported numbers of 61–70 for 60–75 mm SL females.

The mode of egg-laying and question of parental care both remain mysteries to be solved in the future. Dave Bechler has, in our opinion, the most 'green thumb' for maintaining troglobitic Amblyopsids for long periods with high food supplies. But, despite many females developing some visible eggs, Bechler (personal communication) has never seen evidence of reproductive behavior much less egg-laying.

C. Population Size and Population Density

Adaptation to low food supply in caves can perhaps be best seen by comparing the population density per hectare for the best habitats for all the species with the maximum numbers estimated from 4–12 censuses over 1–12 years. For estimates we used a multiplier factor of 1.5–2.5 × from mark-recapture population size calculations in habitats of different complexities and numbers of hiding places. The *C. cornuta* population studied by Ross and Rohde (2003) was in an area of only 50 m^2 and had an estimated 42,000 fish per hectare (2.5 × maximum number of 84 censused)! The *F. agassizii* population studied by Weise (1967) was in only 25 m^2 and had an estimated 80,000 fish per hectare (2.5 × maximum number of 104 censused) while the Mammoth Cave population studied by Poulson was in a 5400 m^2 habitat that had 0 fish per hectare (11 of 12 years) and an estimated 56 fish per hectare one year (2.0 × maximum number of 15 censused). Given the highest caloric cost for metabolic rate and eggs for all species (45 and 4.0 cal per day) it is not surprising that *Forbesichthys* rarely if ever reproduces in caves.

Compared to the spring-cave *Forbesichthys*, all the troglobites had higher densities in caves due to their lower metabolic rate and greater foraging efficiencies. *Typhlichthys* in Shelta Cave, Alabama, were in a 7,800 m^2 of prime habitat and had an estimated 229 fish per hectare (2.5 × maximum of 64 censused) plus as many as 116 crayfish and 30 shrimp per hectare. *Typhlichthys* in Blowing Springs Cave, Tennessee were in an 1100 m^2 area and had an estimated 524 fish per hectare (2.5 × maximum of 23 censused). With much lower metabolic rates both species of *Amblyopsis* had much higher estimated population sizes and much higher densities. *Amblyopsis spelaea* in Upper Twin Cave, Indiana, were in a 3600 m^2 habitat and had an estimated 417 fish per hectare (2.5 × maximum of 130 censused) and in Under the Road Cave, Kentucky in only 300 m^2 of habitat there were an incredible estimated 4199 per hectare (1.5 × maximum of 84 censused). And Pearson and Boston (1995) estimated 2643 *A. spelaea* per hectare in 3702 m^2 of Penitentiary Cave (1.9 × 519 censused). *Amblyopsis rosae* in Logan Cave, Arkansas, in 1100 m^2 of prime habitat had an estimated 1160 fish per hectare (2.5 × maximum 51 censused) and in Cave Springs Cave, Arkansas, in 1300 m^2 of prime habitat had an estimated 2134 fish per hectare (2.5 × maximum of 111 censused). In summary, the estimated maximum density of fish in the 1 - 2 best habitats for each species is as follows. In food-rich surface habitats there were 42,000 *C. cornuta* per hectare and 80,000 *F. agassizii* per hectare. *Forbesichthys* in a cave had only 0–56 fish per hectare <<< 524 and 864 for *T. subterraneus* << 2643 and 4199 for *A. spelaea* >= 932 and 2134 for *A. rosae*.

Despite striking energy economies that have allowed high population densities, there appears to be density-dependent population regulation for the populations that have been most accurately censused over a period of years (Poulson 1969). The first evidence is the irregular appearance of small size classes in all troglobites (Fig. 4 in Poulson 1960). Next, there are slow growth rates when eating predominately copepods. Finally, the density-dependent population regulation inferred for *A. spelaea*, including cannibalism of 8–10 mm fry, suggests energetic limitation on reproduction despite low metabolic rates, efficient foraging, and absence of empty guts at any time.

The data that led to Poulson's (1969) inference of population regulation was based on a nine censuses from 1957–1964 of all sizes of *A. spelaea* in Upper Twin Cave, Indiana. The numbers fluctuated from 84–130 (1.55 ×) but the total mass fluctuated less from 363–465 grams (1.28 ×) and the total metabolic demand (based on routine metabolic rates) fluctuated still less from 153–180 ml O_2 per day (1.17 ×). In addition, the number of incubating females ranged from 0–3 per census and two instances of new hatched young. At the time, Poulson suggested that in years with highest metabolic demand there might be density-dependent cannibalism. An inadvertent manipulative experiment confirmed this hypothesis. About 30 fish of 50–70 mm SL were collected from Under the Road Cave (population 84 fish at the time) for histological and behavioral studies and guts of five fish had small fish ~ 1–2 years of age that had been cannibalized. Two years later, Poulson censused 67 fish including nine fry 9–11 mm SL and there were no cannibalized fish in the guts of the eight large fish collected for anatomical studies.

D. Timing of Reproduction

Troglobitic species in many systems are subject to pronounced seasonality in food availability and water levels (flooding). Accordingly, many species have reproductive cycles synchronized with seasonality. In Amblyopsids, peaks in reproduction occur just after spring floods (Poulson and Smith 1969). Synchronization during this time is adaptive because offspring survival is maximized. Young are produced shortly after spring floods when food availability is still high, yet mortality due to extreme flows is reduced (Poulson and Smith 1969). However, timing of these cues is unpredictable and may occur from late fall into spring. Rises in water level and alkalinity coupled with subtle drops in water temperature may be triggers to reproduction and synchronization of circannian rhythms of reproductive readiness (Poulson 1963, Jegla and Poulson 1970). Synchronization of circannian rhythms of reproductive cycles would allow cave Amblyopsids to maximize reproductive readiness when chances for reproductive success also are at a maximum (Poulson and White 1969).

The reproductive cycle in the epigean Amblyopsid is clear and is probably cued by photoperiod and water temperature. In a detailed year long study at one site Ross and Rohde (2003) showed that *C. cornuta* exhibits a fairly short, spring spawning season that coincides with a fast increase in water temperature from 5 to 15°C. Ovaries begin to enlarge in late fall, show peak development in winter, and all females are spent in April. Throughout the range, reproduction occurs from early March to late-April (Poulson 1963, de Rageot 1992, Jenkins and Burkhead 1994; Ross and Rohde 2003) although a few individuals may spawn in late February. Gravid females with large ova have been collected from 29 January–20 February in the Tar River drainage (Jenkins and Burkhead 1994), and free-living young as small as 9 mm have been collected in mid-May (Rohde and Ross in Jenkins and Burkhead 1994).

The seasonality of reproduction in *F. agassizii* is also clear. In the Illinois populations studied by Weise (1957) and Smith and Welch (1978), most adults presumably spawn underground in late winter based on near disappearance of adults from springs. Ova begin to enlarge in the fall reaching mature size in January when adults move underground (Weise 1957, Poulson 1963). Subterranean spawning is believed to occur from January through April and peaks in February when water levels typically are at their maximum. Fry appear and adults return to the surface by early May. The evidence for seasonality of reproduction in Rich Pond, Kentucky populations is even more clear because Hill sampled biweekly year-round in both the surface and cave habitats. In the cave in winter, both males and females have developed full gonads. Spent males and immatures leave the cave as water levels increase and start to spill out to form the seasonal surface spring run. But spent females and newly hatched young do not exit the cave until 1.5 months later. From this observation, Hill reasonably inferred some parental care though fecundities and egg volumes are much too high for branchial brooding to be possible.

Data on the yearly reproductive cycle for the cave Amblyopsids are comparatively less detailed with *A. spelaea* being the best-documented species. Poulson and Smith (1969) inferred seasonal reproductive cycles for *A. spelaea* based on times of year when gravid females, incubating females with different stages of egg development, and newly hatched and very small young are observed (unpublished data; Eigenmann 1899a). Thus in *A. spelaea*, breeding appears to occur during high water levels from February through April. Females brood eggs in their gill cavities until hatching and hold young until yolk reserves are used up. Eigenmann (1909) showed that this takes 4–5 months. Fry begin to appear in late summer and early fall (e.g., Fig. 4 in Poulson 1960 for Under the Road Cave). The reproductive cycle of *A. rosae* is thought to closely parallel that of *A. spelaea* but the evidence is scant. Gravid females of *A. rosae* have been found from late August through December (Poulson 1963, Adams and Johnson 2001). On one occasion three 10 mm SL fish were observed in mid-July (Adams and Johnson 2001). Except for one other case of an 11 mm SL fish, the smallest fish ever observed are in the mid 20 mm range. The reproductive cycle of *T. subterraneus* appears to be similar but even less evidence is available. Breeding likely occurs in the spring when water levels are at their highest. However, no newly hatched fish have ever been seen or collected and the smallest fish ever observed documented are 15-20 mm (Fig. 4 in Poulson 1960 for Shelta Cave, Alabama) that are possibly nearly a year old. No data on the reproductive cycle of *S. poulsoni* exist.

E.   Growth and Development

Generally, cave organisms exhibit reduced growth rates and delayed development and maturity compared to related surface species. Reduced growth rates represent an adaptive response to low food supplies in cave environments because less energy over a given amount of time is needed (Hüppop 2000). Within the Amblyopsidae, growth and developmental rates decrease with increasing cave adaptation (Poulson 1963).

The fastest growth rates occur in the epigean species. First-year swampfish from North Carolina grow 9.0 mm SL on average from June to January and reach a mean size of 32.6 mm SL by 22 months of age (Ross and Rohde 2003). Slow to no growth occurs during summer, presumably because of lower feeding rates during this time period. An explanation for slow or no growth in the summer might be elevated metabolic rates because of elevated habitat temperature combined with reduced food supplies.

First year spring cavefish grow 10–20 mm per year on average and also exhibit variable growth rates from season to season (Smith and Welch 1978, Hill 1966). Hill (1971) studied squamation and pigmentation development in *F. agassizii*. Scale primordia first appear on the caudal peduncle at around six weeks. By 12 weeks, both squamation and pigmentation pattern are well developed.

Growth rates for troglobitic Amblyopsids are substantially slower with estimates of 1.0–1.25 mm month^{-1} for *T. subterraneus*, 1.0 mm month^{-1} for *A. spelaea*, and 0.9 mm month^{-1} for *A. rosae* based on putative annual scale marks (Poulson 1963 estimated the number of annuli on the scales photographs shown in Figure 8 of Woods and Inger (1957) as 4+ *T. subterraneus*, 3+ to 4 *A. rosae*, and 4+ to 6 *A. spelaea*). Using mark-recapture, Brown (1996) observed growth rates in *A. rosae* at Logan Cave lower than those inferred by Poulson from scale marks (1963). Some fish species do not always show annulus formation because limited feeding precludes annulus development. Thus, it may be reasonable to assume that some cave Amblyopsid populations fail to form annuli in a given year because of low food availability (D. Bechler, personal communication). Therefore, age and growth estimates based on annuli counts are surely underestimates.

Excluding 20 individuals that decreased in size, Brown and Johnson (2001) observed growth rates of 48 *A. rosae* that were recaptured at least once. Growth rates declined with increasing size. Individuals 30–39 mm TL grew, on average, 0.7 mm month^{-1}; 40–49 mm TL, 0.3 mm month^{-1}, and 50+ mm TL, only 0.06 mm month^{-1}. However, some individuals experienced much higher growth rates >1 mm month^{-1}. Observed spurts in growth were not always correlated with increasing food availability, and are influenced by other factors including age, sex, reproduction, and other exogenous factors (Brown 1996, Trajano 2001). One explanation for observed growth spurts is the rare capture of large prey, such as salamander larvae or crayfish (Poulson 2001b). With a highly efficient metabolism, these rare feeding events could lead to significant spurts of growth much higher than average.

Sexual maturity is also delayed in cave Amblyopsids. Both *C. cornuta* and *F. agassizii* reach sexual maturity around 12 months of age. The development of the Y-shaped snout appendage in male *C. cornuta* parallels sexual maturity. Ross and Rohde (2003) documented the development of this appendage noting that it is not discernable by 16 mm SL, but can be readily observed by 23–25 mm SL with full development by 25 mm SL at 13–14 months. On the other hand based on scale mark estimates of growth rate, *T. subterraneus*, *A. rosae*, and *A. spelaea* may take a minimum of 2, 3, and 4 years respectively, to reach sexual maturity. Estimates for the cave species are based on Poulson's (1963) original examination of growth annuli, and, therefore, are very conservative. More reasonable estimates are 4, 6, and 10 years for age at first reproduction of *T. subterraneus*, *rosae*, and *A. spelaea*.

Vent migration is characteristic of the Amblyopsidae and Aphredoderidae (Poulson 1963). Like other developmental stages, time from hatching to completion of vent migration is longest in the troglobitic species (Poulson 1963). Ross and Rohde (2003) have traced migration of the vent in swampfish. Anterior migration of the vent with increasing body size is most rapid from 10–17 mm SL and decreases significantly with very little change ≥ 19 mm SL. Hill (1966) also found that vent migration to the jugular position was complete in 16-18 mm SL *Forbesichthys*. And,

Eigenmann (1909) found that the process took longer for *A. spelaea*. As in all species the vent is located posterior to the pelvic fins in newly hatched young. But in *A. spelaea* 25 mm in length, the vent has only migrated forward in front of the pelvic fins and by 35 mm, it is still positioned just posterior to the pectoral fins.

F. Longevity

Increased longevity of cave organisms compared to their surface relatives is one of several life history adaptations toward a low-r strategy by which cave organisms cope with limited food resources. Prolonged life spans, coupled with a trend from semelparity to iteroparity, increases the chance of population persistence over time, as a population is less likely to be extirpated during times of extremely low food supplies that result in little to no recruitment (Hüppop 2000).

This pattern is evident in the Amblyopsidae as longevity inferred from scale marks increases with increased adaptation to cave environments (Poulson 1963). The shortest life spans are observed in the epigean *C. cornuta* and troglophile *F. agassizii*. Previously, maximum longevity of *C. cornuta* was estimated at 14–15 months (Poulson 1963) with a few two-year-old fish reported (Rhode and Ross 1986). However, a recent study indicates that several swampfish lived up to 22 months with one living 26 months (Ross and Rohde 2003). Spring cavefish are known to live up to three years (Hill 1966, Smith and Welch 1978). It is likely that, as in many short-lived species, death occurs after a single reproductive attempt (semelparity). Therefore, older individuals are those that simply did not acquire enough resources to reproduce at a younger age.

Troglobitic species live considerably longer. Conservatively based on scale marks, southern cavefish were estimated to live 3–4 years (Poulson 1963); however, individuals have been maintained in captivity for over a decade and likely live considerably longer than initial estimates in nature (Noltie and Wicks 2001). Poulson (1963) originally estimated the average longevity of the more troglomorphic species, *A. spelaea* and *A. rosae*, at about 7 and 5 years, respectively. However, Poulson (2001) later questioned his original longevity estimates of the troglobitic species stating they may be off by a factor of 3–4, partly because of the difficulty in determining scale annuli in larger individuals but primarily because of observed growth rates of marked individuals in nature (see above Growth Rates). Accordingly, maximum life spans for *A. spelaea* and *A. rosae* may actually be as long as 24–28 and 15–19 years, respectively. The maximum life span of *S. poulsoni* is unknown, although the largest specimen collected, a 58.3 mm SL female, was estimated to be as old as 8 years (Cooper and Kuehne 1974, USFWS

1982). This estimate was based on the analysis by Poulson of scale marks in a 42 mm SL individual based on the same the criteria used to estimate age in other Amblyopsids. One clearly larger *S. poulsoni* have been observed but not measured in Key Cave (J. Cooper, personal communication) and unquestionably represents an older individual.

G. Metabolic Rates

A trend toward reduction in metabolic rate with increasing cave adaptation exists within the Amblyopsidae (Poulson 1963, 1985). Poulson (1963) found that standard, routine, and active metabolic rates decrease with increasing cave adaptation (Fig. 13). The cause of decreased standard metabolic rate is a combination of a decrease in gill surface area and reduction in the volume and rate of ventilation (Poulson and White 1969, Poulson 2001a). Poulson (2001a) found that the strongest correlates of metabolic rate reduction in Amblyopsids were reduction in ventilation frequency and volume > reduction in brain metabolic rate > reduction

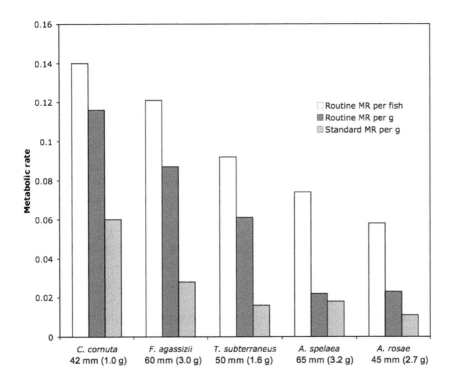

Fig. 13 Standard and routine metabolic rates (MR) in Amblyopsid cavefishes in ml O_2 g^{-1} h^{-1}. Routine metabolic rate for sizes of fish specified are in ml O^2 hr^{-1}.

in gill surface area. No single physiological or morphological trait was in the same rank order as the overall decline in metabolic rate. However, collectively the reductions were in the same order as metabolic rates. To Poulson this is consistent with a general pattern of different pathways to convergent evolution of complex traits. Among Amblyopsid species there were no differences with cave adaptation in histological indices of thyroid activity in adults, although subtle increases in thyroid follicle cell height and colloid vacuolization from the non-growing season (fall/winter) to the growing season (spring/summer) were detected. This was most noticeable within *A. rosae*.

Adams and Johnson (2001) documented a statistically significant effect of body mass on metabolic rate that differed among seasons for a population of *A. rosae* at Logan Cave in Arkansas. A positive relationship was observed during summer and autumn followed by a negative relationship during winter and spring. The authors speculated that seasonal relationships between body mass and metabolic rate may reflect alterations in environmental conditions, such as food availability and ambient dissolved oxygen, however other factors, such as biased sex ratios or seasonal and size susceptibility to handling stress may influence metabolism. But Poulson wonders whether the effects are biologically significant given the small sample sizes and no plausible explanations.

In a study on acclimation of metabolic rate to temperature (next section) the routine metabolic rate of many individuals of each species was measured at 15°C. These data are more extensive than Poulson's (1960) data and include fish from other localities so it is of interest to compare the results. The data are for *C. cornuta* (4 of 23–32 mm SL from Cashie Creek, South Carolina), *F. agassizii* (10 of 27–32 mm from Pine Bluffs, Illinois and 14 of 32–68 mm from Rich Pond, Kentucky), *T. subterraneus* (13 of 29–41 mm from Cave City, Kentucky), *A. spelaea* (13 of 35–70 mm from Sig Chatlet Cave, Kentucky), and *A. rosae* (7 of 36–46 mm from Cave Springs Cave, Arkansas).

Here are minimum and maximum routine metabolic rates ml O_2 g^{-1} hr^{-1} at 15°C with the average and the range among individuals (standardized to a 1 gram fish):

C. cornuta	maximum 0.190 (0.142–0.298) and minimum 0.116 (0.095–0.132)
F. agassizii	maximum 0.137 (0.039–0.396) and minimum 0.087 (0.029–0.092)
T. subterraneus	maximum 0.082 (0.026–0.181) and minimum 0.061 (0.015–0.136)
A. spelaea	maximum 0.037 (0.015–0.093) and minimum 0.023 (0.010–0.023)

A. rosae   maximum 0.031 (0.012–0.070) and minimum 0.026
                  (0.008–0.070)

This decline in routine metabolic rate *C. cornuta* > *F. agassizii* >> *T. subterraneus* > *A. spelaea* = *A. rosae* is really close to that reported by Poulson in (1985).

H.   Tolerance of Temperatures from 5 to 25°C

In these studies of acclimation, data were taken in a torus respirometer. An oxygen electrode was used to measure metabolic rate and an infrared detector was used to measure spontaneous swimming activity.

The results are that metabolic rate and spontaneous activity of all Amblyopsid species acclimate to a wide range of temperatures. This is yet another bit of evidence that belies the constant cave paradigm. In retrospect, it is not surprising that even troglobitic Amblyopsids acclimate since they must deal with floods that bring in prey or allochthonous organic matter that their troglobitic prey need. At the end of winter, rain and snow melt result in cold water entering the cave. In more rare summer floods, the water entering the cave is warm. The Pine Bluffs populations of *F. agassizii* are in springs that only vary from 11–15°C but they acclimate to temperatures of 5, 10, 15, 20, and 25°C the same way as fish from Rich Pond, Kentucky that experience water temperatures of 7–26°C.

The data (Poulson, previously unpublished) show that there is no difference in degree of acclimation among the species and each species shows acclimated metabolic rates and spontaneous activity that are essentially the same from 5 to 25°C. Thus, each species shows almost perfect compensation for temperature. Put another way, metabolic rates and activity are the same in fish acclimated to 5, 10, 15, 20, and 25°C (no significant differences between rates at any temperature for any species). The process of acclimation takes 1–3 weeks depending on the temperature difference. Initially, fish taken form 15 to 5°C show a reduced metabolic rate and activity and fish taken from 15 to 25°C show an elevated metabolic rate and activity.

I.   Starvation Resistance

Poulson (1961, 1963) crudely compared starvation resistance in Amblyopsids and found increasing ability to cope with food scarcity from the epigean *C. cornuta* to the most troglobitic species, *A. rosae*. *Chologaster cornuta* showed the greatest loss in body mass and lowest starvation resistance, expiring after only 45 days (Table 6). On the other hand, *A. rosae* lost only 9% body mass after 20 days and survived over 13 months

Table 6 Effects of starvation on Amblyopsid cavefishes under aphotic conditions (from Poulson 1960). Decreasing standard metabolic rate (SMR) reflects decreased rate of weight loss during starvation and decreased routine metabolic rate (RMR) reflects increased time until spontaneous activity stops during starvation. Data standardized to 45 mm SL fish. Metabolic rates in ml O_2 g^{-1} hr^{-1}.

Species	Habitat	Body mass[1]	SMR	Activity[2]	Minimum – Maximum RMR
C. cornuta	Swamp	34	0.065	46	0.085–0.298
F. agassizii	Spring	24	0.028	116	0.029–0.396
T. subterraneus	Cave	15	0.016	230	0.015–0.181
A. spelaea	Cave	13	0.018	280	0.006–0.093
A. rosae	Cave	9	0.011	400	0.006–0.070

[1] % loss of body mass after 20 days
[2] End of spontaneous activity in days

without food. *Amblyopsis spelaea* can survive up to 20 months without food (Sloan in Eigenmann 1909).

Starvation resistance in cavefish can be achieved by four primary means, acting alone or in concert (Hüppop 2000): 1) consumption of large amounts or a single large prey item at a single feeding event, 2) reduction of activity and metabolic rate, 3) storage of large amounts of fat, and 4) reduction of the digestive absorptive epithelium or reduction of gut length. The first three mechanisms are evident in Amblyopsids. Large prey items occasionally have been observed in the stomach contents of collected fish. Second, metabolic rate decreases within increasing cave adaptation (see Metabolic Rates above and Poulson 1960, 1985, 2001). However, in Poulson's studies, activity of some of the troglobites actually increased over the first six weeks before decreasing later. The observed initial increase in activity has been attributed to searching for food. Lastly, although fat was not directly measured, Poulson (1960) noted that the troglobitic species possessed larger fat stores compared to *C. cornuta* and *F. agassizii*. Thus, fat reserves (though much lower than in *Prietella* and *Astyanax* cavefish), coupled with low metabolic rates explain the increased survival time of troglobitic Amblyopsids under starvation (Hüppop 2000).

However, more detailed studies, like those on other cavefishes (especially by Hüppop 2000), are needed to assess the ability to recover weight after varying amounts of weight loss and to store large amounts of fat quickly when food is available. In addition, it is important to compare the mechanisms of resistance to starvation seen in epigean fish to determine the degree of preadaptation. This has been done for *Astyanax fasciatus* (Hüppop 2000) but not for Amblyopsids. For example, all vertebrates, including humans, when deprived of food first use glycogen, then fat, and as a last resort protein resulting in muscle wasting. Associated with these changes is a reduction in activity and basal metabolic rate.

J. Circadian and Circannian Rhythms

Little information exists about endogenous cycles in cave organisms (Langecker 2000) but some evidence is available for cave Amblyopsids. Some troglobitic species have retained circadian oxygen consumption but have lost circadian activity, as their rhythms cannot be entrained to light-dark cycles (Poulson and Jegla 1969, Poulson and White 1969). This has been viewed as evidence that entrainment of activity cycles to the environment is not maintained by selection in subterranean habitats (Poulson and White 1969) as has been observed in *F. agassizii* (Poulson and Jegla 1969). With regard to endogenous annual (circannian) rhythms, both *A. spelaea* and *A. rosae* exhibit a well-defined yearly reproductive cycle (see above 'Timing of Reproduction'). Because caves do not have reliable seasonal cues (Poulson and Smith 1969), circannian rhythms are adaptive to maximize reproductive readiness when chances for reproductive success also are at a maximum at unpredictable times of the year (Poulson and White 1969).

K. Agonistic Behavior

Agonistic behavior of Amblyopsids has been investigated in detail by Bechler (1980, 1981, 1983). He examined intraspecific dyadic interactions (1 resident and 1 intruder) in 80 liter or 160 liter aquaria with a rock hiding place in each. Only *C. cornuta* showed no agonistic behavior and Poulson suggests that this is because no resources are defendable in its epigean plant and debiris-choked habitat with extremely high food densities.

Among the four other species Bechler documented two submissive acts, "freeze" and "escape", in all species. This is the only study to document "freezing" in hypogean fishes (Parzefall 2000). Poulson has observed freezing by recently hatched *A. spelaea* under threat of cannibalism. Large *Forbesichthys* also cannibalize smaller fish (Hill 1966) and freezing should be an excellent defense since Amblyopsids use only lateral line to detect other fish and prey. Thus it is not surprising that fish that perceived that they are losing in an agonistic encounter 'froze' more often than escaping by fleeing. This allowed them to avoid the most intense kinds of acts.

In order of increasing intensity and decreasing frequency Bechler documented tail beat > chase > attack = head butt > bite and jaw lock (Table 7). The only metric that was in the order of increasing time isolated in caves was total number of agonistic bouts with *F. agassizii* 158 >> *T. subterraneus* 59 = *A. spelaea* 61 >> *A. rosae* 21. Mean duration of bouts had *T. subterraneus* and *A. spelaea* reversed in rank order with *F. agassizii* 53 sec = *A. spelaea* 57 sec >> *T. subterraneus* 26 sec > *A. rosae* 18 sec. Of the six

Table 7 Agonistic behavior in Amblyopsid fishes. From Bechler (1983). In paired interaction tests (1 resident and 1 intruder) *C. cornuta* showed no agonistic behavior. Among the one troglophile and three troglobites, three species showed five to six of six kinds of acts (*F. agassizii* = *A. spelaea* <= *T. subterraneus*) but the most cave-adapted species, *A. rosae*, showed only one of six. See text for interpretation.

Species	C. cornuta	F. agassizii	T. subterraneus	A. spelaea	A. rosae
Habitat	Surface	Spring/Cave	Cave	Cave	Cave
Aggressive acts					
Tail-beat	–	+++	+++	++	++
Head-butt	–	+	+	++	–
Attack	–	+++	+	+	–
Bite	–	+	+	++	–
Chase	–	+	++	++	–
Jaw-lock	–	+(2)	–	+(1)	–
Submissive acts					
Freeze	–	+++	+++	+++	+++
Escape	–	++	++	++	++
Total no. acts	0	158	59	61	21
Mean duration of acts (sec)	0	53	26	57	18

– not observed; + rare; ++ regular; +++ frequent; ++++ always

possible aggressive acts there was less of a trend among species with *F. agassizii*, *T. subterraneus*, and *A. spelaea* showing 5–6 acts and *A. rosae* with showing only tail-beating (Table 7).

One clear difference among species is that *F. agassizii* and *T. subterraneus* always initiated agonistic acts from under or next to rock shelter whereas *A. spelaea* and *A. rosae* patrolled the whole tank and did not set up "stations". This is consonant with the high importance of thigmotaxis to *F. agassizii* and *T. subterraneus* (see Thigmotaxis above). Bechler (personal communication) noted the same prevalence of thigmotaxis with long periods of rest in *F. agassizii* and *T. subterraneus* in a 6.3 m long 4.1 m^2 artificial stream with many rocks. And he noted that both *A. spelaea* and *A. rosae* were active almost all the time and did not set up "stations". To make these observations, Bechler recorded the behavior of four individuals of each species periodically by videotape for 30 days. The four individuals of each species "encountered each other" (within 15 cm) much more infrequently and showed many fewer agonistic interactions per encounter than did the two individuals in aquaria of orders of magnitude smaller volumes. In fact, in aquaria pairs of individuals always were agonistic when encountering each other whereas in the stream the percentages of encounters with any agonistic behavior were 55% *T. subterraneus* > 27% *A. spelaea* = 26% *A. rosae* >= 21% *F. agassizii*. In addition, the bout lengths were shorter and the agonistic acts less intense than in aquaria. The density of four fish in 4.1 m^2 of artificial stream translates to 9,662 per hectare, which is twice the highest we have ever estimated in nature for any cave Amblyopsid. So, it seems to us that in caves encounters would be even less frequent. In fact, neither we or Bechler (personal communication) have ever seen agonistic interactions in the field with hundreds of observations of fish within 15 cm of one another.

Bechler (1983) viewed degenerative evolution as the most likely mechanism for reduction of agonistic behavior in Amblyopsids suggesting conservation of energy due to other adaptations to subterranean life. Thus reduction in metabolic rate and fecundity with increased swimming efficiency reduced selective pressures caused by limited food availability and allowed for reduction in agonistic behavior. He suggested that retention of some levels of agonistic behavior in *A. spelaea* might reflect lower levels of subterranean adaptation or the possibility of defense of prime foraging areas in riffle and pool cave streams. Agonistic behavior has not been observed in *A. spelaea* in nature and Poulson thinks that it is unlikely.

Poulson (this volume) argues that there seems to be no defendable resource in caves and so we should not expect territoriality and associated agonistic behavior to be important. This suggestion is consistent with the lack of any agonistic behavior in *C. cornuta* where there is clearly no defendable resource or food in the dense vegetation in which it lives.

L. Territoriality and Social Groups

Although epigean species are often territorial or form social groups, troglobitic species generally are found in low population densities and usually solitary with a large home range (Langecker 2000). In general, populations of *A. spelaea* and *T. subterraneus* are low in density (but see Poulson 1969 and above TESTS OF HYPOTHESES: FECUNDITY AND EGG SIZE) and individuals are irregularly distributed over suitable habitat (Poulson 1963). Individual fish have large home ranges, cover long distances in search of food, and never defend areas (Poulson 1963, Mohr and Poulson 1966). However, during aggression trials by Bechler (1983), *F. agassizii* and *T. subterraneus* established distinct territories in aquaria with rocks ("stations"), whereas the less sedentary species, *A. spelaea* and *A. rosae*, patrolled the entire tank. Cavefish do not display schooling behavior, although individuals are sometimes observed in close proximity typically around food sources (e.g., underneath a bat roost). Even in caves with the largest populations, 70-80% of fish are found in perhaps only 10-20% of the accessible cave stream (Brown and Johnson 2001, Poulson, unpublished data, Pearson, personal communication).

M. Food Habits

The reported diet of Amblyopsids has been well characterized and was treated in the context of prey detection earlier in this chapter (V. Prey Detection and Availability). All Amblyopsids eat live, moving prey with invertebrates comprising most of the diet. However, cannibalism has been documented in the family and small amounts of nonliving food such as bat guano and detritus have been observed in the stomach contents of some species. These are probably ingested along with live prey but would provide very much lower nutritional benefit per volume than live prey.

Swampfish are likely crepuscular or night feeders (Poulson 1963) principally feeding on amphipods, chironomids, and cladocerans with amphipods comprising the majority of diet (G.B. Pardue and M.T. Huish in Jenkins and Burkhead 1994, Ross and Rohde 2003). Other potential prey items include nematodes, ostracods, ephemeropterans, odonates, plecopterans, megalopterans, coleopterans, trichopterans, and other dipterans. Fifty-three percent of 289 stomachs of day-collected fish examined by Ross and Rohde (2003) were empty from one North Carolina site; however, several fish had noticeable food in the intestines. Prey diversity and percent occurrence and abundance of each organism in the diet is greatest from January–April then substantially decreases in summer and fall (Ross and Rohde 2003). Cannibalism has not been reported in *C. cornuta*.

The diet of *F. agassizii* varies geographically and between cave and surface populations. Illinois populations feed almost exclusively on *Gammarus* amphipods (Forbes and Richardson 1908, Layne and Thompson 1952, Gunning and Lewis 1955, Weise 1957), although amphipods (Weise 1957), insect remnants (Gunning and Lewis 1955), and detritus (Gunning and Lewis 1955) also have been found in stomachs. In Kentucky, surface fish feed principally on chironomids but also copepods, oligochaetes, nematodes, and ostracods (Hill 1969a). On the contrary, individuals of the same population are strongly cannibalistic on younger individuals when in the cave part of the habitat. In caves cannibalism may represent an alternative feeding strategy in response to competition for more typical but extremely rare invertebrate food sources. In rare years when *F. agassizii* occurs in the Mammoth Cave food supply is much greater than usual (Poulson, unpublished data).

The diets of the troglobitic Amblyopsids consists largely of copepods and isopods but rare, larger meals, such as crayfish, salamander larvae, or conspecifics, may result in high growth efficiency and a burst in growth rate (Poulson 2001). An assortment of prey have been reported in stomach contents of *T. subterraneus* including copepods, amphipods, isopods, trichopteran and tendepedid larvae, cladocerans, isopods, and crayfish (Poulson 1960, 1963, Cooper and Beiter 1972); however, copepods are the primary food source accounting for 60–90% of the diet by volume (Poulson 1963). The diet of *A. spelaea* includes copepods, amphipods, isopods, and salamander larvae (Clay 1975). *Amblyopsis spelaea* smaller than 45 mm total length chiefly eat copepods, whereas those in excess of 45 mm may consume amphipods, isopods, and crayfish (Poulson 1963). Larger fish also occasionally cannibalize smaller fish. Few studies have investigated the food habits of *A. rosae*, but gut contents show that the diet of Ozark cavefish comprises the majority of available food sources in the caves they inhabit. Copepods constitute up to 90% of the diet by volume, but isopods, amphipods, cladocerans, ostracods, small crayfish, coleopteran larvae, plecopterans, crickets, and small salamanders also are taken (Poulson 1960, 1963). Bat guano has also been found in stomach contents. It has been suggested that younger individuals are cannibalized. However, Adams and Johnson (2001) observed a presumed parental-young aggregation of *A. rosae* for more than six weeks with no evidence of cannibalism. It remains unknown how long such parent-kin interactions occur and whether kin recognition exists. Based on stomach contents, extent of fat deposits, and annulus formation, *A. rosae* fed less in the fall than other times of the year (Poulson 1960). The diet of *S. poulsoni* has not been examined but likely consists of amphipods, isopods, and small crayfish (Boschung and Mayden 2004).

N. Predators

Troglobitic Amblyopsids are at the top of the food chain in most cave systems that they inhabit, and, therefore, have few natural predators. Epigean fishes may potentially prey on cave Amblyopsids (Brown 1996, Poly and Boucher 1996), as well as crayfish (Romero 1998a), and troglobitic salamanders. Young cavefish are susceptible to cannibalism by larger adults (Poulson 1963), as has been documented in *A. spelaea* and one cave population *of F. agassizii*. Cannibalism may serve as one means to regulate population densities in a food-limited environment (Poulson 1969). Epigean populations of *F. agassizii* likely are occasional prey for other fish, snakes, birds, and mammals (Smith and Welch 1978). At Rich Pond in Kentucky, natural predation is seasonally heavy (J.E. Cooper in Smith and Welch 1978). Epigean predators, such as raccoons, fish, and water snakes, also may be potential threats to cave-inhabiting Amblyopsids; however, no cases of predation on cave-dwelling Amblyopsids by epigean predators have been documented even in cave entrance areas.

O. Parasites and Diseases

Like many cave vertebrates, few parasites have been observed afflicting Amblyopsid hosts. Proteocephalan cestodes have been collected from the pyloric caeca of *A. spelaea* and *F. agassizii*. Three *Proteocephalus poulsoni* were collected from two northern cavefish from Under the Road Cave in Breckinridge Co., Kentucky (Whittaker and Zober 1978). Whittaker and Hill (1968) described *P. chologasteri* from spring cavefish. In southern Illinois, 71 percent of fish examined were parasitized by cestodes and other internal parasites (G. Garoian in Smith and Welch 1978). The acanthocephalan *Neoechinorhynchus cylindratus* has been reported from the intestines of *A. spelaea* (Nickol and Whittaker 1978) but neither frequency nor incidence was reported. In cursory examination of many Amblyopsid guts for food item analyses, Poulson has never seen an obvious parasite.

Reports of external parasites are even fewer. A species of copepod, *Cauloxenus stygius*, is an obligate ectoparasite found on the upper lip of northern cavefish. Little is known about the species. Its distribution is believed to be that of its host but the parasite has rarely been observed and few records exist (Cope 1872, Blatchley 1897, Giovannoli 1933, Lewis 2002ab, Poulson, unpublished data). In other Amblyopsids, small, unidentified leeches also have been reported for *F. agassizii* in Illinois (Smith and Welch 1978). Additional studies using a comparative framework are needed to not only document additional parasitic taxa, but to also compare frequency and density of occurrence between the epigean, troglophilic, and troglobitic Amblyopsids.

Few diseased Amblyopsid cavefishes have been reported in nature. Fournie and Overstreet (1985) reported on an adult *F. agassizii* from Union Co., Illinois, with a retinoblastoma on the right side of the head. This condition may be related to chromosomal abnormalities. At least one other individual collected at the same spring had a similar tumor in appearance and eventually died after the tumor involved the entire head. However, this specimen was not available for histological examination (Bechler, pers. comm. in Fournie and Overstreet 1985).

In troglobitic Amblyopsids, Pearson and Boston (1995) documented a bacterial infection of the fins of *A. spelaea* at Donaldson Cave, Lawrence Co., Indiana. All individuals observed had extremely shortened pectoral and caudal fins with ragged margins. Many fish also had small, red spots scattered over the body. During a recent trip to Donaldson Cave, this condition was not observed (Niemiller, unpublished data). Broken-back syndrome was noted for only three years in perhaps 10% of fish from this site. The likely cause was exposure to pesticide and other chemical contamination since the surface watershed was in row crops and pasture

(Keith and Gray 1979, Keith and Poulson 1981). Gas bubble disease has been documented in recently collected *T. subterraneus* at a spring site in Missouri (Schubert *et al.* 1993). Southern cavefish from Missouri may be particularly susceptible to this disease because of the depths at which individuals reside (Schubert *et al.* 1993, Noltie and Wicks 2001).

There is some logic to believe that caves may be a refuge from many kinds of diseases or parasites so Poulson (this volume) has suggested that we consider this hypothesis carefully and at least keep good records. A corollary is that absence of parasitism and low incidence of disease may have led to evolutionary rudimentation of immune defenses. However, the recovery of many individual *A. spelaea* from fin rot in one cave (W. Pearson, personal communication) may be evidence of perfectly adequate immune responses. It will be worthwhile to compare the Amblyopsids for incidence of and recovery from common aquarium fish diseases.

P. Abundance and Population Sizes

Few studies have attempted to quantify population sizes and relative abundance of Amblyopsids, including *C. cornuta* and *F. agassizii*. The few studies that have attempted to quantify population sizes via techniques such as mark-recapture or survey removal have focused on caves that are known to contain relatively large populations. Other studies for which the most reliable estimates of abundance have been obtained have focused on the species of conservation concern. Additional demographic studies, including long-term censuses, are needed for both epigean and subterranean populations.

In general, the majority of cave Amblyopsids localities yield few fish sightings during single surveys. Although this may be a real reflection of actual abundance in some instances, it is important to realize that the distribution and abundance of these troglobitic species, perhaps with the exception of *S. poulsoni*, likely is greater than currently realized. Localities for which cavefish have been reported represent but a fraction of total available habitat accessible to fish. This fact was clearly illustrated during the Maramec fertilizer pipeline break that resulted in the death of nearly 1,000 southern cavefish and likely many more (see discussion in Noltie and Wicks 2001) from a drainage basin with few records documented previously. The problem with inferring population densities from such fish kill is that we do not know the volume or extent of habitat impacted.

Most observations of cavefish are restricted to caves near the surface and there is some controversy as to whether even the best cavefish caves are sources or sinks (see Poulson, this volume). Poulson believes that populations that we can sample make up the majority of the total number of fish in a cave watershed. Part of his logic is that cavefish will come to be found in the highest available food areas and these are likely to be in shallow caves with allochthonous inputs and a combination of deep pools and shoals. Further we usually find most fish in the highest food supply areas of these caves (see above 'Population Size and Population Density'). He also argues that only few fish will be in deep phreatic habitats with little food input or the most upstream parts of caves inaccessible to humans. He bases this second inference on both Cave Springs Cave, Arkansas, with about 100 *A. rosae* seen on each visit, and Upper Twin Cave, Indiana, with 84–130 *A. spelaea* seen on each visit. As one goes far upstream in both these caves the number of cavefish drops to none as the water gets shallower and faster flowing with no refuges during floods.

Consequently, habitats where we see no fish or only a few at each visit may be population sinks and not sources. Wells and short stream segments encountered in an otherwise dry cave may not be representative of the habitat that most troglobitic Amblyopsids inhabit. To be sure cavefish can disperse through and occupy submerged passages inaccessible to humans but these habitats are probably neither usual for the fish nor optimal. This does not mean, however, that these fish are doomed. They may at least be potential dispersers. They could move long distances given their long lives, low metabolic rates, and foraging efficiencies. In those caves where we always see 6–10 fish they may even be slowly reproducing. If so, these sinks could become a source for re-colonization if some disaster befalls the fish in the best caves or habitats.

Swampfish are reported as generally rare or uncommon throughout their range (Poulson 1963, Cooper and Rohde 1980, Shute *et al.* 1981, Jenkins and Burkhead 1994). However, its reported rarity may be more of a sampling artifact than a reflection of true abundance (Ross and Rohde 2003). Abundance estimates may be biased as most sampling occurs during daylight (Jenkins and Burkhead 1994), whereas *C. cornuta* are most active at night and occur in habitats difficult to seine or electro-fish. In Virginia, the largest series was taken in the Blackwater River drainage via a nonselective ichthyocide (Jenkins and Burkhead 1994). Poulson found that repeated kick-seining at the edges of dense weed beds routinely led to capture of 1–3 swampfish at each attempt for up to six repeats at the same spot. Ross and Rohde's made many collections over a year in a short section of stream and the maximum density estimated from their data is 42,000 per hectare (84×2.5 in 50 m^2).

Historically, *F. agassizii* has been considered rare to uncommon throughout much of its range. Smith and Welch (1978) estimated less than a thousand individuals from eight springs in Union Co., Illinois, and around 40 individuals at Cave Springs Cave in Union Co. However, many hundreds of specimens have been accessioned from a large spring run at Rich Pond in Warren Co., Kentucky. In Tennessee, *F. agassizii* can be locally abundant in ideal habitats (Etnier and Starnes 1993), and, like *C. cornuta*, is easily overlooked because of its nocturnal and reclusive habits (Smith and Welch 1978). Apparent abundance of *F. agassizii* is dependent upon light levels, groundwater velocity, and season (N.M. Welch in Smith 1979). Pine Hills IL spring runs have as many as 80,000 fish per hectare whereas in Mammoth Cave most often had 0 fish per hectare and 50 per hectare only in a year with high food input (Poulson 1969).

We suggest that most localities for cave Amblyopids are population sinks and a very small percentage are population sources. Figure 14 shows that there were <10 fish per survey in 68% of *A. spelaea* caves, 72% of *Typhlichthys* caves, and 87% of *A. rosae* caves. Conversely, there were 30+ fish per survey in only 5% of *A. rosae* caves, 8% of *Typhlichthys* caves, and 14% of *A. spelaea* caves. This pattern is even more accentuated if we consider the 2–3 caves that have 46–87% of all individuals observed for each species.

For *Typhlichthys* the top three caves account for 46% of the 354 fish censused in 39 caves by Niemiller. One population in Putnam County, Tennessee, had 121 fish in a 400 m stream section (Niemiller, personal observation) with an estimated 864 per hectare. In Shelta Cave, Alabama, on one trip there were 64 fish in an area under a gray bat colony (Poulson 1960) with an estimated 229 per hectare (64 × 2.5 in 7000 m^2). In several caves, we routinely see > 40 fish per visit and in two caves these moderate numbers are consistent over a period of years. From most to least numbers per trip, Herring Cave in Rutherford Co, Tennessee had 47, 39, 37, and 32 fish. Blowing Springs Cave in Coffee Co., Tennessee had 52, 37, 31, and 26

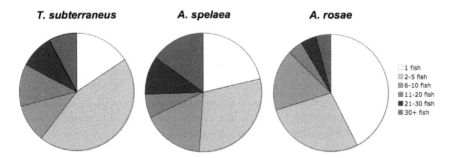

Fig. 14 Portion of cave Amblyopsid localities with the maximum number of observed fish during a single survey correspond to the following cateogories: 1 fish, 2–5 fish, 6–10 fish, 11–20 fish, 21–30 fish, and 30+ fish.

fish including recent surveys by Niemiller and surveys from the 1960s by Poulson.

For *A. spelaea*, the top three caves by number of fish observations account for 87% of the 1591 fish censused in 47 caves, and these caves have had high populations at every survey over the past 50 years (Poulson 1969, Pearson and Boston 1995). During their surveys in the early 1990s, Pearson and Boston (1995) conducted mark-recapture and census studies in several *A. spelaea* caves in Indiana and Kentucky. During a single trip in March 1994, they counted 515 fish at Penitentiary Cave in Breckinridge Co., Kentucky. Over 100 fish were observed during single surveys of two other caves in Kentucky and the Donaldson-Twin Cave complex in Lawrence Co., Indiana. Blue Springs Caverns is also reported to contain a large *A. spelaea* population in excess of 150 individuals (Welch in Keith 1988). Poulson (1969 and VI C this chapter) estimated maximum population densities of *A. spelaea* from 417 per hectare in Upper Twin Cave, Indiana (130×2.5 in 3600 m^2), to 4199 per hectare in Under the Road Cave, Kentucky (84×1.5 in 300 m^2).

For *A. rosae* the top three caves by maximum number of fish per census account for 55% of the 417 fish in 44 caves (G.O. Graening, unpublished data). These three caves have been regularly censused over a period of 20–50 years (Graening unpublished data, Poulson 1960, Brown and Johnson 2001). The top three censuses were 100, 122, and 139 for Cave Springs Cave, Arkansas, 48, 50, and 51 for Logan Cave, Arkansas, and 27, 26, and 33 for Ben Lassiter Cave, Missouri. The densities for the two best caves for the prime habitats with > 80% of all fish in each cave are 2134 per hectare for Cave Springs Cave ($0.8 \times 139 \times 2.5$ in 1300 m^2) and 932 per hectare for Logan Cave ($0.8 \times 51 \times 2.5$ in 1100 m^2).

Speoplatyrhinus poulsoni may be the rarest fish in North America, restricted to Key Cave in Lauderdale Co., Alabama. Alabama cavefish have only been observed in five pools within the cave and is extremely difficult to capture because of difficulty in sampling in deep pools in which it resides. Accordingly, its rarity and difficulty in capture make mark-recapture studies to estimate population size infeasible. Therefore, abundance has been inferred from sight observations during pool surveys. Kuhajda and Mayden (2001) summarized their survey efforts and those of past researchers since the 1970s. Although the numbers of fish observed are comparable from the 1970s to the 1980s and the most recent surveys by Kuhajda and Mayden in the 1990s, abundance is extremely low. Only two of the five pools have been routinely surveyed since the 1970s. A total of 25 survey trips of these two pools have been conducted in the last 30 years (reviewed by Kuhajda and Mayden 2001). Up to nine fish have been observed between these two pools during a single visit with the majority of observations occurring in the mid-1980s. In the most recent surveys that included examination of five pools, Kuhajda and Mayden (2001) observed from 5 to 10 fish with an average of 7.8 fish per survey. Although low in numbers, three different broad size classes were observed indicating that recruitment is occurring. In this cave we can clearly only access a very small amount of habitat and the geology indicates that there is considerable inaccessible habitat.

VII. FORAGING EFFICIENCY AND METABOLIC EFFICIENCY

A. Past and New Studies on Metabolic Efficiency

Poulson (1960, 1963, 1985) has documented a trend of increasing metabolic efficiency from the epigean *C. cornuta*, to the troglophile *F. agassizii*, to *Typhlichthys* and *A. spelaea*, to *A. rosae* to *S. poulsoni*. Earlier in this chapter, we have reviewed the old and new evidence, from eyes and pigment, which corroborates Poulson's early inference that the four troglobites show an increasing period of evolutionary time in caves from *T. subterraneus* to *A. spelaea* to *A. rosae* to *S. poulsoni*. To summarize the original data and new data (see Metabolic Rates) there is a trend of about a five-fold decrease in routine metabolic rates but with a five to ten-fold increase in daily body lengths swimming (Table 2 in Poulson 1985).

Poulson (1985) hypothesized that part of the increased swimming activity is due to longer and inferred bigger relative areas of pectoral and caudal fins that results in greater distance moved per coordinated caudal-pectoral fin stroke. However, only *A. spelaea* has average body lengths moved per fins stroke (0.62–1.25) that is higher than the overlapping body lengths per stroke of *F. agassizii* (0.32–0.72), *T. subterraneus* (0.35–0.67), and *A. rosae* (0.42–0.81).

The spontaneous swimming levels, measured at the same time as metabolic acclimation studies in a 0.5 liter torus respirometer, shows that the lowered metabolic rates of at least *T. subterraneus* and *A. spelaea* is not at the expense of lowered swimming activity. The ranges of swimming activity indices are *C. cornuta* 1–14 <<< *F. agassizii* 19–195 < *T. subterraneus* 70–320 = *A. spelaea* 19–300 >> *A. rosae* 6–55. These studies were done only over about a 20–30-hour period.

A more realistic measure of activity is to follow spontaneous activity in a 0.5-liter torus container with continuously renewed water over a period of 6–12 days. These data show that activity changed in different ways for the Amblyopsid species over the study. In general, over time *Forbesichthys* activity declined drastically (n=2), *Typhlichthys* activity decreased slightly (n=4), *A. spelaea* activity actually increased on average (n=3), and *A. rosae* activity remained the same (n=3). The patterns were consistent within species.

An even more realistic measure of activity is in a 50-liter aquarium. Poulson used an elongate aquarium with a bank of infrared sensors at mid-point to bias for swimming over spontaneous activity. The percent of hours with at least some swimming was measured for two fish of each species for 7 days. The percent of hours swimming were as follows: *C. cornuta* 51 & 53% = *F. agassizii* 52 & 58% < *T. subterraneus* 76 & 86% < *A. spelaea* 92 & 99% = *A. rosae* 86 & 100%. Along with data on swimming speed, this gave estimates of body lengths swum per day while foraging (Table 2 in Poulson 1985) of: *C. cornuta* 1000 body lengths per day < *F. agassizii* 1854 << *T. subterraneus* 5454 << *A. spelaea* 10980 > *A. rosae* 8082.

And a still more realistic measure is provided by Bechler (personal communication) for fish in a 6.3 × 1.5 m artificial stream with rock shelters. Using videotape he followed four individuals at a time periodically for 30 days. The average cm moved per 15 minutes was *F. agassizii* 141 << *A. rosae* 408 < *T. subterraneus* 647 << *A. spelaea* 1274.

B.   Efficiency of Searching for and Capturing Prey

Poulson and Michael Barnett (unpublished data from 1967–1968) studied the searching patterns of Amblyopsids and the efficiency of finding and eating two types of prey at different densities. The prey were ten large cladoceran (*Daphnia magna*) and or five small isopods (*Lirceus*) in a 10-liter aquarium or one cladoceran or one isopod in a 100-liter aquarium. Fish sample sizes were 3–5 individuals of each species.

Poulson made *a priori* predictions of capture efficiency using simulated foraging of average-sized fish for each species (Table 8). He calculated the volumes of water that foraging fish could search in the following way. The width and depth (mm) across which prey could be detected was predicted based on head dimensions, plus projection of neuromast ridges and cupulae above head skin plus distance at which neuromast cupula move to moving prey. This was multiplied by two for head width and for head depth. These two sensitivity distances (horizontal and vertical) were multiplied to get a sensitivity area in mm^2 and this area was multiplied by body lengths swum per minute while foraging length to get a volume of sensitivity in mm^3 per minute. The swimming speed used was that observed when ten *Daphnia* or five *Lirceus* were first introduced to the aquarium with one fish.

Table 8 Predictions of foraging effectiveness. We start by calculating the surface area (in mm^2) of sensitivity to detection of prey around the head. Then, we convert this to volume (in mm^3) searched per time by multiplying surface area by body lengths swum per minute. Compare to Table 9 Time to capture prey and Table 9 % of hours some swimming in 7 days.

Species	C. cornuta	F. agassizii	T. subterraneus	A. spelaea	A. rosae
Habitat	Swamp	Spring	Cave	Cave	Cave
1. Body length (mm)	42	60	50	65	45
2. Neuromast cup + cupula length (mm)	0.3	0.66	0.84	1.14	0.96
3. Distance (mm) at which cupula moves to moving prey	10	16	40	35	30
4. Head width +2 + 3 × head depth + 2 + 3 (mm^2)	283	646	2536	1944	2076
5. Body length × min^{-1} × Body length (mm)	105	222	425	423	320
6. Volume searched (4 × 5) × min^{-1} × 10^{-3} = ml min^{-1}	40	143	1078	822	664
7. Time (min) to search a 10-l aquarium	249	70	9.3	12.2	15.1
8. Time (min) to search a 100-l aquarium	2487	699	93	122	151

The volume of sensitivity per minute in mm^3 was converted to cm^3 to compare to the average volume occupied by different densities of the two prey types in the 10,000 cm^3 oor 100,000 cm^3 aquarium to get a measure of how long it would take for the fish to encounter one prey if it did not repeat searching any volume. This gave predictions of how the species would differ. This prediction was tested for each species by comparing the times to capture of each prey until all were gone. In tests with multiple prey of each type, data were time at which successive prey were captured along with swimming speed in body lengths per minute and turning rate (right or left by 90 degrees) in turns per minute.

With increasing troglomorphy among the Amblyopsids, there is a trend of increasing head size, increased projection of the neuromast cups and cupulae above the head, and increased distance at which cupulae move in response to moving prey. This trend is striking when comparing *Forbesichthys* to any troglobite but the differences among the troglobites are subtle (Table 8; Table 2 in Poulson 1985). There was no trend in swimming speed except that *Forbesichthys* swam slower than any of the troglobites and *Typhlichthys* was the fastest swimming species.

The same relative differences among species were predicted for volume potentially searched in cm^3 per minute (73 cm^3 per minute for *Forbesichthys* <<< 1300 cm^3 for *Typhlichthys* > 864 cm^3 for *A. spelaea* >= 839 cm^3 for *A. rosae*). Thus, all of the 10,000 cm^3 test aquarium could be searched, if the fish used no overlapping paths, in 137 minutes for *Forbesichthys* >>> 8 minutes for *Typhlichthys* < 12 minutes for *A. spelaea* = 12 minutes for *A. rosae*. How does this compare to the time taken to actually find and eat different densities and kinds of prey? The answer is not well. There were unexpected differences in time to capture ten vs. one *Daphnia magna* and the species best at finding and eating *Lirceus* isopods was not the same species best at foraging for *Daphnia* cladocerans (Table 9).

A single *Daphnia* is the foraging challenge that most closely mimics the extremely low prey densities of zooplankton in caves where the maximum ever recorded is ~ 5 × 10^{-2} per 10,000 cm^3! With this test, *Forbesichthys* did worse than predicted as it took 34–2100 minutes to find and eat the single *Daphnia*. *Amblyopsis rosae* was like the other troglobites and better

Table 9 Comparison of computed volume searched (Table 8) with time to capture one isopod and one water flea in a 100-l aquarium and first isopod of five total and first water flea of ten total in a 10-l aquarium. Ranges are reported in parentheses where available.

Species	*C. cornuta*	*F. agassizii*	*T. subterraneus*	*A. spelaea*	*A. rosae*
Habitat	Swamp	Spring	Cave	Cave	Cave
Time to search 100-l aquarium (min)	2487	699	93	122	151
Time to catch one water flea in 100-l (min)	2160	300	48	25	18
Time to catch one isopod in 100-l (min)	620	50 (45–74)	6.2 (1.5–12)	2.8 (1.5–4)	14 (4–36)
Time to catch first of 10 water fleas in 10-l (min)	na	6.8 ± 9.1	1.5 ± 0.5	5.6 ± 5.4	11.0 ± 2.0
Time to catch first of 5 isopods in 10-l (min)	na	4.4 (3.7–5.0)	3.6 (2.2–5.0)	1.8 (0.5–4.0)	9.0 (7.0–11.0)
Hrs (%) swimming over one week	52	55	81	96	93
Body lengths × hr^{-1} in 100-l with no rocks	1000	1854	5454	10980	8000
Body lengths × hr^{-1} in 6.3 m stream w/rocks (4 fish over 30 days)	na	94	557	784	362

than predicted as it took 3–42 minutes to find and eat the single *Daphnia*. What accounts for these departures from prediction? It is not differences in searching speed or turning rate.

All the Amblyopsid species showed behavioral changes as soon as prey were introduced though no controls were used adding inanimate objects. That they were reacting to live prey is shown by sustained increases in swimming speed. *Forbesichthys* increased its speed 20% from 3.8 to 6.1 body lengths per minute. *Typhlichthys* increased its speed 30% from 6.6 to 8.6. *Amblyopsis spelaea* increased its speed 49% from 4.1 to 6.1 and *A. rosae* increased its speed 51% from 4.7 to 7.7 body lengths per minute. All species increased their turning rates by three-fold which helped keep them in the same area where they had caught a prey when multiple prey were in the test aquarium. This behavior would be even more adaptive in caves with much lower densities and much higher patchiness of prey.

When a fish detects a living prey there are also similarities among the species. If a large food item like an isopod or white worm is dropped into an aquarium. Most individuals of all species move toward the area of disturbance (Eigenmann 1909, Hill 1966, Barnett and Poulson, unpublished data). Also when a fish approaches a prey on the bottom, it will pause and back up if the prey stops moving. Then it will "jerk and grab" as the prey starts moving again. All species use a lunge and grab motion as they capture a prey item; none use gape and suck to catch prey.

We have especially detailed observations of the changes in swimming rate, turning rate, and general behavior of *A. spelaea* while foraging for and eating ten *Daphnia magna*. In Michael Barnett's words the fish initially show "lots of turning and jerking' and one fish was "furious at four near misses". After the fish had reduced the number of *Daphnia* to 4–6 their swimming and turning rates increased from 0–3 to 4–18 body lengths per minute and turning increased from 0–2 to 2–5 turns per minute. These rates slowed some when they had reduced prey number to 2-4 and the fish seemed "much less frustrated" if they missed a first strike. One interpretation is that they were becoming "full" since the interval between captures increased.

The relative success of Amblyopsid species (Table 9) was different with ten *Daphnia* in a 10-liter aquarium than with only one in a 100-liter aquarium and this more closely matched the predictions from simulated volume searched per time. Though *Forbesichthys* did better and *A. rosae* worse than predicted, all species were effective foragers. They showed statistically but not ecologically significant differences. The times to catch the first prey ranged in minutes from 1.5 ± 0.5 SD *T. subterraneus* to 11.0 ± 2.0 SD *A. rosae*, with *T. subterraneus* < *A. spelaea* = *F. agassizii* < *A. rosae*. The relative differences were the same for the 5th and 10th prey captured though *F. agassizii* and *T. subterraneus* showed no decline in intervals between prey capture and *A. spelaea* and *A. rosae* capture intervals increased after the 5th prey was captured. Poulson believes that *T. subterraneus* was best at capturing *Daphnia* once detected because its neuromast cupulae are knobbed at the tip and seem to be especially sensitive to slight water movements. We cannot explain why *A. rosae* was the slowest at eating *Daphnia* since it was the fastest by far at finding and eating a single *Daphnia* (above).

Interestingly, *A. spelaea* was fastest at finding and eating isopods whether with five or one per test aquarium. We think this is because isopods are its principal prey in nature (see Food Habits) and its search patterns in nature suggest they are concentrating on finding isopods (see Mohr and Poulson 1966). For sample sizes of three for each species, the range in minutes for capture of the first of five isopods was 0.5–4 for *A. spelaea* to 7–11 for *A. rosae* with *A. spelaea* < *T. subterraneus* = *F. agassizii* < *A. rosae*. For capture of the third of five isopods, *A. spelaea* was still fastest but now *Forbesichthys* was slower than *A. rosae*: *A. spelaea* <= *T. subterraneus* < *A. rosae* < *F. agassizii*. These differences were accentuated with capture times of a single isopod. Times to capture were 2–4 minutes for *A. spelaea* and 7–74 minutes for *F. agassizii* with *A. spelaea* < *T. subterraneus* << *A. rosae* << *F. agassizii*. Part of the explanation is that both *A. spelaea* and *T. subterraneus* search almost only at the bottom where the isopods are located, whereas *F. agassizii* and *A. rosae* continue to search the entire tank. This is a more effective strategy for finding *Daphnia* that swim throughout the water column.

Poulson's perspective from these foraging experiments is that his predictions based on morphological troglomorphy missed some interesting differences in effectiveness when foraging for different prey types. These differences seem to be due partly to the types of prey most common in each species habitat. We have just mentioned this for *A. spelaea*, which lives in caves that have many isopods. We predict that *Forbesichthys* from Pine Hills will outperform even the troglobites with amphipods as prey because it eats only amphipods in the field. Also, *Forbesichthys* from Rich Pond eats mainly chironomid larvae and oligochaetes and so will also be effective foraging for amphipods with their strong thigmotaxis and rapid grabbing during prey capture. We have watched them eat white worms in the lab and they are especially effective. With worms, midge larvae, and amphipods the thigmotaxis used in prey capture complements lateral line detection of prey. We also predict that *A. rosae* and *Typhlichthys* may be most effective at foraging for copepods that are their main prey. This should also be true for small *A. spelaea* < 40 mm SL that eat mainly copepods and few isopods in the field. We think *Typhlichthys* will do the best with copepod prey because our lab observations show that it is especially sensitive to slight water movement, even 'excitedly' moving towards a previously quiescent isopod when it started to wave its antennae!

C. Hypothesized Tradeoffs of Living in Caves

Despite the negative (–) tradeoff of low food supply in caves, it is clear that *Forbesichthys* spawns in caves. Hill's (1966) study of *Forbesichthys* was much more complete than Weise's (1957) though both provide excellent laboratory observations of feeding and behavior. In both sites, the fish reproduce in the cave since adults disappear from the springs for a few months starting in late fall and then adults and small young appear outside in the spring. Unlike Weise, Hill was able to study the fish in both the spring run and in the cave and the contrasts provide important insights into the tradeoffs of the cave and spring run for feeding, refuge, and reproduction.

The Rich Pond Cave may be a refuge from parasitism and predation for newly hatched young but it has 3–4 orders of magnitude less food that Hill sampled outside and the new hatched fish of 6 mm SL stop growing and die if they do not leave the cave. If they leave the cave, they grow to about 35 mm SL their first year eating > 80% chironomid larvae by volume. When fish then overwinter in the cave their guts are 99.9% empty. When fish leave the cave the following spring, they grow to about 50mm SL with a diet of approximately 20% *Forbesichthys* and 78% chironomid larvae. When fish return to the cave again, about 82% of their guts have *Forbesichthys* and the rest are empty. This indicates that the spring cavefish is especially food-limited in caves and may not be a good model for the way other Amblyopsids became isolated in caves.

Forbesichthys may only be a marginal troglophile because there is no documentation of it reproducing in caves. The best-studied populations emerge nightly or seasonally from food-poor caves to feed in spring runs that have abundant live prey. If the surface spring runs dried up with climatic warming and drying then these populations might well go extinct. So a better habitat type to allow gradual evolution of food finding and metabolic efficiencies would be a sinking stream. These would have reliable allochthonous organic input and so much higher food supplies than in caves that exit to springs. *Forbesichthys* is occasionally found in such caves, in the Mammoth Cave Region, but these populations have not been studied to see if they have incipient adaptations to low food supply.

D. Matching Amblyopsid Adaptations to Cave Type

Poulson has had a bit of an epiphany as he has spent three months reviewing old data, working up old unpublished data, and carefully re-reading what he thinks are the best studies available. He agrees with Noltie and Wicks (2001) that *A. rosae* and *T. subterraneus* occupy very different kinds of caves and karst in Missouri and here extend their suggestions about how cave type relates to all troglobitic Amblyopsid biology.

Typhlichthys occurs over such a wide geographic area with such a great difference in cave and karst type that it is difficult to make generalizations. If the hypothesis of some match of adaptations to cave type is correct then we predict differences between Missouri populations and Kentucky–Tennessee populations. The Missouri *Typhlichthys* are in caves that are far below the surface, have extensive underground watersheds, and have relatively low input of allochthonous organic matter due to often great distances from surface inputs and low currents. One prediction is that they will have more efficient metabolic patterns and foraging with longer lives and larger sizes than reported thus far for *T. subterraneus*. The studied *Typhlichthys* east of the Mississippi River occur in caves that are close to the surface, have small underground watersheds, and often have streams that flood and bring in allochtonous organic matter. For these *Typhlichthys*, we may expect to find local adaptation of populations that have been isolated in caves independently. Local adaptation is especially likely since genetic evidence indicates less gene flow than among cave populations of other Amblyopsid troglobites.

Amblyopsis rosae occur in caves that are shallow with small underground watersheds (Noltie and Wicks 2001) and small rocky streams that never flood severely. The shallow caves often have extensive allochthonous organic input and guano from gray bat maternity colonies in the habitats with the largest populations of fish. Poulson (1985) has previously suggested that this high food supply is related to the relatively short life and relatively frequent reproduction of *A. rosae* compared to other Amblyopsid troglobites. He argued that small maximum size (smallest of all cave Amblyopsids) is associated with less food needed to get the extra energy needed to allow reproduction. Here we extend the connection of high food supply to relatively fast growth rates allowed by an especially energy efficient metabolic pattern. This along with an inferred small clutch size (see Fecundity) allow for more frequent reproduction that is manifested in a fairly even size-frequency distributions in the best habitats.

Amblyopsis spelaea is the largest Amblyopsid and occurs in large streams that have the highest input of allochthonous organic matter and often strong currents. The extreme of input of coarse particulate organic matter is in Penitentiary Cave, Kentucky (Pearson and Boston 1995), with 519 fish censused in 3702 meters of rimstone pools with 10–30 cm of leaf, twig, and acorn fragments. And the extreme for live prey input is in Twin Cave, Indiana, with 130 fish censused in 3600 m^2 of riffles and pools with 70 ml of plankton including hundreds of cladocerans and copepods and 500 ml of fine silt and organic matter in an 800 liter sample during one flood (Scott 1909); this cave also has very high densities of troglobitic isopod prey. Poulson now hypothesizes that even those cave stream *A. spelaea* habitats now predominately silt-bottomed and slow-flowing still have high food supplies and once had higher stream velocities and greater flooding. His reasoning is that the range of *A. spelaea* is just south of the farthest glacial advances where there would have been much greater flooding and faster cave enlargement than at present. He thinks that the dense populations of isopods in many *A. spelaea* caves are explained by cave stream character and that the efficient foraging for isopods by fish > 35 mm SL is associated with isopods as the most frequent and abundant prey larger than copepods. Regular flooding with strong currents has selected for large size and abundant isopods have allowed large size and a large clutch size with branchial incubation. Consonant with this hypothesis is Bechler's observation (personal communication) that in an artificial stream *A. spelaea* was the most active swimmer of the four Amblyopsids studied.

An interesting corollary of large size of *A. spelaea* may be that it allowed syntopic coexistence with the much smaller *Typhlichthys* in the Mammoth Cave System (Poulson 1992). Recall that *T. subterraneus* is in the more food-rich small shaft drain streams and is replaced by *A. spelaea* as you census downstream to the larger and more food-poor base level streams. These *A. spelaea* in the Mammoth Cave System grow to much larger sizes (115–130 mm SL) than farther north (80–90 mm SL) and this may reflect character displacement. As a final note, Poulson thinks that he missed the significance of large size in *A. spelaea* because he always reported traits at the same 45 mm SL size. This was a kind of control when comparing troglomorphic trends (Table 2 in Poulson 1985). Poulson (2001b) has discussed the + and − tradeoffs of large body size for cavefish and for *A. spelaea* the positives are much greater than the negatives.

What kinds of specializations to the character of Key Cave might we look for in *S. poulsoni*? Key Cave is the only locality for *S. poulsoni* and its character may have changed due to anthropogenic impacts (see below Threats). Another complication is that we cannot access much of its habitat even using SCUBA. Nonetheless, Key Cave is a maze system developed at and well below the water table. It does not flood and the only allochtonous organic input seems to be bat guano. The predominant prey is undoubtedly copepods though a large fish might be able to eat a very small crayfish. The maximum size estimated for visual sightings is around 60 mm SL. They are apparently very slow growing based on a conservative scale mark age estimate of 5 years for a 42 mm SL fish. If Alabama cavefish are like other Amblyopsids, we can multiply this estimate by 2–3 to get a probable age of 10–15 years. Also, since adults of other Amblyopsids slow down greatly in growth rate a 60 mm SL fish (estimated size of largest observed fish) may be 20–30 years old. This suggests extremely low available food supply and is consonant with estimates that its metabolic rate is even lower than for *A. rosae*. It is also consonant with the extreme neoteny inferred from its huge relative head size and unbranched fin rays in adults. These extremes may represent both adaptation to its very low food supply and the inference that it has been isolated in caves the longest of all the troglobitic Amblyopsids based on pigment and optic system rudimentation (Table 4; Table 2 in Poulson 1985). Whether we will be allowed to test the predicted genetic, physiological, and behavioral correlates of these morphological trends is problematic because the species is listed as 'critically endangered'. Poulson would at least love to see his namesake alive before he dies (but see Poulson, 2009a)!

VIII. CONSERVATION

The conservation status of subterranean fishes has received increasing attention in the past few years. According to Proudlove (2006), 63 of the 104 known species of subterranean fishes are listed by the International Union for the Conservation of Nature and Natural Resources (IUCN 1996, 2000). All four troglobitic Amblyopsids are included on the list. Three are considered Vulnerable and the other, *S. poulsoni*, is considered Critically Endangered. Two species, *A. rosae* and *S. poulsoni*, are listed by the United States Fish and Wildlife Service (USFWS) under the Endangered Species Act.

In this section, we review the conservation status of Amblyopsid fishes including the two nontroglobitic species, examine the major threats facing each species, and conservation measures that have either been implemented or proposed. This section largely follows that of Proudlove's (2006) chapter titled *"The Conservation Status of Subterranean Fishes."* We encourage readers seeking more information about the conservation of other species of troglobitic fishes to examine the works of Proudlove (2006) and the 'Threatened Fishes of the World' series in the journal *Environmental Biology of Fishes*.

A. Conservation Status

All troglobitic Amblyopsids are considered vulnerable or endangered across their respective distributions (Table 10). In general, there is trend towards increasing threat status with increased cave adaptation. The non-troglobitic species are apparently secure throughout their respective ranges, although disjunct populations of *F. agassizii* in southeast Missouri are listed as endangered (Missouri Natural Heritage Program 2008). *Typhlichthys subterraneus* (as currently recognized) is the most widely distributed and least cave-adapted (Poulson 1963) of the cave Amblyopsids. As such, it is considered the most secure (although it is afforded protection in several states) and is considered endangered only in Georgia where it ranges only into the extreme northwest corner of the state. *Amblyopsis spelaea* is afforded protection at the state level and is a species of concern by USFWS (USFWS 1996). *Amblyopsis rosae* is listed as endangered throughout its range and was listed as threatened by USFWS in 1984 with a recovery plan written in 1989 (USFWS 1989). The most cave-adapted Amblyopsid, *S. poulsoni*, has been described as "the rarest American cavefish and one of the rarest of all freshwater fish" (USFWS 1996) and is found only in Key Cave, Lauderdale Co., Alabama. Because of its extremely limited distribution and suspected low population size, *S. poulsoni* is designated as Critically Endangered by IUCN and was listed as threatened by USFWS in 1977 and later elevated to endangered in 1988.

B. Threats

Proudlove (2006) listed five broad threats that subterranean fishes can face. This list includes: (1) habitat degradation, (2) hydrological manipulations, (3) environmental pollution, (4) overexploitation, and (5) impacts of introduced aquatic animals. Many of the threats discussed below are interrelated because of their wide range of potential effects. For example, dam construction can result in direct destruction and degradation of cavefish habitat, alter hydrological patterns, and allow surface species to colonize and either compete or prey on existing cavefish populations. Here we generally follow the broad classification of threats listed by Proudlove (2006) and review the threats to Amblyopsid populations and focus the majority of our discussion on three troglobitic species, *A. rosae*, *A. spelaea*, and *S. poulsoni*, for which aspects of conservation have been most thoroughly examined. We focus on the first four of Proudlove's list as little work has investigated the effects of introduced species on cavefish populations.

Table 10 Conservation status of each Amblyopsid species throughout its respective distribution

Listing	C. cornuta	F. agassizii	T. subterraneus	A. spelaea	A. rosae	S. poulsoni
Federal Listing	Not listed	Not listed	Not listed	Not listed	Threatened	Endangered
Global Status	G5	G4G5	G4	G4	G3	G1
IUCN Red List Category	Not listed	Not listed	Vulnerable	Vulnerable	Vulnerable	Critically Endangered
Alabama			Protected (S3)			Endangered, Protected (S1)
Arkansas			Inventory Element (S1)		Endangered (S1)	
Georgia	Not listed (SU)		Endangered (S1)			
Illinois		Not listed (S1S2)				
Indiana				Endangered (S1)		
Kentucky		Not listed (S4S5)	Special Concern (S2S3)	Special Concern (S3)		
Missouri		Endangered (S1)	Not listed (S2S3)		Endangered (S2)	
North Carolina	Not listed (S4)					
Oklahoma					Endangered (S1)	
South Carolina	Not listed (SNR)					
Tennessee		Not listed (S4)	Deemed in Need of Management (S3)			
Virginia	Not listed (S3)					

Sources: Alabama Natural Heritage Program 2007, Illinois Endangered Species Protection Board 2006, Indiana Department of Natural Resources 2007, Kentucky State Nature Preserves Commission 2005, Missouri Natural Heritage Program 2008, Oklahoma Department of Wildlife Conservation 2008, Romero and Conner 2007, Withers et al. 2004.

Habitat Degradation and Alteration

Habitat degradation and alteration can result from the direct destruction or manipulation of habitat during quarrying and mining operations, highway construction, and urban development. Keith (1988) reported two *A. spelaea* localities were either partly or completely destroyed by quarrying operations. However, the majority of habitat degradation and alteration threats are indirect resulting in loss of habitat because of siltation, sedimentation, and alteration of hydrological flow patterns and levels. Many caves in the eastern Interior Low Plateau have massive silt banks along streams that are likely associated with farming that began in the 1800s (Poulson, personal observation). However, some caves contain cavefish populations that are found entirely on silt substrates and have high population sizes and frequent reproduction. Likewise, on a much longer time scale, huge changes in habitat composition and food availability must have occurred with glacial cycles during the Pleistocene. Studies are needed to assess the actual rather than the potential impacts on cavefish populations from increased siltation and sedimentation.

Land development within cave recharge zones can alter surface runoff patterns or even block or destroy major recharge points. This can result in dramatic habitat alteration because of increased or decreased water volume, water velocity, sedimentation, or stream scouring depending on local hydrological patterns. In forested areas, increased erosion and production of sediment because of logging can result in increased siltation and sedimentation or the complete blockage of a cave passage. For surface Amblyopsids, loss of forested areas can cause the decline or loss of local populations. The installation of an electric transmission line has been cited as the cause of population decline at a *C. cornuta* locality in Virginia (de Rageot 1992). Removal of the surrounding forest caused increased insolation and drying of aquatic habitat.

Reduced input of surface runoff in recharge zones could have dramatic impacts on reproduction of Amblyopsid cavefishes. Cavefish are believed to rely on increased flow and small temperature changes associated with cave flood events during winter and spring to coordinate reproduction and spawning (Poulson 1963, 1969). Because most populations of cavefish are reportedly small (*S. poulsoni* in particular), the importance of successful timing of sexual maturation and spawning is magnified (Kuhajda 2004). Reductions in surface runoff may disrupt the environmental cues necessary for successful reproduction leading to greater susceptible to extirpation.

Impoundments have resulted in many problems and threats for cavefish populations. A primary example is the construction of Lock and Dam #6 constructed on the Green River below Mammoth Cave in 1906. Although the Green River naturally back-floods into the cave system, flood levels have increased since dam construction (Lisowski and Poulson 1981). The Styx and Echo River areas in Mammoth Cave experienced an apparent decline in cave biota, including cavefish, from the late 1800s to the 1920s (Elliott 2000). From the 1950s to the 1970s, cavefish were still observed but were only large in size and low in abundance. The present rarity of *A. spelaea* in the Echo and Styx River parts of the system likely is related to flooding and silting associated with deforestation and construction of Lock and Dam #6 and other impoundments along the Green River and the Nolin River, a tributary to the Green River, in the 1970s (Poulson 1969, 1996, Elliott 2000). In addition to alteration and loss of habitat, river impoundments may have resulted in decrease in within cave downstream transport of particulate organic matter by hydraulic damming when releases from the upstream Green River dam keep the river level at Mammoth Cave from declining completely after flood peaks. This exacerbates the decline of cave biota in the base-level stream of Mammoth Cave (Poulson 1996). This is discussed in more detail by Poulson (1992).

Hydrological Manipulations

Hydrological manipulations can include underground water removal for human consumption, irrigation, or industry. However, some hydrological manipulations, such as impoundments or increased surface runoff, can raise water tables and alter habitats (see above). Lowering of the water table resulting from direct human consumption, irrigation, or industrial use may threaten cavefish populations. Because *S. poulsoni* lives in a zone of seasonal oscillation of the water table (Trajano 2001), lowering of the water table (and drainage alterations) may isolate fish in these pools exposing them to decreased oxygen levels if decomposable organic mater is present and increased concentrations of contaminants and death if pools dry out during the dry season (Kuhajda 2004). Accordingly, lowering of the water table because of a proposed industrial park for the city of Florence, Alabama, within the recharge area of Key Cave has been cited as a threat to *S. poulsoni* (USFWS 1977, Kuhajda and Mayden 2001).

Groundwater Pollution

Groundwater pollution has been listed as factor negatively affecting populations for all cave Amblyopsids. This threat includes eutrophication and contamination from agricultural and industrial runoff containing pesticides, fertilizers, and heavy metals, sewage effluent, spills and illegal dumping of hazard materials, and thermally altered runoff. Although few studies have examined the direct effects of groundwater pollution on cavefish populations in detail, several studies implicate this threat in population declines.

Groundwater pollution may be acute in nature, such as a toxic spill resulting in a large impulse of contaminants, or chronic occurring over several months to years (Proudlove 2001). Both forms have been attributed to cavefish declines or extirpations from cave systems. Nearly 1,000 dead or dying *T. subterraneus* were expelled from Meramec Spring in Missouri after a fertilizer pipeline rupture in November 1981 caused acute, catastrophic deoxygenation of groundwater (Vandike 1984, Crunkilton 1985). In contrast, several decades worth of gross pollution by decomposable organic matter (creamery waste) and heavy metal contamination (electroplating waste) is the suspected cause of the apparent extirpation of *T. subterraneus* along with other cave life at Hidden River Cave in Kentucky (Lewis 1996). However, *T. subterraneus* and other cave life have re-colonized areas previously affected from far upstream refuges.

Several sources have been implicated in heavy metal and hazardous chemical contamination of groundwater throughout the distributions of cave Amblyopsids. These sources include drilling related to oil and gas development, inactive wells, industrial effluents, accidental spills, underground storage tank leaks, sinkhole dumping, and runoff from roads (road salt) and agricultural fields (pesticides and herbicides) (Keith 1988, Pearson and Boston 1995, Kuhajda and Mayden 2001, Kuhajda 2004). Heavy metal runoff from a local landfill may threaten populations of *T. subterraneus* in Pulaski Co., Kentucky (Tercafs 1992). Increased contaminant levels may cause increased susceptibility to disease. Pesticides were attributed as the cause of "broken back syndrome" that affected perhaps 10 percent of a population of *A. spelaea* in Indiana (Keith and Poulson 1981).

At least four of these threats, industrial effluents, underground storage tank leaks and sinkhole dumping, have been connected to the decline of *T. subterraneus* and other cave life from Hidden River Cave (Pearson and Boston 1995, Lewis 1996). Organic enrichment from sewage treatment plant effluents and septic tank leaks also have been implicated at Hidden River Cave and other caves with Amblyopsids. Organic enrichment can increase nutrients in an otherwise low-nutrient environment and drastically alter food web dynamics, increase risk of disease, and dramatically decrease dissolved oxygen levels. Brown *et al.* (1998) attributed a 30% decline in a population of *A. rosae* to increased levels of inorganic and organic compounds.

Groundwater pollution is a major threat to *S. poulsoni* (Kuhajda 2004) and has been considered as one of the factors likely to cause the decline of the species. Herbicide and pesticide application to cotton fields in the recharge zone of Key Cave have been shown to have direct access to into the cave. Moreover, a sewage sludge disposal operation also lies within the recharge zone of the cave (Aley 1986). Thermal pollution in the form of thermally altered runoff may alter reproductive cycles of Alabama cavefish (USFWS 1982, Kuhajda 2004) and potentially reduce or eliminate recruitment. These and other threats associated with either the current or potential alteration and degradation of the Key Cave recharge area and aquatic habitat, along with the perceived small population size and extremely small distribution have resulted in the listing of *S. poulsoni* as 'Critically Endangered' because of its high risk of extinction (Hilton-Taylor 2000). These threats have been addressed by Kuhajda (2004) and outlined in detail in three different versions of the USFWS Recovery Plan for *S. poulsoni* (USFWS 1982, 1985, 1990).

Collection and Cave Visitation

The collection of cavefish, illegal or otherwise, for the aquarium trade or scientific purposes may pose a threat to all cave Amblyopsids. Because of their uniqueness to hobbyists and the ease at which individuals can be captured, cave Amblyopsids can be easily exploited. Over-collection of fish can reduce or even eliminate local populations. The rarity of *A. spelaea* in the Echo River and River Styx sections of Mammoth Cave system and its presumed absence from adjacent caves to the north have led some to speculate that the species was either introduced or decimated during the 1800s when it was sold as a novelty (Poulson 1968, Elliott 2000). Over-collection by both the scientific community and amateurs is thought be a concern for populations of *A. rosae* (Culver 1986, USFWS 1989).

Commercial exploitation of caves can either alter or even destroy considerable amounts of cavefish habitat. Commercial caves increase human traffic and disturbance in addition to increased light levels. At least five populations of *A. spelaea* are indirectly affected by commercial cave tours including Blue Springs Caverns and Upper Twin Cave in Indiana and Hidden River Cave and Mammoth Cave in Kentucky (Pearson and Boston 1995). However, the exact impacts and long-term effects of commercial cave operations remains to be examined. At Upper Twin Cave, no differences in apparent abundance of *A. spelaea* exist between times when tours are conducted and times when tours are not in operation (Poulson, personal observation). And the continued abundance of fish in Upper Twin Cave since Eigenmann's studies suggests that pole boat tours in the downstream part of the cave, starting in the 1950s, have not compromised the populations.

Human disturbance caused by increased traffic is more of a concern than commercial exploitation. The activities of even the most cautious caver may have serious impacts on cave organisms in shallow, silt-bottomed streams. Disturbance caused by substantial cave visitation may alter breeding of cavefish populations, disturb food sources, and unknowingly stress individual fish by increasing fish activity. However, no evidence has been obtained for any of the above.

Increased disturbance caused by human disturbance is thought to negatively affect grey bat (*Myotis grisescens*) colonies by increasing bat mortality or the eventual abandonment of a cave. If bat colonies are extirpated, cavefish populations may lose an important source of food and nutrients (USFWS 1989). This could be a very serious threat given that the caves with the largest *A. rosae* populations are also grey bat maternity colony caves. However, more studies are needed to assess the impacts and lasting effects of intense cave visitation on cave fauna. The huge decline in the rich aquatic fauna of Shelta Cave, Alabama, including *Typhlichthys*, has been attributed to the loss of a major grey bat roost (Hobbs, personal communication).

Mechanisms of Impact: Community Signatures

Poulson (1992) has hypothesized community signatures both as early warning signs and to identify the four most common kinds of pollution impact. These are meant to supplement Indices of Biological Integrity (IBIs) developed by William Pearson and his students to detect any changes in aquatic cave communities over time. As with IBIs, community signatures require un-impacted baselines. And impacted caves serve as inadvertent experiments to test the community signature hypotheses. Poulson (2009b) proposes that aquatic cave communities will show different responses to siltation, eutrophication, acute toxicity, and chronic toxicity. These signatures are based both on first principles of ecology and toxicology and the differences in life history and metabolic rate of different aquatic cave fauna. The obvious caveat for using community signatures is that several kinds of pollution may occur in the same cave. In these cases, there should be extremely low IBIs. Historically, Hidden River Cave, Kentucky, lost all its troglobites in the main stream. The normal species were replaced by sheets of colonial sewage bacteria and by dense mats of sludge worms that were not killed by the massive inputs of decomposable organic matter with attendant low dissolved oxygen and/or high concentrations of heavy metals.

Differences are expected with chronic toxicity at low levels and acute toxicity at high levels. Heavy metals, like lead and mercury, and organic compounds, like chlorinated hydrocarbons, are usually toxic because they are not naturally occurring and so organisms have not evolved physiological ways of sequestering or detoxifying them. As with herbicides used on weeds as a model, the species with fastest metabolic and growth rate will be most quickly and seriously affected by a pulse of heavy metals or pesticides. Also, species with low metabolic and growth rates may not be affected by pulsed toxicity but are at particular risk with chronic low levels of toxins due to continued bioaccumulation over a long lifetime. If they are also top predators they are in double jeopardy due to biomagnification of toxins along food chains. With acute toxicity their prey may be killed.

Aquatic cave communities impacted by acute toxicity may be missing the organisms at the beginning of the food chain. Troglophiles at the base of food chains will be more affected than troglobites because the troglophiles have the highest growth and metabolic rates. If there is a pulse of toxin input, as in a railroad car derailing or truck accident on a highway near a sinkhole, then *Forbesichthys* is most at risk and *A. rosae* and *Speoplatyrhinus* at least risk.

The community signature for low-level toxic pollution will be different with opposite vulnerabilities than for high-level toxic pollution. Among Amblyopsids, troglobites will be most affected due to both biomagnification and bioaccumulation. Size frequency distributions will show that the largest and oldest fish are missing or under-represented compared to smaller fish. Even troglobitic crayfish may be at risk if the food supply is low and since longevities are documented to be as long as a century in food-poor caves! Short-lived species like copepods and isopods may be unaffected even if they are troglobitic. They may even increase in density if predatory fish and crayfish are reduced in numbers due to chronic toxicity.

It is difficult to find an unimpacted control cave to provide a comparison for the expected community signature for siltation. The problem is that a great number of caves have had increases in silt levels associated with land clearing for agriculture over the past 200 years. Nonetheless, the expected impact of siltation is homogenization of the stream bottom habitat. At the extremes, silt can cover rock and gravel refuges for isopods and amphipods in riffles and so there will be fewer prey washing into pools and shoals deep enough for fish and crayfish. Silt is also likely to cover or mix with fine particulate organic matter that copepods, isopods, and amphipods graze or ingest. Also, at low levels it may foul the biofilms on rocks that are grazed by everything except predators. Poulson (1992) has provided detailed examples for the lower levels of the Mammoth Cave where siltation is due to the combined effects of downstream and upstream dams on the hydropattern of Green River into which the cave streams flow.

The community signature for organic enrichment (eutrophication) is the most clear since it is sensitive to the level of pollution. The mechanism is that decomposition of organic matter provides increased food at low levels but uses up dissolved oxygen. At low levels of organic enrichment species at the beginning of the food chain increase in numbers and their size frequency distributions are skewed more to smaller size classes with increased rates of reproduction. This is especially true if they are troglophiles. Over time the faster reproducing troglophiles simply outproduce the troglobites even though more energy efficient troglobites cannot be outcompeted. See Poulson (this volume) for an explanation of how this may lead to demographic swamping of troglobites by the faster reproducing troglophiles.

At high levels of organic enrichment, especially if pulsed in time, all the normal aquatic fauna is replaced by species tolerant of extremely low dissolved oxygen. At the extreme there are only stringy mats of colonial sewage bacteria, like *Sphaerotilus*, and tubificid worms. The red tubificids have a hemoglobin that can bind oxygen at very low concentrations and so these worms can be seen as waving mats at the stream edge where a little bit of dissolved oxygen remains.

C. Conservation Measures

Several conservation measures have been proposed or implemented for populations of cave Amblyopsids. Fencing or gating of cave entrances have been proposed or implemented to reduce and control human visitation to sensitive cave ecosystems including Amblyopsid caves. Special bat gates are needed to allow entry and exit by bats but stop human entry. Bat Conservation International and The National Speleological Society have been leaders in the improvement and installation of such bats on an increasing number of bat caves.

Protection of cave surface and subsurface watersheds is probably the most important intervention for cavefish caves. Thomas Aley (Ozark Underground Laboratory, Protem, Missouri) is one of the best practioners of state-of-the-art water tracing that is critical to delineating cave watersheds. Among others his studies have led to the protection of watersheds of at least Key Cave, Alabama, the only locality for *Speoplatyrhinus poulsoni*, and for the best *Amblyopsis rosae* cave, Cave Springs Cave in Arkansas. Watershed protection has included establishing preserves as well as institution of best land management practices around sinkholes and sinking creeks that includes reforestation. In other cases water tracing has identified the source of pollutants and so allowed legal action that remedied the situation. Hidden River Cave in Kentucky is one example.

We suggest that what we have called source caves deserve complete protection of their watersheds. Recall that a few caves for each species have the vast majority of all individuals ever censused. To us, attention to protecting these caves is a number one priority for the near future.

Despite the fact that the population of *S. poulsoni* appears to have remained stable over the past 30 years, perturbations within the recharge basin could alter the status of this species (Kuhajda and Mayden 2001). Accordingly, USFWS purchased 1060 acres within the recharge basin in January 1997 and established the Key Cave National Wildlife Refuge. Likewise, the Logan Cave population of *A. rosae* is protected by the 123-acre Logan Cave National Wildlife Refuge, and another population is protected by the 40-acre Ozark Cavefish National Wildlife Refuge.

Introduction of cavefish to new localities or to caves that were historic localities is worth considering. Until we learn to breed Amblyopsids the only source for introductions is existing caves with thriving populations. To protect genetic integrity these source caves should only be in watersheds that include the recipient cave. In the case of *Speoplatyrhinus*, there is only one cave so spreading of risk of extinction by introductions to adjacent caves will require very careful consideration.

Acknowledgements

First we thank the editors of this book, especially Eleonora Trajano, for inviting our participation. Second, we thank a number of friends and colleagues who have allowed us to cite unpublished observations, answered questions, and discussed ideas while we have been writing this chapter. In alphabetical order these are Ginny Adams, Tom Aley, Claude Baker, Richard Borowsky, Mike Barnett (deceased), Tom Barr, David Bechler, Ken Christiansen, Cheryl Coombs, John ('cOOp') Cooper, Dave Culver, Bill Elliott, Ben Fitzpatrick, Beep Hobbs, Bill Jeffery, Jim Keith, Bob Kuehne (deceased), Bernie Kuhajda, Brian Miller, Tom Near, Bill Pearson, Al Romero, "Leo" Trajano, Ted Uyeno, and Fred Whittaker. Third, we thank John ('cOOp') Cooper, Dante Fenolio, Bernie Kuhajda, Rick Mayden, Al Romero, and Uland Thomas for use of fish photographs and drawings. Fifth, we thank authors of what we consider to have been especially thorough, insightful, groundbreaking, and helpful research papers, reviews or dissertations concerning one or more species in the Amblyopsidae. These include Dave Bechler, Zach Brown and Jim Johnson, Carl Eigenmann (deceased), Michael Fine, Loren Hill, Kathrin Hüppop, Bill Pearson, Doug Noltie and Carol Wicks, Fernandus Payne (deceased), Graham Proudlove, Steve Ross and Fred Rohde, Al Romero, Dave Swofford, John Weise, Norbert Welch, and Loren Woods and Bob Inger. Last we thank all of our many colleagues over the years who have been helpful in the field and with discussions about cave biology. You know who you are and we thank you one and all.

References

Adams, G.L. and J.E. Johnson. 2001. Metabolic rate and natural history of Ozark cavefish, *Amblyopsis rosae*, in Logan Cave, Arkansas. *Environmental Biology of Fishes* 62: 97-105.

Agassiz, J.L.R. 1847. Plan for an investigation of the embryology, anatomy and effect of light on the blind-fish of the Mammoth Cave, *Amblyopsis spelaeus. Proceedings of the American Academy of Arts and Sciences* 1: 1-180.

Agassiz, J.L.R. 1853. Recent researches of Prof. Agassiz. *American Journal of Science and Arts* 16: 134.

Alabama Natural Heritage Program. 2007. Alabama inventory list: The rare, threatened and endangered plants and animals of Alabama. Alabama Natural Heritage Program, Montgomery, Alabama, 55 pp.

Aley, T. 1986. Hydrogeologic investigations for a proposed landfill near Florence, Alabama. A contract study and report for Waste Contractors, Inc.

Aley, T and C. Aley. 1979. Prevention of adverse impacts on endangered, threatened, and rare animal species in Benton and Washington counties, Arkansas. Phase 1. Report, Ozark Underground Laboratory, Protem, Missouri. 35 pp.

Armstrong, J.G. 1969. A study of the spring dwelling fishes of the southern bend of the Tennessee River. M.Sc. Thesis, University of Alabama, Tuscaloosa, Alabama. 60 pp.

Armstrong, J.G. and J.D. Williams. 1971. Cave and spring fishes of the southern bend of the Tennessee River. *Journal of the Tennessee Academy of Science* 46: 107-115.

Aumiller, S.R. and D.B. Noltie. 2003. Chemoreceptive responses of the southern cavefish *Typhlichthys subterraneus* Girard, 1860 (Pisces, Amblyopsidae) to conspecifics and prey. *Subterranean Biology* 1: 79-92.

Barr, T.C. and J. Holsinger. 1985. Speciation in cave faunas. *Annual Review of Ecology and Systematics* 16: 313-337.

Bailey, V. 1933. Cave life of Kentucky, mainly in the Mammoth Cave region. *American Midland Naturalist* 14: 385-635.

Balon, E.K. 1975. Reproductive guilds of fishes: A proposal and definition. *Journal of the Fisheries Research Board of Canada* 32: 821-864.

Bechler, D.L. 1976. *Typhlichthys subterraneus* Girard (Pisces: Amblyopsidae) in the Jackson Plain of Tennessee. *National Speleological Society Bulletin* 38: 39-40.

Bechler, D.L. 1980. The evolution of agonistic behavior in Amblyopsid fishes. Ph.D. dissertation. Saint Louis University, Saint Louis, Missouri. 160 pp.

Bechler, D.L. 1981. Agonistic behavior in the Amblyopsidae, the spring, cave, and swamp fishes. *Proceedings of the International Congress of Speleology* 8: 68-69.

Bechler, D.L. 1983. The evolution of agonistic behavior in Amblyopsid fishes. *Behavioral Ecology and Sociobiology* 12: 35-42.

Bergstrom, D.E. 1997. The phylogeny and historical biogeography of Missouri's *Amblyopsis rosae* (Ozark cavefish) and *Typhlichthys subterraneus* (southern cavefish). M.S. Thesis. University of Missouri, Columbia, Missouri.

Bergstrom, D.E., D.B. Noltie and T.P. Holtsford. 1995. Ozark cavefish genetics: The phylogeny of Missouri's Ozark cavefish (*Amblyopsis rosae*) and southern cavefish (*Typhlichthys subterraneus*). Final report to Missouri Department of Conservation.

Blatchley, W.S. 1897. Indiana caves and their fauna. Report of the Indiana State Geologist for 1896, pp. 120-212.

Blaxter, J.H.S. 1965. The feeding of herring larvae and their ecology in relation to feeding. *California Cooperative Oceanic Fisheries Investigations, Reports* 10: 79-88.

Bonaparte, C.L. 1846. Catalogo metodico dei pesci europei. *Atti Scienz. Ital. Settima Adunanza* (1845): 1-95.

Boschung, H.T. and R.L. Mayden. 2004. *Fishes of Alabama*. Smithsonian Institution Press, Washington.

Branson, B.A. 1991. The Mammoth Cave blindfish. *Tropical Fish Hobbyist* 40: 35, 39-40.

Breder, C.M. and D.R. Rosen. 1966. *Modes of Reproduction in Fishes*. Natural History Press, New York.

Brown, A.V. and L.D. Willis. 1984. Cavefish (*Amblyopsis rosae*) in Arkansas: Populations, incidence, habitat requirements and mortality factors. Final Report, Federal Aid Project E-1-6, Arkansas Game and Fish Commission, Little Rock, Arkansas. 61 pp.

Brown, A.V. and C.S. Todd. 1987. Status review of the threatened Ozark cavefish (*Amblyopsis rosae*). *Proceedings of the Arkansas Academy of Science* 41: 99-100.

Brown, A.V., G.O. Graening and P. Vendrell. 1998. Monitoring cavefish population and environmental quality in Cave Springs Cave, Arkansas. Final Report to Arkansas Game and Fish Commission, Little Rock, Arkansas.

Brown, J.Z. 1996. Population dynamics and growth of Ozark cavefish in Logan Cave National Wildlife Refuge, Benton County, Arkansas. M.S. Thesis, University of Arkansas, Fayetteville, Arkansas. 105 pp.

Brown, J.Z. and J.E. Johnson. 2001. Population biology and growth of Ozark cavefishes in Logan Cave National Wildlife Refuge, Arkansas. *Environmental Biology of Fishes* 62: 161-169.

Burr, B.M. and M.L. Warren. 1986. A distributional atlas of Kentucky fishes. Kentucky Nature Preserves Commission Scientific and Technical Series 4. 399 pp.

Charlton, H.H. 1933. The optic tectum and its related fiber tracts in blind fishes. A. *Troglichthys rosae* and *Typhlichthys eigenmanni*. *Journal of Comparative Neurology* 57: 285-325.

Clay, W.M. 1975. *The Fishes of Kentucky*. Kentucky Department of Fish and Wildlife Resources, Frankfurt, Kentucky.

Cooper, J.E. and D.P. Beiter. 1972. The southern cavefish, *Typhlichthys subterraneus* (Pisces, Amblyopsidae), in the eastern Mississippian Plateau of Kentucky. *Copeia* 1972: 879-881.

Cooper, J.E. and R.A. Kuehne. 1974. *Speoplatyrhinus poulsoni*, a new genus and species of subterranean fish from Alabama. *Copeia* 1974: 486-493.

Cooper, J.E. and F.C. Rohde. 1980. *Chologaster cornuta* Agassiz. In: *Atlas of North American Freshwater Fishes*, D.S. Lee, C.R. Gilbert, C.H. Hocutt, R.E. Jenkins, D.E. McAllister and J.R. Stauffer Jr. (eds.), North Carolina State Museum of Natural History, Raleigh, North Carolina, p. 481.

Cope, E.D. 1871. Life in the Wyandotte Cave. *The Annals and Magazine of Natural History* 8: 368-370.

Cope, E.D. 1872. On the Wyandotte Cave and its fauna. *American Naturalist* 6: 109-116.

Cox, U.O. 1905. A revision of the cave fishes of North America. *Appendix to the Report of the Commission for Fisheries for 1904*: 377-393.

Crunkilton, R. 1985. Subterranean contamination of Maramec Spring by ammonium nitrate and urea fertilizer and its implication on rare cave biota. *Missouri Speleologist* 25: 151-158.

Culver, D.C. 1982. *Cave Life: Evolution and Ecology*. Harvard University Press, Cambridge.

Culver, D.C. 1986. Cave faunas. In: *Conservation Biology: The Science of Scarcity and Diversity*, M.E. Soule (ed.). Sinauer Associates, Sunderland, pp. 427-443.

Culver, D.C., T.C. Kane and D.W. Fong. 1995. *Adaptation and Natural Selection in Caves: The Evolution of* Gammarus minus. Harvard University Press, Cambridge and London.

Davidson, R. 1840. An excursion to the Mammoth Cave and the Barrens of Kentucky, with some notices of the early settlement of the state. Published in the same year by both A.T. Skillman and Son, Lexington and Thomas Cowperthwait and Co., Philadelphia.

Dearolf, K. 1956. Survey of North American cave vertebrates. *Proceedings of the Pennsylvania Academy of Science* 30: 201-210.

de Rageot, R. 1992. Notes on the swampfish (*Chologaster cornuta* Agassiz) in the Dismal Swamp of Virginia. *Banisteria* 1: 17-18.

DeKay, J.E. 1842. Zoology of New York or the New York fauna; comprising detailed descriptions of all the animals hitherto observed within the state of New York, with brief notices of those occasionally found near its borders, and accompanied by appropriate illustrations. *Part IV, Fishes*. Printed by W. and A. White and J. Visscher, Albany.

Eigenmann, C.H. 1897. The Amblyopsidae, the blind fish of America. *Report of the British Association for the Advancement of Science* 1897: 685-686.

Eigenmann, C.H. 1898. A new blind fish. *Proceedings of the Indiana Academy of Sciences* 1897: 231.

Eigenmann, C.H. 1899a. A case of convergence. *Science* 9: 280-282.

Eigenmann, C.H. 1899b. The blind fishes of North America. *Popular Science Monthly* 56: 473-486.

Eigenmann, C.H. 1899c. *Blind Fish*. Biological Lectures at The Marine Biological Laboratory, Woods Hole. Athenaeum Press, Boston.

Eigenmann, C.H. 1899d. The eyes of the blind vertebrates of North America. I. The eyes of the Amblyopsidae. *Archiv. fiir. Entwickelungsmechanik* 8: 545-617.

Eigenmann, C.H. 1900. The blind-fishes. *Biological Lectures from the Marine Biological Laboratory of Woods Hole* 2: 113-126.

Eigenmann, C.H. 1905. Divergence and convergence in fishes. *Biological Lectures from the Marine Biological Laboratory of Woods Hole* 8: 59-66.

Eigenmann, C.H. 1909. *Cave Vertebrates of America. A Study in Degenerative Evolution.* Carnegie Institution of Washington, Washington, D.C.

Elliott, W.R. 2000. Conservation of North American cave and karst biota. In: *Ecosystems of the World, Volume 30, Subterranean Ecosystems,* H. Wilkens, D.C. Culver and W.F. Humphreys (eds.), Elsevier, Amsterdam, pp. 665-690.

Etnier, D.A. and W.C. Starnes. 1993. *The Fishes of Tennessee.* The University of Tennessee Press, Knoxville, Tennessee.

Evans, M.H. 1940. *Brain and Body of Fish.* Technical Press, London.

Fine, M.L., M.H. Horn and B. Cox. 1987. *Acanthonus armatus,* a deep-sea teleost fish with a minute brain and large ears. *Proceedings Royal Society of London* B 230: 257-265.

Flint, J. 1820. *Letters from America July 10, 1818 – November 9th 1820.* W. and C. Tate, Edinburgh (Reprinted in 1822).

Forbes, S.A. 1882. The blind cave fishes and their allies. *American Naturalist* 16: 1-5.

Forbes, S.A. and R.E. Richardson. 1908. *The Fishes of Illinois.* Illinois State Laboratory of Natural History, Urbana, Illinois.

Fournie, J.W. and R.M. Overstreet. 1985. Retinoblastoma in the spring cavefish, *Chologaster agassizi* Putnam. *Journal of Fish Diseases* 8: 377-381.

Fuller, T.C. 1958. Rare, blind, white cave fish are found in Kentucky cave by "amateur spelunkers". *Kentucky Happy Hunting Ground* 14: 16-17.

Garman, S. 1889. Cave animals from southwestern Missouri. *Bulletin of the Harvard Museum of Comparative Zoology* 17: 225-240.

Georgia Natural Heritage Program. 2007. Tracking list of special concern animals of Georgia. Georgia Natural Heritage Program, Georgia Department of Natural Resources, Atlanta, 11 pp.

Giovannoli, L. 1933. Invertebrate life of Mammoth and other neighboring caves. *American Midland Naturalist* 14: 600-623.

Girard, C.F. 1859. Ichthyological notices. *Proceedings of the Academy of Natural Sciences of Philadelphia* 1859: 56-68.

Green, S.M. and A. Romero. 1997. Responses to light in two blind cave fishes (*Amblyopsis spelaea* and *Typhlichthys subterraneus*) (Pisces: Amblyopsidae). *Environmental Biology of Fishes* 50: 167-174.

Greenwood, P.H., D.E. Rosen, S.H. Weitzman and G.S. Myers. 1966. Phyletic studies of teleostean fishes, with a provisional classification of living forms. *Bulletin of the American Museum of Natural History* 131: 339-456.

Gunning, G.E. and W.M. Lewis. 1955. The fish population of a spring-fed swamp in the Mississippi bottoms of southern Illinois. *Ecology* 36: 552-558.

Hempel, G. 1965. On the importance of larval survival for the population dynamics of marine food fish. *California Cooperative Oceanic Fisheries Investigations Reports.* 10: 13-23.

Hill, L.G. 1966. Studies on the biology of the spring cavefish, *Chologaster agassizi* Putnam. PhD Dissertation, Department of Biology, University of Louisville. 104 pp.

Hill, L.G. 1968. Oxygen preference in the spring cavefish, *Chologaster agassizi*. *American Fisheries Society Transaction* 97: 448-454.

Hill, L.G. 1969a. Feeding and food habits of the spring cavefish, *Chologaster agassizi*. *American Midland Naturalist* 82: 110-116.

Hill, L.G. 1969b. Distributional and population isolation of the spring cavefish, *Chologaster agassizi* (Pisces: Amblyopsidae). *Proceedings of the Oklahoma Academy of Science* 48: 32-36.

Hill, L.G. 1971. Scale development and patterns of squamation on the spring cavefish, *Chologaster agassizi* (Amblyopsidae). *Proceedings of the Oklahoma Academy of Science* 51: 13-14.

Holsinger, J.R. 2000. Ecological derivation, colonization, and speciation. In: *Ecosystems of the World, Volume 30, Subterranean Ecosystems*, H. Wilkens, D.C. Culver and W.F. Humphreys (eds.), Elsevier, Oxford, pp. 399-415.

Hubbs, C.L. 1938. Fishes from the caves of Yucatan. *Carnegie Institution of Washington Publication* 491: 261-295.

Hüppop, K. 2000. How do cave animals cope with the food scarcity in caves? In: *Ecosystems of the World, Volume 30, Subterranean Ecosystems*. H. Wilkens, D.C. Culver and W.F. Humphreys (eds.), Elsevier, Oxford, pp. 159-188.

Hilton-Taylor, C. 2000. 2000 IUCN Red List of Threatened Species. IUCN, Gland, Switzerland and Cambridge, United Kingdom.

Hunter, J.R. 1972. Swimming and feeding behavior of larval anchovy *Engraulis mordax*. *Fishery Bulletin* 70: 821-838.

Hutchings, J.A. 1991. Fitness consequences of variation in egg size and food abundance in brook trout *Salvelinus fontinalis*. *Evolution* 45: 1162-1168.

Illinois Endangered Species Protection Board. 2006. Endangered and threatened species list. Illinois Endangered Species Protection Board, Springfield, Illinois. Available online at http://dnr.state.il.us/espb/datelist.htm.

Indiana Department of Natural Resources. 2007. Indiana's species of greatest conservation need. Division of Fish and Wildlife, Indiana Department of Natural Resources, Indianapolis, Indiana. 2 pp.

IUCN 1996. 1996 IUCN Red List of Threatened Animals. IUCN, Gland and Cambridge.

IUCN 2000. 2000 IUCN Red List of Threatened Species. IUCN, Gland and Cambridge.

Jegla, T.C. and T.L. Poulson. 1970. Circadian rhythms, I: Reproduction in cave crayfish *Orconectes pellucidus inermis*. *Comparative Biochemical Physiology* 33: 347-355.

Jenio, F. 1980. Predation of freshwater gammarids (Crustacea: Amphipoda). *Proceedings of the West Virginia Academy of Science* 51: 67-73.

Jenkins, R.E. and N.M. Burkhead. 1994. *Freshwater Fishes of Virginia*. American Fisheries Society, Bethesda, Maryland, 1079 pp.

Jones, G.M. and K.E. Spells. 1963. A theoretical and comparative study of the functional dependence of the semicircular canal upon its physical dimensions. *Proceedings Royal Society of London* B 157: 403-419.

Jones, S.R. and C.A. Taber. 1985. A range revision for western populations of the southern cavefish *Typhlichthys subterraneus* (Amblyopsidae). *American Midland Naturalist* 113: 413-415.

Jordan, D.S. 1889. Descriptions of fourteen species of fresh-water fishes collected by the U.S. Fish Commission in the summer of 1888. *Proceedings of the United States National Museum* 11: 351-362.

Jordan, D.S. 1929. *Forbesichthys* for *Forbesella*. *Science* 70: 68.

Jordan, D.S. and B.W. Evermann. 1927. New genera and species of North American fishes. *Proceedings of the California Academy of Science* 16: 501-507.

Kalayil, P.K. and W.M. Clay. 1976. Immunological characteristics and relationships of tissue antigens in Amblyopsid fishes. *Federation Proceedings* 35: 751.

Keith, J.H. 1988. Distribution of northern cavefish, *Amblyopsis spelaea* DeKay, in Indiana and Kentucky and recommendations for its protection. *Natural Areas Journal* 8: 69-79.

Keith, J.H. and L.M. Gray. 1979. A preliminary study of the occurrence of broken-back syndrome in the northern cavefish (*Amblyopsis spelaea*) at Spring Mill State Park, Mitchell, Indiana. *Proceedings of the Indiana Academy of Sciences* 88: 163.

Keith, J.H. and T.L. Poulson. 1981. Broken-back syndrome in *Amblyopsis spelaea*, Donaldson-Twin Caves, Indiana. Cave Research Foundation 1979 Annual Report: 45-48.

Kentucky State Nature Preserves Commission. 2005. Rare and extirpated biota of Kentucky. Kentucky State Nature Preserves Commission, Frankfort, Kentucky. 19 pp.

Kuhajda, B.R. 2004. The impact of the proposed Eddie Frost Commerce Park on *Speoplatyrhinus poulsoni*, the Alabama cavefish, a federally endangered species restricted to Key Cave, Lauderdale County, Alabama. *Endangered Species Update* 21: 57.

Kuhajda, B.R. and R.L. Mayden. 2001. Status of the federally endangered Alabama cavefish, *Speoplatyrhinus poulsoni* (Amblyopsidae), in Key Cave and surrounding caves, Alabama. *Environmental Biology of Fishes* 62: 215-222.

Langecker, T.G. 2000. The effects of continuous darkness on cave ecology and cavernicolous evolution. In: *Ecosystems of the World, Volume 30, Subterranean Ecosystems*, H. Wilkens, D.C. Culver and W.F. Humphreys (eds.), Elsevier, Amsterdam, pp. 135-158.

Layne, J.N. and D.H. Thompson. 1952. Recent collections of the Amblyopsid fish *Chologaster papillifera* in Illinois. *Copeia* 1952: 39-40.

Lewis, J.J. 1996. The devastation and recovery of caves affected by industrialization. *Proceedings of the 1995 National Cave Management Symposium*: 214-227.

Lewis, J.J. 1998. The subterranean fauna of the Blue River area. Final Report, The Natural Conservancy, 266 pp.

Lewis, J.J. 2002a. Conservation assessment for northern cavefish copepod (*Cauloxenus stygius*). USDA Forest Service, Eastern Region, Report. 10 pp.

Lewis, J.J. 2002b. Conservation assessment for southern cavefish (*Typhlichthys subterraneus*). USDA Forest Service, Eastern Region, Report. 9 pp.

Lewis, J.J. and T. Sollman. 1998. Groundwater monitoring in significant aquatic caves that lie beneath impending residential developments in the Blue River basin of southern Indiana. Final Report, U.S. Fish and Wildlife Service. 89 pp.

Lisowski, E.A. and T.L. Poulson. 1981. Impacts of Lock and Dam Six on base level ecosystems in Mammoth Cave. Cave Research Foundation 1981 Annual Report: 48-54.

Mayden, R.L. and F.B. Cross. 1983. Reevaluation of Oklahoma records of the southern cavefish, *Typhlichthys subterraneus* Girard (Amblyopsidae). *Southwestern Naturalist* 28: 471-473.

McAllister, D.E. 1968. The evolution of branchiostegals and associated opercular, gular, and hyoid bones and the classification of teleostome fishes, living and fossil. *Bulletin of the National Museum of Canada* 221: 1-239.

McDonald, E.F. and W.L. Pflieger. 1979. The spring cavefish, *Chologaster agassizi* (Pisces: Amblyopsidae), in southeastern Missouri. *American Midland Naturalist* 102: 194-196.

McNulty, J.A. 1978a. The pineal organ of the troglophilic fish, *Chologaster agassizi*: An ultrastructural study. *Journal of Neural Transmission* 43: 47-71.

McNulty, J.A. 1978b. Fine structure of the pineal organ in the troglobytic fish, *Typhlichthys subterraneus* (Pisces: Amblyopsidae). *Cell and Tissue Research* 195: 535-545.

Means, M.L. and J.E. Johnson. 1995. Movement of threatened Ozark cavefish in Logan Cave National Wildlife Refuge, Arkansas. *Southwestern Naturalist* 40: 308–313.

Missouri Natural Heritage Program. 2008. Missouri species and communities of conservation concern checklist. Missouri Department of Conservation, Jefferson City, Missouri, 55 pp.

Mohr, C.E. and T.L. Poulson. 1966. *The Life of the Cave*. McGraw-Hill, New York.

Montgomery, J.C., S. Coombs and C.F. Baker. 2001. The mechanosensory lateral line system of the hypogean form of *Astyanax mexicanus*. *Environmental Biology of Fishes* 62: 87-96.

Murray, A.M. and M.V.H. Wilson. 1999. Contributions of fossils to the phylogenetic relationships of the percopsiform fishes (Teleostei: Paracanthopterygii): Order restored. In: *Mesozoic Fishes 2—Systematics and Fossil Record*. G. Arratia and H.-P. Schultz (eds.), Verlag Dr. Friedrich Pfeil, Munich, Germany, pp. 397-411.

Nelson, J.S. 1994. *Fishes of the World*. Third Edition. John Wiley Sons, New York.

Nelson, J.S. 2006. *Fishes of the World*. Fourth Edition. John Wiley Sons, New York.

Nickol, B.B. and F.B. Whittaker. 1978. *Neoechinorhynchus cylindratus* (Acanthocephala) from the troglodytic fish, *Amblyopsis spelaea*, in Kentucky. *Proceedings of the Helminthological Society of Washington* 45: 136-137.

Niemiller, M.L. and B.M. Fitzpatrick. 2008. Phylogenetics of the southern cavefish, *Typhlichthys subterraneus*: Implications for conservation and management. *Proceedings of the 2007 National Cave and Karst Management Symposium* (In Press).

Noltie, D.B. and C.B. Wicks. 2001. How hydrogeology has shaped the ecology of Missouri's Ozark cavefish, *Amblyopsis rosae*, and southern cavefish, *Typhlichthys subterraneus*: Insights on the sightless from understanding the underground. *Environmental Biology of Fishes* 62: 171-194.

Oklahoma Department of Wildlife Conservation. 2008. Oklahoma's endangered and threatened species and species of special concern. Natural Resources Section, Oklahoma Department of Wildlife Conservation, Oklahoma City. Available online at http://www.wildlifedepartment.com/endanger.htm.

Page, L.M. and B.M. Burr. 1991. *A Field Guide to Freshwater Fishes of North America North of Mexico*. Houghton Mifflin, Boston.

Parenti, L.R. 2006. *Typhlichthys eigenmanni* Charlton, 1933, an available name for a blind cavefish (Teleostei: Amblyopsidae), differentiated on the basis of characters of the central nervous system. *Zootaxa* 1374: 55-59.

Parzefall, J. 2000. Ecological role of aggressiveness in the dark. In: *Ecosystems of the World, Volume 30, Subterranean Ecosystems*. H. Wilkens, D.C. Culver and W.F. Humphreys (eds.), Elsevier, Amsterdam, pp. 221-228.

Patterson, C. 1981. The development of the North American fish fauna—a problem of historical biogeography. In: *The Evolving Biosphere*. P.L. Forey (ed.), Cambridge University Press, Cambridge, pp. 265-281.

Patterson, C. and D.E. Rosen. 1989. The Paracanthopterygii revisited: Order and disorder. In: *Papers on the Systematics of Gadiform Fishes*. D.M. Cohen (ed.), Science Seriees, Natural History Museum Los Angeles Co., No. 32, pp. 5-36.

Payne, R. 1907. The reactions of the blind fish, *Amblyopsis spelaeus*, to light. *Biological Bulletin* 13: 317-323.

Pearson, W.D. and C.H. Boston. 1995. Distribution and status of the northern cavefish, *Amblyopsis spelaea*. Final report, Nongame and Endangered Wildlife Program, Indiana Department of Natural Resources, Indianapolis.

Poly, W.J. and C.E. Boucher. 1996. Nontroglobitic fishes in caves: Their abnormalities, ecological classification and importance. *American Midland Naturalist* 136: 187-198.

Poly, W.J. and G.S. Proudlove. 2004. Family Amblyopsidae Bonaparte 1846—cavefishes. *California Academy of Sciences Annotated Checklists of Fishes* 26. 7 pp.

Poulson, T.L. 1960. Cave adaptation in Amblyopsid fishes. PhD Dissertation, Department of Zoology, University of Michigan, Ann Arbor. University Microfilms 61-2787.

Poulson, T.L. 1963. Cave adaptation in Amblyopsid fishes. *American Midland Naturalist* 70: 257-290.

Poulson, T.L. 1967. Comparison of terrestrial cave communities. Cave Research Foundation Annual Report 1967: 39.

Poulson, T.L. 1968. Aquatic cave communities. Cave Research Foundation Annual Report 1968: 16-18.

Poulson, T.L. 1969. Population size, density, and regulation in cave fishes. *Actes of the 4th International Congress of Speleology*, Ljubljana, Yugoslavia 4-5: 189-192.

Poulson, T.L. 1985. Evolutionary reduction by neutral mutations: Plausibility arguments and data from Amblyopsid fishes and Linyphiid spiders. *Bulletin of the National Speleological Society* 47: 109-117.

Poulson, T.L. 1992. The Mammoth Cave ecosystem. In: *The Natural History of Biospeleology*. A. Camacho (ed.), Museo Nacional de Ciencias Naturales, Madrid, pp. 569-611.

Poulson, T.L. 1996. Research aimed at management problems should be hypothesis-driven: Case studies in the Mammoth Cave Region. In: *Proceedings of the 1995 National Cave Management Symposium, Spring Mill State Park, Mitchell, Indiana*, G.T. Rea (ed.). Indiana Karst Conservancy, Indianapolis, Indiana, pp. 267-273.

Poulson, T.L. 2001a. Morphological and physiological correlates of evolutionary reduction of metabolic rate among Amblyopsid cave fishes. *Environmental Biology of Fishes* 62: 239-249.

Poulson, T.L. 2001b. Adaptations of cave fishes with some comparisons to deep-sea fishes. *Environmental Biology of Fishes* 62: 345-364.

Poulson, T.L. 2005. Food sources. In: *Encyclopedia of Caves*, D.C. Culver and W.B. White (eds.), Elsevier, Amsterdam, pp. 255-263.

Poulson, T.L. 2009a. New studies of *Speoplatyrhinus poulsoni* (Pisces: Amblyopsidae). *Proceedings of the 15th International Congress of Speleology* 3: 1337-1342.

Poulson, T.L. 2009b. Field metrics for cave stream bio-integrity. *Proceedings of the 15th International Congress of Speleology* 3: 1343-1348.

Poulson, T.L. and P.M. Smith. 1969. The basis for seasonal growth and reproduction in aquatic cave organisms. *Actes of the 4th International Congress of Speleology*, Ljubljana, Yugoslavia 4-5: 197-201.

Poulson, T.L. and T.C. Jegla. 1969. Circadian rhythms in cave animals. *Actes of the 4th International Congress of Speleology*, Ljubljana, Yugoslavia 4-5: 193-195.

Poulson, T.L. and W.B. White. 1969. The cave environment. *Science* 165: 971-981.

Poulson, T.L. and K.H. Lavoie. 2000. The trophic basis of subterranean ecosystems. In: *Ecosystems of the World, Volume 30, Subterranean Ecosystems*. H. Wilkens, D.C. Culver and W.F. Humphreys (eds.), Elsevier, Amsterdam, pp. 118-136.

Protas, M.E., C. Hersey, D. Kochanek, Y. Zhou, H. Wilkens, W.R. Jeffery, L.I. Zon, R. Borowsky and C.J. Tabin. 2006. Genetic analysis of cavefish reveals molecular convergence in the evolution of albinism. *Nature Genetics* 38: 107-111.

Proudlove, G.S. 2001. The conservation status of hypogean fishes. *Environmental Biology of Fishes* 62: 201-213.

Proudlove, G.S. 2006. *Subterranean Fishes of the World*. International Society for Subterranean Biology, Moulis.

Putnam, F.W. 1872. The blind fishes of the Mammoth Cave and their allies. *American Naturalist* 6: 6-30.

Rohde, F.C., R.G. Arndt, D.G. Lindquist and J.F. Parnell. 1994. *Freshwater Fishes of the Carolinas, Virginia, Maryland, and Delaware.* University of North Carolina Press, Chapel Hill, North Carolina.

Romero, A. 1998a. Threatened fishes of the world: *Amblyopsis rosae* (Eigenmann, 1898) (Amblyopsidae). *Environmental Biology of Fishes* 52: 434.

Romero, A. 1998b. Threatened fishes of the world: *Typhlichthys subterraneus* Girard, 1860 (Amblyopsidae). *Environmental Biology of Fishes* 53: 74.

Romero, A. 1998c. Threatened fishes of the world: *Speoplatyrhinus poulsoni* Cooper and Kuehne, 1974 (Amblyopsidae). *Environmental Biology of Fishes* 53: 293-94.

Romero, A. 2001. Scientists prefer them blind: The history of hypogean fish research. *Environmental Biology of Fishes* 62: 43-71.

Romero, A. 2002. Between the first blind cave fish and the last of the Mohicans: The scientific romanticism of James E. DeKay. *Journal of Spelean History* 36: 19-29.

Romero, A. and L. Bennis. 1998. Threatened fishes of the world: *Amblyopsis spelaea* DeKay, 1842 (Amblyopsidae). *Environmental Biology of Fishes* 51: 420.

Romero, A. and M. Conner. 2007. Status report for the southern cavefish, *Typhlichthys subterraneus*, in Arkansas. Final report to the Arkansas Game and Fish Commission, Little Rock, Arkansas.

Rosen, D.E. 1962. Comments on the relationships of the North American cave fishes of the family Amblyopsidae. *American Museum Novitates* 2109: 1-35.

Ross, S.W. and F.C. Rohde. 2003. Life history of the swampfish from a North Carolina stream. *Southeastern Naturalist* 2: 105-120.

Schubert, A.L.S., C.D. Nielsen and D.B. Noltie. 1993. Habitat use and gas bubble disease in southern cavefish (*Typhlichthys subterraneus*). *International Journal of Speleology* 22: 131-143.

Scott, W. 1909. An ecological study of the plankton of Shawnee Cave. *Biological Bulletin* 17: 386-402.

Shute, J.R., P.W. Shute and D.G. Lindquist. 1981. Fishes of the Waccamaw River Drainage. *Brimleyana* 6: 1-24.

Smith, P.W. 1979. *The Fishes of Illinois.* University of Illinois Press, Urbana, Illinios.

Smith, P.W. and N.M. Welch. 1978. A summary of the life history and distribution of the spring cavefish, *Chologaster agassizi*, Putnam, with population estimates for the species in southern Illinois. *Illinois Natural History Survey Biological Notes* 104: 1-8.

Smith, W.L. and W.C. Wheeler. 2006. Venom evolution widespread in fishes: A road map for the bioprospecting of piscine venoms. *Journal of Heredity* 97: 206-217.

Storer, D.H. 1846. Synopsis of the fishes of North America. *Memoirs of the American Academy* 2: 436.

Swofford, D.L. 1982. Genetic variability, population differentiation, and biochemical relationships in the family Amblyopsidae. M.S. Thesis. Eastern Kentucky University, Richmond, Kentucky.

Swofford, D.L., B.A. Branson and G. Sievert. 1980. Genetic differentiation of cavefish populations (Amblyopsidae*). Isozyme Bulletin* 13: 109-110.

Tellkampf, T.G. 1844. Ueber den blinden Fisch der Mammuth-Hohle in Kentucky, mit Bemerkungen ueber einige undere in dieser Hohle lebenden Thiere. *Archiv fur Anatomie, Physiologie und wissenschaftliche Medicin* 4: 381-394.

Tellkampf, T.G. 1845. Memoirs on the blind-fishes and some other animals living in the Mammoth Cave in Kentucky. *New York Journal of Medical and Collateral Sciences* 5: 84-93.

Tercafs, R. 1992. The protection of the subterranean environment. Conservation principles and management tools. In: *The Natural History of Biospeleology,* A.I. Camacho (ed.), Monografias del Museo Nacional de Ciencias Naturales, Madrid, pp. 481-524.

Trajano, E. 2001. Ecology of subterranean fishes: An overview. *Environmental Biology of Fishes* 62: 133-160.

Trajano, E. 2007. The challenge of estimating the age of subterranean lineages: examples from Brazil. *Acta Carsologica* 36: 191-198.

United States Fish and Wildlife Service (USFWS). 1977. Final threatened status and critical habitat for five species of southeastern fishes. *Federal Register* 42: 45526-45530.

United States Fish and Wildlife Service (USFWS). 1982. Recovery plan for the Alabama cavefish, *Speoplatyrhinus poulsoni* Cooper and Kuehne, 1974. U.S. Department of Interior, U.S. Fish and Wildlife Service, Atlanta, Georgia.

United States Fish and Wildlife Service (USFWS). 1985. Revised recovery plan for the Alabama cavefish, *Speoplatyrhinus poulsoni* Cooper and Kuehne, 1974. U.S. Department of Interior, U.S. Fish and Wildlife Service, Atlanta, Georgia.

United States Fish and Wildlife Service (USFWS). 1989. Ozark cavefish recovery plan. U.S. Department of Interior, U.S. Fish and Wildlife Service, Atlanta, Georgia.

United States Fish and Wildlife Service (USFWS). 1990. Alabama cavefish, *Speoplatyrhinus poulsoni* Cooper and Kuehne, 1974 (second revision) recovery plan. U.S. Department of Interior, U.S. Fish and Wildlife Service, Atlanta, Georgia.

United States Fish and Wildlife Service (USFWS). 1996a. Proposed establishment of Key Cave National Wildlife Refuge, Lauderdale County, Alabama. U.S. Department of Interior, U.S. Fish and Wildlife Service, Atlanta, Georgia.

United States Fish and Wildlife Service (USFWS). 1996b. Region 3 decision on the status recommendation for the northern cavefish (*Amblyopsis spelaea*). U.S. Department of Interior, U.S. Fish and Wildlife Service, Bloomington, Indiana.

Vandike, J.E. 1981. The effects of the November 1981 liquid fertilizer pipeline break on groundwater in Phelps County, Missouri. Report for Water Resources Data and Research, Missouri Department of Natural Resources, Division of Geology and Land Survey, Rolla. 24 pp.

Vandike, J.E. 1984. Hydrogeologic aspects of the November 1981 liquid fertilizer pipeline break on groundwater in the Meramec Spring recharge area, Phelps County, Missouri. *Proceedings of the 1984 National Cave Management Symposium, Missouri Speleology* 25: 93-101.

Verrier, M.L. 1929. Observations sur le comportement d'un poisson cavernicole *Typhlichthys osborni* Eigenmann. *Bulletin of the National Museum of Natural History Paris* (Series 2) 1: 82-84.

Weber, A. 2000. Fish and amphibia. In: *Ecosystems of the World, Volume 30, Subterranean Ecosystems*, H. Wilkens, D.C. Culver and W.F. Humphreys (eds.), Elsevier, Amsterdam, pp. 109-132.

Weise, J.G. 1957. The spring cave-fish, *Chologaster papilliferus*, in Illinois. *Ecology* 38: 195-204.

Whittaker, F.H. and L.G. Hill. 1968. *Proteocephalus chologaster* sp. n. (Cestoda: Proteocephalidae) from the spring cavefish *Chologaster agassizi* Putnam, 1782 [error = Putnam 1872] (Pisces: Amblyopsidae) of Kentucky. *Proceedings of the Helminthological Society of Washington* 35: 15-18.

Whittaker, F.H. and S.J. Zober. 1978. *Proteocephalus poulsoni* sp. n. (Cestoda: Proteocephalidae) from the northern cavefish *Amblyopsis spelaea* DeKay, 1842 (Pisces: Amblyopsidae) of Kentucky. *Folia Parasitologica* 25: 277-280.

Wiley, E.O., G.D. Johnson and W.W. Dimmick. 2000. The interrelationships of acanthomorph fishes: A total evidence approach using molecular and morphological data. *Biochemical Systematics and Ecology* 28: 319-350.

Wilkens, H. 1988. Evolution and genetics of epigean and cave *Astyanax mexicanus* (Characidae, Pisces): Support for the neutral mutation theory. *Evolutionary Biology* 23: 271-367.

Willis, L.D. and A.V. Brown. 1985. Distribution and habitat requirements of the Ozark cavefish, *Amblyopsis rosae*. *American Midland Naturalist* 114: 311-317.

Withers, D.I., K. Condict and R. McCoy. 2004. A guide to the rare animals of Tennessee. Division of Natural Heritage, Tennessee Department of Environment and Conservation, Nashville, Tennessee. 16 pp.

Woods, L.P. and R.F. Inger. 1957. The cave, spring, and swamp fishes of the family Amblyopsidae of central and eastern United States. *American Midland Naturalist* 58: 232-256.

Wyman, J. 1843a. Description of a blind-fish from a cave in Kentucky. *American Journal of Science and Arts* 45: 94-96.

Wyman, J. 1843b. Description of a blind-fish from a cave in Kentucky. *Annals and Magazine of Natural History* 12: 298-299.

Wyman, J. 1851. Account of dissections of the blind fishes (*Amblyopsis spelaeus*) from the Mammoth Cave, Kentucky. *Proceedings of the Boston Society of Natural History* 3: 349, 375.

Wyman, J. 1854a. The eyes and organs of hearing in *Amblyopsis spelaeus*. *Proceedings of the Boston Society of Natural History* 4: 149-151.

Wyman, J. 1854b. On the eye and the organ of hearing in the blind fishes (*Amblyopsis spelaeus* DeKay) of the Mammoth Cave. *Proceedings of the Boston Society of Natural History* 4: 395-396.

Wyman, J. 1872. Notes and drawings of the rudimentary eyes, brain and tactile organs of *Amblyopsis spelaeus*, in Putnam (1872). *American Naturalist* 6: 6-30.

CHAPTER 8

Subterranean Fishes of Mexico (*Poecilia mexicana*, Poeciliidae)

Martin Plath[1] and Michael Tobler[2,3]

[1]Institute of Ecology, Evolution and Diversity, Department of Ecology and
Evolution, J.W. Goethe University Frankfurt, Siesmayerstrasse 70-72,
D-60054 Frankfurt am Main, Germany
E-mail: mplath@bio.uni-frankfurt.de
[2]Texas A&M University, Department of Fish and Wildlife Sciences, 2258
TAMU, College Station, TX 77843-2258 USA
[3]University of Zürich, Institute of Zoology, Winterthurerstrasse 190
CH-8057 Zürich, Switzerland

INTRODUCTION

The Atlantic molly (*P. mexicana*) is a widespread freshwater fish living along
the Atlantic coast of Mexico and Central America (Miller 2005). It inhabits
coastal lagoons, estuaries, lowland ponds and rivers up to highland streams
(Bussing 1998, Miller 2005). Mollies are small fish less than 10 centimeters
long, and belong to the family Poeciliidae (livebearers). Poeciliids give
birth to fully developed young, which are instantly independent. Males
use a modified anal fin, the so-called gonopodium, to transfer sperm
bundles (spermatozeugmata) to the female during copulation (Rosen and
Bailey 1963, Meffe and Snelson 1989).

In the southern Mexican Cueva del Azufre system, *P. mexicana* has
colonized at least two caves. Cavernicolous *P. mexicana* have been referred
to as the cave molly (Gordon and Rosen 1962, Parzefall 2001). Cave mollies
are the only known poeciliids that naturally inhabit subterranean habitats
(Proudlove 2006). The only other known poeciliid species occurring in a
cave is *Gambusia affinis* introduced into the 'Spunnulate' dolinas system of
Torre Castiglione in southeastern Italy (Camassa 2001). Various aspects of

the evolutionary ecology of *P. mexicana* from the Cueva del Azufre system have been studied thoroughly during the past decades. Today, the cave molly ranks among the best-studied cavefishes worldwide along with the cave form(s) of the Mexican tetra *Astyanax mexicanus* (see chapters 5 and 6 in this volume). This chapter provides a synopsis particularly over some recent advances in the understanding of the evolutionary ecology of cavernicolous *P. mexicana*.

HISTORY AND OVERVIEW

The cave molly has always been tightly connected to the local native culture. The Cueva del Azufre (also known as Cueva de las Sardinas and Cueva Villa Luz) is sacred to the Rain God of local Zoque Indians. Once a year around Easter (towards the end of the dry season), the Zoque penetrate into the cave, and poison and capture cave mollies with the aid of rotenone- and deguelin-containing barbasco root (*Lonchocarpus* sp., Fabaceae). The fish are then cooked and eaten. During this act, which is accompanied by prayers, the dead fish act as messengers to the Rain God, bringing on the first rains of the rainy season.

Scientific research in the cave and the recognition of the uniqueness of its inhabitants, commenced in the late 19th century. According to an unpublished manuscript of the recently deceased American ichthyologist Robert Rush Miller, the first scientific collections of fish in the Cueva del Azufre date back to 1896, when A. Dugas sent specimens to the US National Museum. M.W. Stirling, Chief of the Bureau of American Ethnology (Smithsonian Institution), and his collaborators researched the cave and collected cave mollies several times between 1944 and 1948. The first map of the cave and a scientific description of the cave molly – not as a distinct species but a divergent population of *Poecilia sphenops*[1] – were published by Gordon and Rosen in 1962. A second, more detailed map was later published by Hose and Pisarowicz (1999).

Ever since the 1960s, several research groups have worked in the Cueva del Azufre, specifically on the fish inhabiting the cave. This research was spearheaded by a group from the University of Hamburg led by Jakob Parzefall. Parzefall pioneered maintaining and breeding the cave molly in captivity and published numerous scientific studies on this fish. He established the cave molly as a model system to study the evolution of cave animals, with special emphasis on behavioral adaptations to darkness (Parzefall 1993, 2001).

---

[1]Later research indicated that, based on the shape of their teeth, the molly populations in and around the cave are *P. mexicana* not *P. sphenops* (Miller, R.R. 1983. Checklist and key to the mollies of Mexico (Pisces, Poeciliidae, *Poecilia*, subgenus *Mollienesia*). *Copeia*: 817-822.

The pioneering work on the cave molly was primarily based on laboratory studies. In the course of time, this led to a somewhat simplifying view in two directions: (1) The predominant view of the Cueva del Azufre system was that it consists of two distinct habitat types, the Cueva del Azufre and normal surface habitats. (2) The absence of light was assumed to be the most important selective factor. More recent work led to a redefinition of the Cueva del Azufre system to amend these views by including additional habitats in comparative analyses and by recognizing the importance of hydrogen sulfide (H_2S), which is present in some habitats of this system, as an important selective factor. The discovery of a non-sulfidic cave (Cueva Luna Azufre) and independent sulfidic springs in surface habitats (El Azufre) in 2006 allowed extending the comparative approach of studying the ecology and evolution of *P. mexicana* under divergent environmental conditions. The inclusion of a wider array of populations – especially from adjacent surface habitats – and the recognition of a multitude of both abiotic and biotic environmental factors as potential sources of natural selection allowed for a better understanding of the ecological and evolutionary divergence in cave-inhabiting populations of *P. mexicana*.

In this chapter, we review the current knowledge on the evolutionary ecology of *P. mexicana* in the Cueva del Azufre system. We first compare various aspects of the abiotic environments in which populations of *P. mexicana* occur and discuss divergent traits occurring in the different populations as well as their potential adaptive significance. We then examine differences in the biotic environment of the different habitat types, discuss the consequences for the ecology of different populations, and highlight again the evolutionary responses of *P. mexicana*. By the end of this chapter, we synthesize the current knowledge and scrutinize support for the idea that parapatric ecological speciation due to local adaptation to divergent environmental conditions may be occurring in this system. Throughout, we make an attempt to disentangle the effects of different selective forces – primarily darkness (i.e., evolutionary trends that may be shared also with other cave fishes), toxic hydrogen sulfide, and selective pressures stemming from ecological differences among habitat types – on the evolution of different populations of *P. mexicana*. Most information currently available considers the behavior, ecology and evolution of the cave molly population from the sulfidic Cueva del Azufre, but in some cases, fish from the recently discovered Cueva Luna Azufre have already been included in recent research projects, and – where available – first results are presented.

AN INTRODUCTION TO THE CUEVA DEL AZUFRE SYSTEM

The Cueva del Azufre system is located near the village of Tapijulapa in the state of Tabasco, Mexico (Figure 1). All watercourses in the study area eventually drain into the Río Oxolotan, which is part of the Río Grijalva drainage system (Tobler *et al.* 2006b). The Cueva del Azufre system provides an unparalleled 'natural experiment' with two abiotic environmental factors – the presence or absence of light and H_2S – occurring in a fully 2×2 factorial design. All habitat types are connected and located within a perimeter of roughly 3 km. Most importantly, all of the habitat types are inhabited by *P. mexicana* (Tobler *et al.* 2006b, Tobler *et al.* 2008a). The four habitat types typically recognized for comparative analyses include:

- The Cueva del Azufre itself is a **sulfidic cave** structured into 13 different chambers, the nomenclature of which follows Gordon and Rosen (1962; Figure 2). The front chambers obtain some dim light through breaks in the ceiling, whereas the rearmost cave chambers are completely dark. The cave is drained by a creek fed by a number of springs throughout the cave, most of which contain high levels of dissolved H_2S (Tobler *et al.* 2006b, Tobler *et al.* 2008b).

- The recently discovered Cueva Luna Azufre is – despite its name – a **non-sulfidic cave** (Figure 3). The creek in the Cueva Luna Azufre is also fed by springs, but none of these contain H_2S (Tobler *et al.* 2008a). The Cueva Luna Azufre is substantially smaller than the Cueva del Azufre. Although the two caves are in close proximity, they are located within different hills that are separated by a surface valley.

- The El Azufre is a **sulfidic surface habitat**. The creek originates in the hills southwest of the two caves and is fed by multiple sulfidic as well as non-sulfidic springs. The El Azufre flows through the valley that separates the two caves. Both caves drain into the El Azufre, which eventually joins the Río Oxolotan (Figure 1).

- Fish from several **non-sulfidic surface habitats** in the vicinity of the divergent habitats are used for comparative analyses (Figure 1). In earlier studies, primarily fish from the Río Teapao (about 25 km south of Tapijulapa) were used for comparison with cave mollies (Parzefall, 2001). Our recent research has concentrated on more proximate surface populations from the nearest large rivers, namely the Río Oxolotan and the Río Amatan, as well as some of their tributaries that are similar in size and structure to the El Azufre (Tobler *et al.* 2008b). *Poecilia mexicana* from non-sulfidic surface habitats represent the ancestral form that resembles the fish that must originally have colonized the sulfidic and cave habitats.

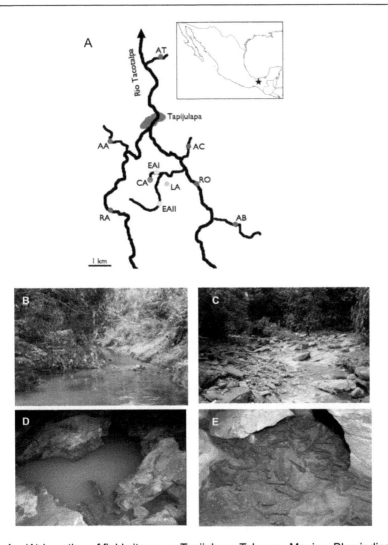

Fig. 1 (A) Location of field sites near Tapijulapa, Tabasco, Mexico. Blue indicates non-sulfidic surface sites (AT Arroyo Tres, AA Arroyo Tacubaya, RA Río Amatan, AC Arroyo Cristal, RO Río Oxolotan, AB Arroyo Bonita), yellow indicates sulfidic surface sites (EA I, II El Azufre, sites I and II, red: the sulfidic Cueva del Azufre, brown: the non-sulfidic Cueva Luna Azufre. The inserted figure shows the location of the Cueva del Azufre system in Mexico. (B-E) Habitat types inhabited by *P. mexicana* in the Cueva del Azufre system. (B) non-sulfidic surface habitat (Arroyo Bonita), (C) sulfur sources in the springhead area of El Azufre, (D) Cueva del Azufre. (E) *P. mexicana* occur at high abundance inside the Cueva del Azufre.

Color image of this figure appears in the color plate section at the end of the book.

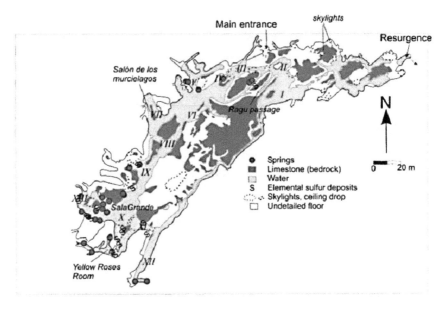

Fig. 2   Simplified map of the Cueva del Azufre. Simplified by Laura Rosales with permission of Bob Richards from Hose and Pisarowicz (1999); chambers (I-XIII) approximate location from Gordon and Rosen (1962).

Like in other caves around the world, the Cueva del Azufre and the Cueva Luna Azufre are characterized by the absence of light and consequently by the reduction of fluctuations in temperature. H_2S, however, although also present in other cave systems (Sarbu *et al.* 1996, Macalady *et al.* 2006), is less known as environmental factor influencing organisms, and some biologically relevant aspects of this chemical are thus introduced here briefly.

H_2S as an Environmental Factor

Already in the first description of the Cueva del Azufre, Gordon and Rosen (1962) reported the presence of toxic hydrogen sulfide (H_2S) in the Cueva del Azufre as a faint odor noticeable throughout the cave, but the effects of H_2S only received marginal attention in early studies on cave mollies, and previous studies have mainly focused on darkness as the driving evolutionary force in this system.

Because of its lipid solubility, H_2S freely penetrates biological membranes and readily invades organisms (Reiffenstein *et al.* 1992). Like cyanide, sulfide is an inhibitor of the cytochrome c oxidase blocking the electron transport in aerobic respiration, thereby hampering the function

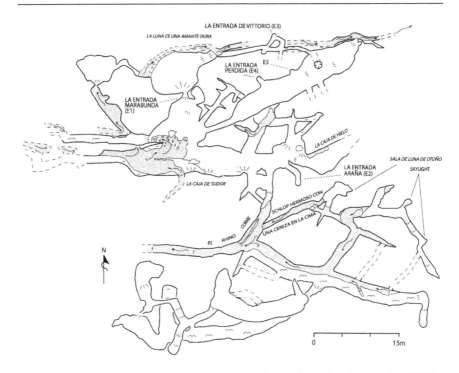

Fig. 3 Simplified map of the Cueva Luna Azufre Azufre (after Pisarowicz 2005).

of mitochondria and the production of ATP (Lovatt Evans 1967, Nicholls 1975, Petersen 1977, National Research Council 1979). H_2S is also able to modify oxygen transport proteins, such as globins in vertebrates, with which it forms non-functional sulfhemoglobin and sulfmyoglobin (Carrico *et al.* 1978, Park *et al.* 1986). It further can inhibit about 20 other enzymes and – at least in mammals – has neurotoxic effects that arrest the central respiratory drive (Bagarinao 1992, Reiffenstein *et al.* 1992). Due to its biochemical effects, H_2S is highly toxic for aerobic organisms even in micromolar amounts (Torrans and Clemens 1982, Bagarinao 1992, Grieshaber and Völkel 1998). It has been shown to play a role in natural fish kills (Bagarinao and Lantin-Olaguer 1999, Luther *et al.* 2004) and to aggravate mortality due to hypoxia and low pH (Bagarinao and Lantin-Olaguer 1999). Lower concentrations of H_2S may have a chronic toxic effect (National Research Council 1979).

H_2S is highly reactive at room temperature and is spontaneously oxidized to polysulfides, elemental sulfur, thiosulfate, sulfite, and sulfate in water (Cline and Richards 1969, Chen and Morris 1972). This reaction is biologically relevant because the presence of H_2S leads to and aggravates hypoxia in aquatic systems, resulting in a distinct inverse correlation in the concentrations of H_2S and oxygen (Bagarinao 1992).

In a systematic survey of abiotic environmental conditions inside the Cueva del Azufre and the El Azufre, high concentrations of H_2S were found, reaching more than 300 μM in spring areas (Tobler *et al.* 2006b, Tobler *et al.* 2008b). Further away from the spring areas, the concentrations are lower (10-80 μM/l) because of the oxidation of sulfide and the emission from the water body to the atmosphere; these concentrations, however, still have to be considered as highly toxic. H_2S concentrations in the Cueva del Azufre system are high compared to other systems with naturally containing H_2S, such as deep sea hydrothermal vents (Van Dover 2000, Nybakken 2001). Sulfide concentrations in the Cueva del Azufre system seem to vary to some extent over time and are likely dependent on the discharge of the springs relative to the precipitation in the area (Tobler *et al.* 2006b). Some sulfidic surface springs, for example, only carry water during the rainy season (authors, personal observation).

Although the water in sulfidic habitats of the Cueva del Azufre system still contains low concentrations of oxygen, the conditions are considered hypoxic. This is different from systems like swamps and tidal pools, in which H_2S originates biogenically from sulfate reduction of bacteria (Jorgensen and Fenchel 1974, Jorgensen 1982, 1984). Here, anoxic conditions are a prerequisite for the production of H_2S.

DIVERGENCE OF *P. MEXICANA* ALONG ABIOTIC GRADIENTS

Genetic Differentiation Among Habitats

All habitat types in the Cueva del Azufre system are physically connected. In theory, *P. mexicana* should be able to move freely within and between habitat types. Molecular genetic analyses, using microsatellites as well as cytochrome b sequences, gave insights not only into the phylogenetic relationship among mollies from the different habitat types, but also on migration patterns and genetic differentiation due to differences in abiotic conditions.

Cytochrome b sequence data in *P. mexicana* from the Cueva del Azufre system suggests that fish from the different habitat types are closely related to each other (Tobler *et al.* 2008b). There is extensive haplotype-sharing among habitat types and private alleles have only diverged by a few point mutations among habitat types. The current data suggest that the divergent habitats (including the sulfidic surface habitats and the two caves) were originally colonized by fish from adjacent non-sulfidic surface habitats. Both caves were likely colonized independently from the El Azufre.

Likewise, population genetic analyses based on nuclear markers (10 unlinked microsatellite loci) indicated that *P. mexicana* in different habitat types are genetically distinct from one another and that gene-flow between habitats is low despite their spatial proximity (Plath *et al.* 2007a, Tobler *et al.* 2008b). Thus, neither are *P. mexicana* populations in the divergent habitats sink populations that persist only through continuous immigration of individuals from normal epigean habitats, nor are *P. mexicana* in the Cueva del Azufre system panmictic. The current data suggest that each habitat type harbors a genetically distinct population (Figure 4). Thus, *P. mexicana* did not only diverge into cave and surface populations, but the presence of H_2S also seems to restrict gene flow. Genetically detected migration events were more common between sites of the same habitat type than between sites differing in at least one of the two abiotic environmental parameters (Figure 5).

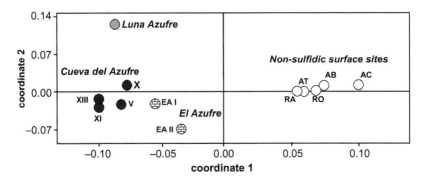

Fig. 4 Genetic differentiation in the Cueva del Azufre system based on microsatellite data (10 unlinked loci). The principal component analysis is based on F_{st} values among populations (GenAlEx). Note the gap between the "sulfur plateau" cluster (left) and all surrounding populations from non-sulfidic surface sites (right). Data from Tobler *et al.* (2008b).

Cavernicolous *P. mexicana* consequently represent one of the few known systems in which cave-dwellers coexist parapatrically with their recent ancestors, to which they are still connected by low rates of gene-flow (see also Schilthuizen *et al.* 2005). The system thus provides a unique and rare opportunity not only to study ecological and evolutionary processes associated with cave colonization, but it is an emergent model system to investigate mechanisms leading to local adaptation and reproductive

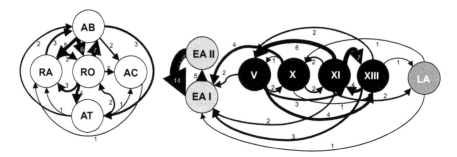

Fig. 5 Population assignment using microsatellite data (10 unlinked loci). Given are numbers of 1st generation migrants and the direction of migration events, as calculated from GENECLASS. For abbreviations (population code) see Figure 1. Note that genetically detected "migrants" do not necessarily represent actual migration events.

isolation due to divergent natural selection, i.e., ecological speciation (Schluter 2000, Rundle and Nosil 2005). Both of these aspects will be discussed in more detail below.

Phenotypic Differentiation and Adaptation Along Abiotic Gradients

Poecilia mexicana from different habitat types did not only diverge genetically, but also phenotypically, for example in terms of their morphology and their behavior. We first discuss divergent traits that are likely driven by and adaptive to differences in the abiotic environmental conditions among habitat types. Below we will also look at traits that likely evolved in response to different ecological conditions among habitats.

Each habitat type in the Cueva del Azufre system harbors *P. mexicana* with distinct morphological features (Tobler *et al.* 2008a, Tobler *et al.* 2008b). Most importantly, there is independent variation explained along the axis surface-cave as well as along the non-sulfide-sulfide gradient (Figure 6). The observed morphological differences among *P. mexicana* from different habitat types are not caused by environmentally induced phenotypic variation alone, since laboratory stocks maintained under identical conditions retained the distinct morphology of wild-caught fish from the respective habitat types for multiple generations, indicating that morphological traits have a heritable basis (Peters and Peters 1968, Tobler *et al.* 2008b). Likewise, divergent behavioral traits of *P. mexicana* from different habitat types have been shown to be heritable (Plath *et al.* 2004c, Plath *et al.* 2006c, Plath 2008, Plath and Schlupp 2008).

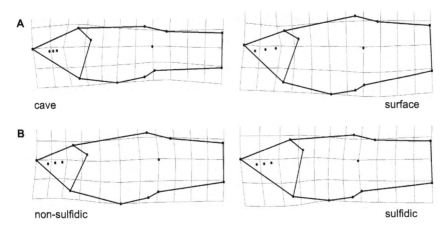

Fig. 6 Morphological divergence due to (A) darkness, and (B) hydrogen sulfide. Cartesian transformation grids depicting the effects of both selection pressures (deformations are exaggerated three times). Note the smaller eye size and more slender body in cave dwelling *P. mexicana*, and the larger head (and larger gills, respectively) in mollies from sulfidic habitats.

Differentiation in Caves

Like cave-dwelling species in other taxonomic groups, *P. mexicana* have diverged in multiple traits that are commonly associated with cave evolution, both in terms of regression of traits that are no longer under selection in darkness as well as in the evolution of novel traits. Compared to fish from surface habitats, *P. mexicana* from the two caves are characterized by a reduction in eye size (Gordon and Rosen 1962, Walters and Walters 1965, Peters *et al.* 1973, Plath *et al.* 2007a, Tobler *et al.* 2008b). Relatively large eyes were found in surface habitats, smaller eyes in the anterior cave chambers (V, X), and the smallest eyes were detected in the innermost cave chamber XIII (Plath *et al.* 2007a). This pattern shows some congruence with a previously proposed morphocline in eye size (Parzefall 1970, Peters *et al.* 1973, Parzefall 2001). However, the morphological and genetic data did not support the proposed mechanism for this morphocline, namely its maintenance by migration from both directions into the middle cave chambers. This would have led to an increased variance in eye size in the middle cave chambers, which was not found (Plath *et al.* 2007a).

Unlike the eyes of other cave dwellers (Porter and Crandall 2003), those of cavernicolous *P. mexicana* are still functional, and cave mollies readily show behavioral responses similar to conspecifics from surface populations to visual stimuli (Parzefall 2001, Körner *et al.* 2006, Parzefall *et al.* 2007). A study of visual pigments in cave mollies as well as various

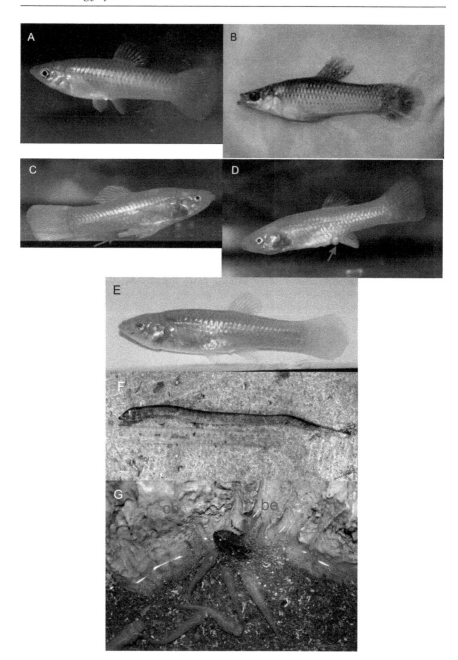

Fig. 7 Morphological divergence in *P. mexicana*. (A) Female and (B) male from a non-sulfidic surface site (Arroyo Cristal). Note the bright coloration of

Fig. 7 Contd.

related *Poecilia* species/populations using microspectrophotometry indicated that all species/populations exhibit four cone types including UV-sensitive cones as described for other poeciliid species (Körner *et al.* 2006). Differences in the absorption characteristics between species/populations were small and, most importantly, the maximum absorbance wavelength (λ_{max}) values of the visual pigments of the cave form of *P. mexicana* are almost unchanged, although stabilizing selection on visual pigments is not evident in this population.

Besides the reduction of eye size, cave populations of *P. mexicana* also have reduced body pigmentation (Figures 7A-E, Gordon and Rosen 1962, Peters *et al.* 1973, Tobler *et al.* 2008a) and have more slender bodies than conspecifics from epigean habitats (Figure 6, Tobler *et al.* 2008b) a trait commonly observed in cave organisms (Langecker 2000). The reduction in body height is not associated to poor nutritional condition of cave populations (Tobler 2008), but could be driven by divergent predatory regimes or resource scarcity (see below).

Poecilia mexicana from the Cueva del Azufre have also evolved a series of constructive traits most of which seem to improve communication and orientation in darkness. So far, only individuals from the Cueva del Azufre but not the Cueva Luna Azufre have been investigated, so that it is unclear whether the newly discovered population exhibits similar traits. Female cave mollies exhibit a distinct genital pad that is absent in epigean fish (Figure 7D). Supposedly, this pad secrets chemical signals that play a role during reproduction (Walters and Walters 1965, Zeiske 1968, Parzefall 1970, 1973) and are detected by males by the increased number of taste buds on the head (Parzefall 1970). Also, cave mollies exhibit a hyper-developed cephalic lateral line system (Walters and Walters 1965; Parzefall 2001). The mechano-sensory lateral line system has been documented to mediate spatial orientation in other cave fishes (Abdel-Latif *et al.* 1990, Burt de Perera 2004). Last but not least, a part of the otolith morphology – the suculus – diverged in cavernicolous *P. mexicana* (Schulz-Mirbach

Fig. 7 Contd..

large,dominant males in (B). (C) Male from chamber XI of the Cueva del Azufre, "golden morph". Arrow shows the gonopodium, the male copulatory organ. (D) female from chamber XI with a pronounced genital pad (arrow) around the genital pore. (E) Cave molly female from chamber V, "sunken eye type". (F) The role of *Ophisternon aenigmaticum* as a molly predator in the Cueva del Azufre remains to be examined. This dead individual was found in El Azufre and had probably been flushed out of the cave. The ruler is 30 cm long. (G) Inside the Cueva del Azufre, cave mollies face predation by giant water bugs (*Belostoma* sp., be). ch: chironomid larvae, one of the food sources of cave mollies.

Color image of this figure appears in the color plate section at the end of the book.

et al. 2008). The sulcus is a furrow-like structure on the inner face of an otolith and is attached to the sensory epithelium. Otoliths from surface- and cave-dwelling populations show pronounced differences in the depth of the sulcus. Fish from surface habitats have otoliths with a flat sulcus (Figure $8A_{1-2}$), whereas cave mollies possess otoliths characterized by a deeper sulcus (Figure $8B_{1-2}$). In addition, otolith weight in fish from cave chamber X is significantly higher than in fish from El Azufre. Greater otolith weight in cave mollies (relative to the body size of the fish) may be an effect of the increased depth of the sulcus. This difference in the depth of the sulcus may reflect a greater hair cell area for sensory cells. As otoliths are static position receptors, these differences may allow for a better position perception in absence of visual cues (see Poulson 1963 for divergent otolith morphology in troglobitic amblyopsids).

The modification of sensory structures in the cave molly is also reflected in its behavior. Cave mollies from the Cueva del Azufre have evolved the ability for communication in darkness, a trait that is lacking in surface populations. While surface-dwelling *P. mexicana* predominantly rely on visual signals, e.g., during mate choice, cave mollies are able to assess mate quality in darkness (Plath *et al.* 2004c, 2006c). Most importantly, the ability for non-visual communication is heritable and largely independent of ontogenetic effects, i.e. whether fish were reared in light or permanent darkness. This indicates a sensory shift in the detection of traits relevant for mate choice as an evolutionary response to life in darkness. Using visual cues for mate choice is the ancestral state in *P. mexicana*, and the

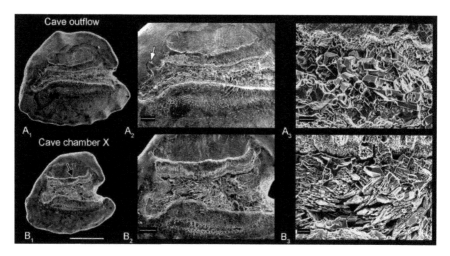

Fig. 8 Otoliths of *P. mexicana* from the sulfidic El Azufre at the surface (A), and from cave chamber X of the Cueva del Azufre (B) (from Schulz-Mirbach *et al.* 2008).

use of non-visual cues has evolved as a novel trait in the cave population. Our data indicate that the cave molly population from the independently colonized Cueva Luna Azufre has not evolved the ability for comparable communication in darkness and, in that respect, are more similar to surface fish than the population from the Cueva del Azufre (Tobler *et al.* 2008c).

In summary, cave mollies are characterized by a number of regressive and constructive traits when compared to surface populations. Although currently far less understood, mollies from sulfidic habitats also diverged from mollies in non-sulfidic habitats, irrespective of whether they live in a cave or a surface habitat.

Adaptations to Sulfidic Habitats

Individuals from sulfidic habitats are characterized by an increase in head size (Figure 6), which is correlated with an increase in gill filament length (Tobler *et al.* 2008b). An increase in gill size facilitates oxygen uptake in hypoxic environments (Graham 2005) and has been documented in various fish species living in other types of hypoxic environments (Chapman *et al.* 1999, Chapman *et al.* 2000, Chapman and Hulen 2001, Timmerman and Chapman 2004).

This highlights the importance of respiratory adaptations facilitating efficient oxygen acquisition for survival in sulfidic habitats. Sulfide detoxification in organisms capable of tolerating high and sustained concentrations of H_2S is primarily achieved through its oxidation to less toxic sulfur compounds and subsequent excretion through the kidneys (Curtis *et al.* 1972, Bagarinao 1992, Ip *et al.* 2004). Due to the hypoxic conditions in sulfidic habitats, oxygen available for respiration is generally limited, but at the same time oxygen is required for coping with the toxic effects of H_2S. Organisms inhabiting sulfidic habitats should therefore be under selection for maximizing oxygen uptake. While some fish species rely on air-breathing using specialized breathing organs or derived respiratory epithels in the stomach or guts to do so (Bagarinao and Vetter 1989, Brauner *et al.* 1995, Affonso and Rantin 2005), *P. mexicana* from the Cueva del Azufre rely on aquatic surface respiration (ASR), where fish exploit the more oxygen-rich air-water interface using their gills (Plath *et al.* 2007c).

Indeed access to the water surface and the possibility to perform ASR critically influence survival in sulfidic water even over short periods of time. When exposed to sulfidic water, cave mollies invest more than 75% of their time budget performing ASR. An experiment in the natural habitat revealed that access to the water surface is the major predictor of the survival even over short periods of time (24 h) (Plath *et al.* 2007c).

Physiological adaptations of *P. mexicana* to detoxify H_2S remain to be studied in detail. Invertebrates living in sulfidic environments often harbor symbiotic sulfide-oxidizing bacteria (which detoxify H_2S) on their respiratory organs or as endosymbionts, or show unusually structured (tube-like) mitochondria in response to exposure to H_2S (Bagarinao 1992, Grieshaber and Völkel 1998, Van Dover 2000). A pilot study using electron-microscopic techniques, however, did not reveal any symbiotic bacteria on or in the gills of cave mollies, and mitochondria are normally shaped (Figure 9; Giere *et al.*, unpublished data).

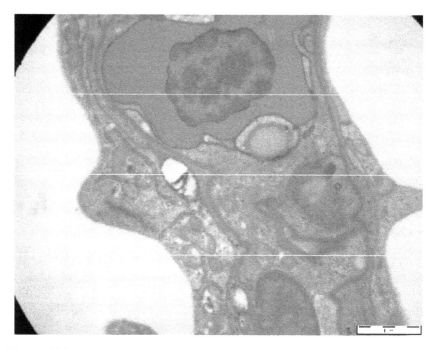

Fig. 9 EM detail of a gill filament of *P. mexicana* from the sulfidic Cueva del Azufre showing regular mitochondria and nuclei. The bar equals 1 micrometer (Giere *et al.* unpublished data).

There is evidence though that *P. mexicana* from sulfidic habitats have physiological adaptations absent in conspecifics from non-sulfidic habitats. While fish from sulfidic habitats can sustain in sulfidic water, *P. mexicana* from non-sulfidic habitats lost motion control within seconds after exposure to H_2S (Tobler *et al.* 2008d). Similar results were obtained when re-examining data from Peters *et al.* (1973) on sulfide tolerances of laboratory-reared animals (Figure 10), suggesting that differences in sulfide tolerance in *P. mexicana* is at least in part genetic.

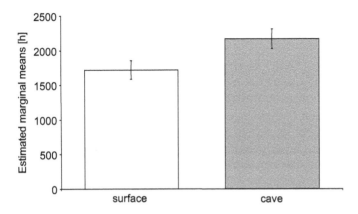

Fig. 10 Molly populations differ in their tolerance to H_2S. We reanalyzed data published by Peters *et al.* (1973). Sulfide tolerance was measured in fish from non-sulfidic (surface) habitats and the sulfidic Cueva del Azufre. All fish used were maintained in the laboratory in non-sulfidic water prior to the study. We reanalyzed the data using a General Linear Model (GLM) with time sustained in sulfidic water as a dependent variable, population as an independent variable and H_2S concentration as a covariate. H_2S concentration did not have an effect ($F = 0.563$, $P = 0.463$) but there were significant differences among populations ($F = 5.168$, $P = 0.035$). In the figure, the estimated marginal means ($\pm$ SD) from the GLM are shown. Overall, *P. mexicana* from non-sulfidic surface populations (left) had a lower sulfide tolerance than conspecifics from the Cueva del Azufre (right).

ECOLOGICAL DIFFERENCES AMONG HABITATS

The different habitat types in the Cueva del Azufre system differ in more than abiotic environmental conditions. The energy basis among habitats likely differs qualitatively and quantitatively. Due to the lack of light, photosynthetic primary production is absent in cave habitats (Poulson and Lavoie 2000) and likely impaired in sulfidic surface habitats, since H_2S is also toxic for most phototrophic primary producers (Bagarinao 1992). On the other hand, the presence of sulfide allows for chemoautotrophic primary production through sulfide oxidizing bacteria (Sarbu *et al.* 1996, Hose *et al.* 2000, Colaco *et al.* 2002). Allochthonous input probably occurs in all habitat types, but is dominated by leaf litter and terrestrial insects in surface habitats (Allan 1995) and bat guano in the case of the two caves (cf. Poulson and Lavoie 2000).

The food web of the ecosystem of the Cueva del Azufre appears to rely to some extent on *in situ* chemoautotrophic bacterial primary production (Langecker *et al.* 1996, Hose *et al.* 2000), whereas most other caves studied so far mainly rely on energy input from the neighboring surface habitats. Ecosystems relying on chemoautotrophic primary production are known from deep-sea hydrothermal vents and cold seeps (Peek *et al.* 1998, Van Dover 2000). A sulfurous chemoautotrophic cave-ecosystem, the Movile cave, has been described from Romania (Sarbu *et al.* 1996). There, only invertebrates – many of them endemic to the cave – use this unusual food base. The Cueva del Azufre is the only known ecosystem in which a vertebrate, *P. mexicana*, inhabits a chemoautotrophic cave.

The different habitat types also differ dramatically in the composition of ecological communities. The current knowledge on the biodiversity in the different aquatic habitats is still fragmentary and data are mostly restricted to fish. In non-sulfidic surface habitats, diverse fish communities can be found, which – as is typically for the wider region – are dominated by cichlid and poeciliid species (Table 1, Tobler *et al.* 2006b). By contrast, the diversity of fishes is heavily reduced in the divergent habitats, and *P. mexicana* occurs as the single dominant species. *Heterandria bimaculata* and *Xiphophorus hellerii* (Poeciliidae), *Astyanax aeneus* (Characidae) as well as *"Cichlasoma" salvini* and *Thorichtys helleri* (Cichlidae) occur in areas of the El Azufre where H_2S in not measurable (interestingly, *P. mexicana* is basically absent in the El Azufre as soon as other fish species are present). Furthermore, a facultatively air-breathing synbranchid eel (*Ophisternon aenigmaticum*) occurs in sulfidic surface habitats and has occasionally been reported in the Cueva del Azufre (Parzefall, personal communication; Figure 7F). In the Cueva Luna Azufre, *P. mexicana* has so far been the only recorded fish.

Other taxonomic groups, such as snails and other aquatic invertebrates, were only surveyed in the Cueva del Azufre, but a comparative analysis of different habitat types remains to be carried out. However, it is likely that the invertebrate diversity follows a similar pattern with reduced species diversity in the sulfidic and cave habitats. A reduced species diversity and dominance of a few specialists have been documented from other caves (Gibert and Deharveng 2002) and other sulfidic habitats (Tobler *et al.* 2008d and unpublished data).

Interspecific Interactions

Predation

Poecilia mexicana in non-sulfidic surface habitats are primarily exposed to piscivorous fishes and birds as predators. Piscivorous birds are also present in sulfidic surface habitats, but neither bird or fish predators occur in the two caves (Tobler *et al.* 2006b, Tobler *et al.* 2007a). However, in the Cueva del Azufre (but not the Cueva Luna Azufre), an aquatic heteropteran (*Belostoma* sp.) occurs at high densities (over 1 individual/

Table 1 Fish communities in the Cueva del Azufre system. Table after Tobler *et al.* (2006b) updated with additional data from 2007 and 2008.

	Cueva del Azufre	Cueva Luna Azufre	El Azufre	Non-sulfidic surface habitats
Characidae				
Astyanax aeneus (Günther 1860)			x[1]	x
Brycon guatemalensis Regan 1908				x
Ariidae				
Potamarius nelsoni (Evermann & Goldsborough 1902)				x
Pimelodidae				
Rhamdia guatemalensis (Günther 1864)				x
Rhamdia laticauda (Kner 1858)				x
Loricariidae				
Pterygoplichthys cf. *disjunctivus* (Weber, 1991)				x[2]
Batrachoididae				
Batrachoides goldmani Evermann & Goldsborough 1902				x
Atherinopsidae				
Atherinella alvarezi (Díaz Pardo 1972)				x
Belonidae				
Strongylura hubbsi Collette 1974				x
Poeciliidae				
Heterophallus milleri Radda 1987				x
Heterandria bimaculata (Heckel 1848)			x[1]	x
Poecilia mexicana Steindachner 1863	x	x	x	x
Priapella chamulae Schartl, Meyer & Wilde 2006				x
Xiphophorus hellerii Heckel 1848			x[1]	x
Synbranchidae				
Ophisternon aenigmaticum Rosen & Greenwood 1976	x[3]		x	
Centropomidae				
Centropomus undecimalis (Bloch 1792)				x
Cichlidae				
'Cichlasoma' salvini (Günther 1862)			x[1]	x
Oreochromis cf. *aureus* (Steindachner 1864)				x[2]
Paraneetroplus gibbiceps (Steindachner 1864)				x
Theraps lentiginosus (Steindachner 1864)				x
Thorichthys helleri (Steindachner 1864)			x[1]	x
Vieja bifasciata (Steindachner 1864)				x
Vieja intermedia (Günther 1862)				x
Eleotridae				
Gobiomorus dormitor Lacépède 1800				x

[1]Only in non-sulfidic microhabitats
[2]Introduced species
[3]Parzefall, personal communication

m² in cave chamber V) and seems to be the predominant predator of *P. mexicana* (Tobler *et al.* 2007b). *Belostoma* are large aquatic hemipterans (Figure 7G). As generalist feeders, they usually prey on aquatic insects, snails, amphibians, and fish (Menke 1979). Belostomatids are sit-and-wait predators that catch bypassing prey items with their raptorial forelegs that are strongly incrassate, with the femora often grooved to accept the tibiae. Upon capture, *Belostoma* inject toxins causing prey paralysis and digestive enzymes causing tissue necrosis (Swart and Felgenhauer 2003).

Several experiments assessing the prey choice of *Belostoma* gave insights into the potential role of predation on the evolution of cave mollies. *Belostoma* exhibit a prey preference for larger cave mollies over smaller fish (Plath *et al.* 2003b, Tobler *et al.* 2007b) as well as for males over females of the same size (Tobler *et al.* 2008e). These preferences parallel the prey preference of other predators (such as birds and fish) in surface habitats that also preferentially attack large and male prey items (Endler 1987, Trexler *et al.* 1994). In other poeciliids, differences in susceptibility to predation among sexes as well as among size classes have been documented to have profound impacts on various traits, such as life history strategies, sex ratios and size distributions. In cave mollies, the higher susceptibility to predation of large individuals, for example, has been hypothesized to maintain male size polymorphisms in the Cueva del Azufre population. Here, large males are favored by female choice (see below) but more likely to be preyed upon by *Belostoma*, which gives an advantage in survival to small males (Figure 11). It remains to be studied how strong the contribution of size- and sex-specific predation by *Belostoma* is. Given that sex ratios are more equal in cave mollies than in surface habitats (Table 2), the impact of predation by *Belostoma* on sex ratios is probably much weaker than predation by herons and predatory fishes on the same species in surface habitats.

Parasites

Differences in the ecological communities as well as the abiotic environmental conditions among habitats have been hypothesized to cause differences in parasite infection in *P. mexicana*. Parasites may become locally extinct either directly, through selection by adverse environmental conditions (such as H_2S) on free-living parasite stages, or indirectly, through selection on other host species involved in a parasite's life cycle (Figure 12, Tobler *et al.* 2007a). This hypothesis was tested by comparing the infection rates of the digenean trematode *Uvulifer* sp. among several populations of *P. mexicana* (Tobler *et al.* 2007a). The cercariae of *Uvulifer* sp. infect fish by penetrating their skin, causing the production of black cysts ("black spots") in which the host encapsulates the parasite (Spellman and Johnson 1987,

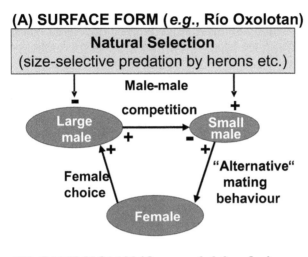

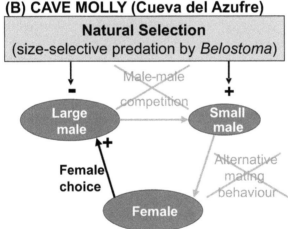

Fig. 11 Direction of sexual and natural selection on male body size (i.e., size at maturation) in mollies from (A) non-sulfidic surface sites and (B) the Cueva del Azufre. (+) indicates an advantage for this male class due to the respective selection factor, while (–) indicates an evolutionary disadvantage. After Plath *et al.* (2004a).

Bush *et al.* 2001). Consistent with the predictions, we found populations from sulfidic habitats to be significantly less parasitized by *Uvulifer* sp. than populations from non-sulfidic waters, and cave mollies showed no signs of this parasite at all. A subsequent, yet unpublished study, however, indicated that this pattern in not consistent across different parasite species. For example, *P. mexicana* from the Cueva del Azufre show a high

Table 2 Sex ratios in several populations of *P. mexicana* in Mexico. The rationale behind female-biased sex ratios in the recently discovered non-sulfidic Cueva Luna Azufre despite the apparent absence of predators in that cave remains to be studied.

Habitat	Sampling site	Year	Sex ratio (males/females)	n (adult)
Surface,	Río Purificacion, stream	1998	0.07	49
non-sulfidic	Río Purificacion, pool	1998	0.08	68
	Río Purificacion, stream	2002	0.50	81
	Río Purificacion, stream (Barretal)	2002	0.08	13
	Río Mante tributary	2002	0.63	31
	Río Mante, Ciudad Mante	2002	0.46	19
	Buena Vista (Río Tuxpan)	2002	0.38	95
	Arroyo Cristal	2004	0.28	100
	Arroyo Cristal	2006	0.36	19
	Arroyo Bonita	2006	0.27	19
	Río Amatan	2007	0.38	66
Sulfur	El Azufre	1998	2.00	313
creek	El Azufre	2002	1.98	173
	El Azufre, Clear Creek[1]	2002	0.94	35
	El Azufre	2004	0.99	100
Cave	Cueva del Azufre, chambers			
	III-V	2004	1.33	191
	X	1998	1.09	148
	XI	2004	0.56	74
	XIII	1998	0.49	138
	XIII	2002	1.65	45
	XIII	2004	0.56	145
	Cueva Luna Azufre	2007	0.19	31

[1] A small freshwater tributary of the sulfidic El Azufre.

prevalence of *Dactylogyrus* sp. (a monogenean gill parasite) infections, a parasite that is far less frequent in other populations. Overall, it cannot be concluded that *P. mexicana* from divergent habitats are generally less parasitized, but the parasite communities clearly differ among habitat types both in species composition as well as the frequency of occurrence of different parasite species (Tobler *et al.*, unpublished data).

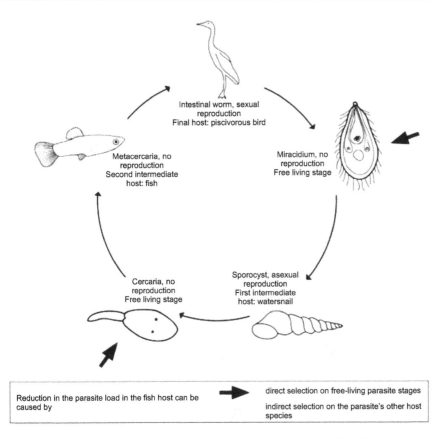

Fig. 12 Life cycle of *Uvulifer* sp., a typical trematode parasite of *Poecilia mexicana*. Arrows indicate pathways by which extreme environmental conditions (H_2S and/or darkness) can affect the prevalence of the parasite directly or indirectly, i.e., due to selection against further hosts in its life cycle. From Tobler *et al.* (2007a).

Trophic Ecology

Resource Use and Trophic Morphology

Poecilia mexicana in non-sulfidic surface habitats primarily fed on detritus and algae, which is consistent with previous studies investigating the food habits of closely related species and *P. mexicana* in other parts of its range (Darnell 1962, Winemiller 1993, Kramer and Bryant 1995, Bussing 1998, Miller 2005). The most pronounced difference in diet was observed between non-sulfidic surface habitats and the other habitat types, where invertebrates were consumed by *P. mexicana* (Tobler 2008). Especially in the cave habitats, invertebrates made up the majority of the gut content.

A substantial amount of this invertebrate diet (about 50%), however, stemmed from bat guano, not living invertebrates. This finding is also consistent with a previous study that investigated the gut content of *P. mexicana* from the sulfidic cave only (Langecker *et al.* 1996). Consequently, the colonisation of extreme habitats in *P. mexicana* was accompanied by a shift in resource use.

Likely, both a change in resource availability and differences in the competitive regime (i.e., competitive release) were driving the shift in resource use of *P. mexicana*. Indeed, specimens collected in different habitat types did not only differ in what they fed on, but also the diversity of food items present in their intestines; fish from cave habitats exhibited a significantly higher diversity (i.e., a higher individual trophic niche width). It is unlikely that this increase in the trophic niche width was caused by a higher diversity in resources available, since there is ample evidence that caves (and sulfidic habitats for that matter) exhibit a reduced species diversity compared to surface habitats (Poulson and White 1969, Gibert and Deharveng 2002). Niche expansion as a response to competitive release (MacArthur *et al.* 1972, Schluter and McPhail 1992, Begon *et al.* 1996) or as a response to resource scarcity (MacArthur and Pianka 1966, Schoener 1971, Fenolio *et al.* 2006) may drive the increase in the diversity of resources used. Both scenarios, however, are not necessarily consistent with *P. mexicana* from sulfidic surface habitats having a low individual dietary niche width. At least in terms of fishes, the communities in the sulfidic surface habitats are comparatively poor compared to those in the cave habitats (Tobler *et al.* 2006b, Tobler *et al.* 2008, Riesch *et al.* 2009a).

The major difference in trophic morphology among populations living in different habitat types was found between *P. mexicana* from non-sulfidic surface habitats and those from extreme habitats, which parallels the major dietary shift with incorporation of invertebrates into the diet. Intestinal tract lengths in fishes are typically correlated with the amount of plant material ingested (Kramer and Bryant 1995), and *P. mexicana* from divergent habitats (which consume less plant material) had shorter intestinal tracts (Tobler 2008, Tobler *et al.* 2009). They were also characterized by wider and thicker jaws, which may be advantageous in handling larger (Wainwright 1996) and/ or more evasive prey items (Hulsey and Garcia de Leon 2005, Higham *et al.* 2007).

Differentiation in skull morphology among populations inhabiting different habitat types is less pronounced than the differentiation in general body shape, which seems to be driven predominantly by abiotic environmental factors, i.e., the lack of light (eye size reduction) as well as the presence of H_2S and hypoxia (increase in head and gill size). Thus, differences in jaw morphology (Tobler 2008) are not simply explained by a correlated response to selection on other characteristics of the head (Chapman *et al.* 2000), but may actually be adaptive to differential use of resources among populations. The finding that skull morphology is at least partly determined by genetics supports the idea that evolutionary divergence in skull morphology by natural selection is possible.

Energy Limitation

Early studies examining body condition in this system (by comparing length-weight regressions) found *P. mexicana* from normal surface habitats to exhibit the highest body condition, while cave populations had the lowest, and specimens from sulfidic surface habitats were intermediate (Tobler *et al.* 2006b, Plath *et al.* 2007c, Tobler *et al.* 2008a). Morphological differences among populations, however, affected these results since cave population are generally more slender bodied. In fact, *P. mexicana* from sulfidic surface habitats have equally low amounts of storage lipids as fish from the cave populations (Tobler 2008).

Fish from cave and sulfidic habitats may have a low condition for different reasons. *Poecilia mexicana* from the non-sulfidic cave exhibited low amounts of storage fats likely because resources are scarce. Caves relying on energy input from surface habitats are known to be energy limited (Streever 1996, Hüppop 2000, Poulson and Lavoie 2000). Bat guano is thought to be the trophic base of cave food webs whenever bats are present and provide an energy-rich food base (Culver 1982, Willis and Brown 1985), but recent work indicates that this is not necessarily the case (Graening and Brown 2003).

Sulfidic habitats in turn have been suggested to be resource-rich due to the presence of chemoautotrophic bacterial primary production (Langecker *et al.* 1996). The paradox of fish with low body condition living in an apparently resource-rich environment may be explained in two (not mutually exclusive) ways. (1) Although resource-rich, sulfidic habitats may lack particular nutrients for fish or provide an imbalanced diet, which may negatively affect condition (Jeyasingh 2007). (2) Coping with the toxic environment may be energetically costly. Although the physiological mechanisms of sulphide-tolerance are not well understood in *P. mexicana*, detoxifying H_2S has been shown to be energetically costly under hypoxic conditions in the mudskipper, *Boleophthalmus boddaerti* (Ip *et al.* 2004). Short-term survival of *P. mexicana* in sulfidic water is directly dependent on energy-availability and possibility to perform ASR (see above) (Plath *et al.* 2007c). ASR itself is physiologically costly and constrains an individual's energy budget, leaving less time for foraging (Kramer 1983, Weber and Kramer 1983, Chapman and Chapman 1993, Tobler *et al.* 2009). *Poecilia mexicana* in sulfidic habitats thus seem to be living in a resource-rich habitat but paying a cost for coping with the toxic conditions. The high resource availability in these habitats may, in fact, be one of the factors making life in such extreme environments possible at all.

Low energy availability and/or high energetic costs of sustaining in a stressful environment likely are strong selective factors acting on *P. mexicana* from divergent habitat types. In fact, a multitude of traits, ranging from shift in life history strategies to the reduction of costly behavioral traits, may have evolved as a response to energy limitation in the divergent habitats.

Life History Modifications

Life history traits, such as female fecundity, can evolve rapidly in response to environmental changes (Reznick *et al.* 1990, Reznick *et al.* 2001). Some recent studies highlight that life history modifications also occurred in *P. mexicana* from the Cueva del Azufre system. In a pilot study, female fecundity (number of ripe oocytes and developing embryos) was assessed in wild-caught *P. mexicana* from the Cueva del Azufre and a non-sulfidic surface creek (Arroyo Cristal) as well as in laboratory-reared individuals from the Cueva del Azufre and the Río Oxolotan, which were reared in either light or darkness (Riesch *et al.* 2009b). Compared to mollies from surface habitats, cave molly females exhibited a dramatically reduced fecundity, suggesting an evolutionary shift in energy allocation by cave molly females (Figure 13). On the other hand, oocytes are larger. Differences between populations were found to be stronger than those described for guppies (*Poecilia reticulata*) from different habitat types in Trinidad, where divergent predatory regimes appear to be the main driver in life history divergence (Reznick and Endler 1982).

Even in the laboratory setting, the fecundity of cave mollies was generally lower compared to that of surface-dwelling mollies. In both populations fecundity was slightly higher under the respective "natural" light conditions (light for the Río Oxolotan population and darkness in the case of cave mollies) (Riesch *et al.* 2009b, in press-a). Our results suggest a genetic component to the reduction in female fecundity in the cave molly, because in the laboratory cave mollies were reared in the absence of toxic H_2S and the overall effect was largely independent of the light treatment. The results from a recent study including Luna Azufre fish (Riesch *et al.* in press-b) corroborate these findings: extremophile populations generally have fewer, but larger eggs.

Differences in egg size among populations are also accompanied by differences in maternal nutrient provisioning after conception

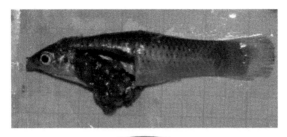

Non-sulfidic surface sites

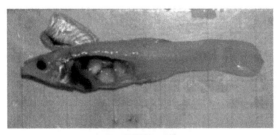

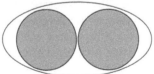

Cueva del Azufre

Fig. 13 Life history trait evolution in *P. mexicana*. Assuming that the volume of the ovary constrains the number of oocytes a female can produce, *P. mexicana* females have two options: they can either produce numerous small oocytes (above, female from Arroyo Cristal) or few, but large oocytes containing more yolk (below, cave molly female from cave chamber X of the Cueva del Azufre (compare Trexler and DeAngelis, 2003).
Color image of this figure appears in the color plate section at the end of the book.

(matrotrophy). The extent of maternal provisioning can be estimated by comparing the weight of neonates to the weight of ova at conception. In purely lecitotrophic species (i.e. species without matrotrophy), embryos loose weight during gestation due to metabolic losses, but weight loss should be minimal or even absent in species with matrotrophy. In *P. mexicana*, the weight of neonates as a percentage of the weight of the egg at fertilization was approximately 58% in fish from non-sulfidic surface sites, 86% in the Cueva Luna Azufre population, and 57% in the fish from the Cueva del Azufre. Our data indicate that this double-provisioning of nutrients (via more yolk and more matrotrophy) leads to larger body size of the neonates in the cave populations compared to fish from non-sulfidic surface sites (Riesch *et al.* in press-a, in press-b).

Female reproductive investment is usually considered much more energy demanding than male investment into spermatogenesis (Bateman 1948, Trivers 1972); however, also sperm production is not cost-neutral (Nakatsuru and Kramer 1982). Thus, male sperm stores in five populations of *P. mexicana* from naturally sulfidic and non-sulfidic habitats were examined (Franssen *et al.* 2008). Pronounced differences among populations in the number of sperm stripped per male were also reflected by differences in gonad weights. The largest sperm stores were detected in males from non-sulfidic surface creeks (Arroyo Cristal and Arroyo Tres), while males from sulfidic surface waters (El Azufre) had smaller sperm stores. Inside the sulfur cave (Cueva del Azufre), males were collected in two cave chambers, and male sperm stores were similarly low. Hence, the extreme environmental conditions in sulfidic habitats appear to constrain male sperm production.

In the laboratory male sperm production of cave mollies (Cueva del Azufre) and surface mollies (Río Oxolotan) was studied in response to nutrient availability (Franssen *et al.* 2008). Males of both populations were maintained both under a high and a low food treatment. Surface dwelling and cave molly males produced comparable amounts of sperm when fed ad libitum, suggesting that unlike female fecundity male sperm production is to a large extent a plastic response of the males to adverse environmental conditions. Interestingly, less sperm was stripped from surface males after one week of food deprivation, while no such decrease was found in the case of cave molly males. This may point towards an increased starvation resistance in cave mollies.

BEHAVIORAL ECOLOGY

Reduction of Costly Behavioral Traits

Anti-predator Behaviors

Shoaling behavior protects fishes from avian and piscine predation, but at the same time costs of group living arise, e.g. due to increased food competition (Krause and Ruxton 2002). Cave mollies (both caves) live in an environment in which avian and piscine predators are lacking, and they are energy limited. Altogether, this should favor the reduction of shoaling. When shoaling behavior (i.e. the tendency to associate with a stimulus shoal) was compared among surface dwelling populations of (*Poecilia mexicana*) and the two cave forms, shoaling was found to be reduced in both cave forms compared with surface dwelling mollies (Plath and Schlupp 2008). An analysis of shoaling in population hybrids revealed that this reduction is heritable (Parzefall 1993).

Also a plastic response of shoaling behavior to starvation was found in surface fish (Río Purificación population); individuals that had been food deprived for one week prior to the tests spent less time shoaling than well-fed ones, probably because food competition is more relevant to starved fish (Plath and Schlupp 2008). Moreover, starved fish are probably physically weaker, and hence, inferior in food competition (Krause *et al.* 1999). From the latter result it can be concluded that sustained malnutrition over evolutionary times may indeed have selected for a genetic reduction of shoaling behavior in cave mollies in an environment with absence of avian and piscine predation.

Shoaling behavior often comes along with 'conspecific cueing' (Kiester 1979, Wolf 1985), where phenotypically similar individuals, such as members of the same species, attempt to form homogenous groups, because individuals in uniform shoals are worse targets to predators. Heterospecific fishes, such as swordtails (*Xiphophorus hellerii*) are common in all non-sulfidic waters around the Cueva del Azufre and the Cueva Luna Azufre (Table 1) and occur in some non-sulfidic parts of the El Azufre, but are absent in the caves. When surface-dwelling *P. mexicana* females (Río Oxolotan) were given an opportunity to associate with a conspecific or a swordtail female, they strongly preferred the conspecific female when visual cues were available to the female (Riesch *et al.* 2006b). No association preference was observed when only non-visual cues were provided. Females from El Azufre showed only a slightly weaker response. In contrast, cave-dwelling females showed no preference under all testing conditions, suggesting that the preference to associate with conspecifics was reduced. These results were corroborated by a subsequent study using conspecific vs. guppy (*P. reticulata*) females as stimulus fish (Tobler *et al.* 2006a).

Aggression

Parzefall (1974) demonstrated that aggressive male interactions occur only in surface populations, but not in the cave molly. In surface-dwelling Atlantic mollies the fish normally form shoals with a size dependent rank-order in males. When females become receptive and ready to be inseminated, large dark-colored males vigorously defend such females. Aggression is most pronounced when a male of the same size tries to approach such a female, but similar-sized males will generally show intense fighting when they first meet, i.e., before a rank-order is established. The aggressive repertoire consists of seven different behavior patterns (aggressive fin erection, tail lifting, s-positioning, tail beating, ramming, biting and circling; Parzefall 1974, 2001). Small, subdominant males with more cryptic body coloration typically hide in a head down position and try to sneak copulations (Parzefall 1969).

In the cave molly (Cueva del Azufre), all patterns of aggressive behavior are reduced. This reduction was found to increase gradually from the animals originating from the front chambers of the Cueva del Azufre to those found in the rear cave chambers (Figure 8 in Parzefall 2001). When cave specimens were tested with males of either their own or the epigean population, the reduction of aggressive behavior remained unchanged (Figure 9 in Parzefall 2001). In all tests, the mean number of aggressive behaviors of the fish from the innermost cave pool (XIII) was about zero. In mixed tests the majority of the cave molly males answered the initial attacks of the epigean males with sexual patterns (following, nipping and copulation), only a few reacted defensively for a short time. No cave molly male ever attacked.

When Parzefall (1974, 1979) tested hybrids between cave and epigean fish, the frequency distribution for aggressive fin erection and s-position revealed a genetically based reduction of the aggressive behavior in the cave molly. On the basis of quantitative genetic data, Parzefall (1979) concluded that a closely linked polygenic system exhibiting additive gene interaction exists and controls the aggressive behavior in *P. mexicana*. Given that males do not fight for access to females, intrasexual selection via aggressive male-male competition breaks down (Plath *et al.* 2004a; Figure 11).

When comparing the aggressive behavior of cave mollies with that of other cave-dwellers, reduced aggression seems to be the exception rather than the rule. For example, in the cave form of the Mexican tetra *Astyanax mexicanus*, males and females may defend small feeding territories and use signals effective in close body contact to communicate with competitors. Biting, circling and tail-beating have been observed, but no fin-spreading, snake-swimming and ramming like in the epigean relatives (Parzefall and Trajano, this volume). It seems that the aggressive behavior of these cave fish evolved as an adaptation to the lightless habitat after the reduction of the visually mediated aggression of the epigean ancestor (see chapter 4 by Parzefall and Trajano). Also other cave fishes, like the Somalian cyprinid *Phreatichthys andruzii*, can be highly aggressive (authors, personal observation).

The reduction of aggressive behavior in the cave molly has previously been interpreted as an adaptation to darkness, but this argumentation failed to explain why other cave fishes can be highly aggressive (Parzefall 1993). Therefore, it seems reasonable to assume that the extreme environmental conditions in the Cueva del Azufre act as a strong selection factor, favoring the reduction of any energy-demanding behaviors.

Male Mating Behavior

Male poeciliids from normoxic surface habitats usually spend most of their time either defending females or attempting to mate (Parzefall 1969, Houde 1997). *Poecilia mexicana* males show more sexual behavior than wanted by the females, leading to a sexual conflict (Plath *et al.* 2003b). Indeed, *P. mexicana* females from all populations examined so far (Río Oxolotan, El Azufre and Cueva del Azufre), tend to avoid the presence of a male when given a choice to associate with a male or a female stimulus fish in a simultaneous, dichotomous choice situation (Plath *et al.* 2001, Plath and Tobler 2007). Only during few days of receptivity during their monthly sexual cycle, cave molly females appear to be attracted by males (Plath *et al.* 2001).

Nevertheless, poeciliid males also attempt to mate with non-receptive females, i.e., they show coercive mating. Sexual harassment by males has been reported from several livebearing fishes (Poeciliidae) and has been shown to inflict costs to females (Plath *et al.* 2007b). For example, poeciliid females have reduced feeding opportunities when accompanied by a male, because females need to dedicate attention to avoiding male copulation attempts (Plath *et al.* 2007b). When comparing male sexual activity, and female feeding time reduction by male harassment in Río Oxolotan, El Azufre and Cueva del Azufre fish (Plath *et al.* 2003b), we could demonstrate that (a) unlike surface mollies, cave molly and El Azufre males show low sexual activity even in the laboratory (i.e. in light and in the absence of H_2S), (b) females do not suffer feeding time reductions from male harassment in those populations, and (c) small Río Oxolotan males show especially high rates of sexual behavior (probably to compensate for their disadvantage in female choice, Figure 11), but this pattern breaks down in the cave molly and the El Azufre population (see Plath *et al.* 2004b, Plath *et al.* 2005b, Plath *et al.* 2006a for further laboratory and field studies).

After the discovery of the second cave form of *P. mexicana* from the non-sulfidic Cueva Luna Azufre, male sexual behavior, and potential costs of male harassment in terms of reduced feeding rates of the females in the presence of a male, could also be examined in the latter cave form, and was compared to previous data from two populations inhabiting clear-water surface sites (Río Oxolotan and Tampico in central Mexico), the sulfidic El Azufre, and the Cueva del Azufre (Plath 2008). In all populations with at least one physicochemical stressor present (H_2S or absence of light), a reduction in male sexual activity was recorded. While females from both habitats without physicochemical stressors spent less time feeding around males, no such costs of male harassment were detected in populations from extreme habitats, including the Cueva Luna Azufre cave fish. This study corroborated the interpretation that energy-limitation is indeed the driving force selecting for reduced male sexual activity, since fish from all habitats with presence of physicochemical stressors were found to show a low body condition factor under natural conditions (Tobler 2008). As revealed by the analysis of male sperm production (see above), those males need to allocate relatively more energy to somatic maintenance, and thus allocate relatively less energy into reproduction.

Another study examined the question of whether low male sexual activity, lack of sexual harassment and lack of size-dependent mating behavior in the cave molly (Cueva del Azufre population) is phenotypically plastic (Plath *et al.* 2004b). When cave molly males were sexually deprived for one week, male sexually activity increased slightly, but still no costs of harassment for females were detected. Also when female feeding motivation was manipulated by food deprivation or when a combination of both treatments was used, females did not suffer from male harassment. The number of sexual behaviors was not correlated with male body size in any experiment, indicating that even after sexual deprivation small cave molly males do not switch to the pattern of size-dependent ("alternative") mating behavior known in surface-dwelling *P. mexicana*. Also when another male, i.e., a rival, could interact with the same female did small males not change their behavior, and still showed less sexual behavior than large males (Riesch *et al.* 2006a).

When plasticity in male sexual behavior in response to sexual deprivation was observed in the El Azufre population (Plath *et al.* 2006a), costs of male harassment for females were also not detectible following the described treatments. However, although El Azufre males also maintained relatively low sexual activity after sexual deprivation, mating behavior was found to be negatively correlated with male body size. In conclusion, reduced sexual activity has a strong genetic component in *P. mexicana* from sulfidic habitats, and the genetic basis for size-dependent mating behavior is lacking at least in the cave molly (Figure 11), while it still has a phenotypically plastic component in the El Azufre population.

Altogether, these results suggest a disruption in the antagonistic coevolution between male persistence traits (harassment) and female resistance traits (vigilance and flight behavior) in cave mollies. It seems that female vigilance behavior is reduced compared to surface fish, because males show extremely low sexual activity under natural conditions (Plath *et al.* 2005b), and even experimentally increased male sexual activity no longer elicits a response in cave molly females.

Sexual Selection

Size Preferences

The cave molly is one of very few cave fishes in which the evolution of mating preferences has been investigated so far (but see Plath *et al.* 2006b for mating preferences in cave tetras). One male trait females are commonly responsive to throughout the Animal Kingdom is large male body size (Ryan and Keddy-Hector 1992, Andersson 1994). Like most poeciliids, epigean *P. mexicana* females exhibit a strong mating preference for large bodied males (Plath *et al.* 2004b). Mating preferences for large males are visually mediated in surface populations, but cave molly females from the Cueva del Azufre have evolved the ability to discriminate between males of different size using non-visual cues. The ancestral visually mediated preference has not been lost and can be observed when females are tested under light conditions (Plath *et al.* 2003a, Plath *et al.* 2004c, Plath *et al.* 2004d). Interestingly, cave mollies from the independently colonized Cueva Luna Azufre have not evolved the ability to perform mate choice in darkness (Tobler *et al.* 2008c).

The differences in the ability to discriminate between different sized males in darkness had consequences for male traits under selection (male size at maturity). Pronounced male size polymorphisms are characteristic for epigean populations as well as the cave molly population from the Cueva del Azufre. Costs of growing large and being preferentially preyed upon by predators are balanced in large males by a mating advantage mediated by female choice (Figure 11). In the Cueva Luna Azufre, however, large males appear to be absent. The lack of the ability in females to discriminate between large and small males might have led to a reduction of advantages of being large, such that growth in males has been counter-selected (Tobler *et al.* 2008c).

Mating preferences for body size cannot only be found in females, but also in males. Males likely prefer to mate with large females because their fecundity – which is a direct function of body size – is higher. Like females, male cave mollies from the Cueva del Azufre population have evolved the ability to determine female size in darkness, while epigean populations are unable to do so. Male mating preferences are clear-cut both in simultaneous (Plath *et al.* 2006) and sequential (full contact) mate choice experiments (Plath 2008). Males from the Cueva Luna Azufre have so far not been tested for their ability to discriminate between differently sized females in darkness.

Indicator Value of Nutritional State

Theory predicts that female preferences may evolve relative to the 'indicator value' of the male trait (Harvey and Bradbury 1991, Kirkpatrick and Ryan 1991, Andersson 1994, Grether 2000). Traits that reliably reflect male body condition could function as indicators of the male's (in)ability to cope with its environment, such as environmental fluctuation, food shortage, or other stressful conditions. One such male trait is its nutritional state, namely, the extension of its abdomen. If males are unable to gather enough food, they will inevitably look malnourished. Female *P. mexicana* from the different populations clearly differed in mating preferences for well nourished vs. starved males. River-dwelling females never showed a clear preference, but sulfur creek females did so in the visual cues treatment. Cave molly females strongly preferred to associate with the well-fed male both when visual and non-visual communication was possible, i.e., they showed this preference also in darkness (Plath *et al.* 2005a).

As discussed before, fish from both the sulfidic surface and cave habitats are energy limited, whereas mollies from non-sulfidic rivers have access to sufficient food (Plath *et al.* 2005a). The weak response in the river population reflects that good male nutritional state has a low 'indicator value' in such habitats (basically, every male is well fed), such that females do not base they mate choice on this trait. By contrast, a male that shows a good nutritional state despite the environmental harshness in its habitat will possibly provide indirect benefits to the female's offspring (in the form of "good genes"). In all populations studied, there was variation in male nutritional state (Plath *et al.* 2005a). Hence, under natural conditions, there is a potential for benefits to females by female choice, because males differ in "quality".

SYNTHESIS

Insights into the Origin of Cave Organisms

Active Colonization

Originally, all cave organisms invariantly descended from epigean ancestors that colonized subterranean habitats and subsequently diverged into distinct evolutionary lineages. Cave colonization has long been viewed as a passive process, where organisms were accidentally washed into the underground and got trapped (Wilkens 1979, Langecker 1989). According to Romero and Green (2005), this notion predominantly derived from the – faulty – logic that caves provide a harsh environment that organisms would try to avoid if possible. In fact, most cave habitats are continuous with adjacent epigean habitats, and no permanent physical barriers prevent organisms from returning to their original habitats (e.g., Romero *et al.* 2002, Reis *et al.* 2006). Alternatively, cave colonization by epigean organisms may be active and advantageous, and potential advantages include environmental stability, exploitation of unoccupied niches, as well as a reduction in competition and predation (Romero and Green 2005). Empirical evidence for active colonization, however, has been scarce to date.

Negative phototactic behavior, which has been reported as a potential mechanism for active cave colonization, does not explain cave colonization in *P. mexicana* as cave forms – just like their epigean ancestors – consistently exhibit positive phototactic behavior under various light conditions (Parzefall *et al.* 2007). Thus, phototactic behavior alone neither explains the initial colonization of subterranean habitats by *P. mexicana*, nor the restricted gene flow among populations.

The comparative analysis of the biotic environment of the different habitat types as well as the ecology of *P. mexicana* in the Cueva del Azufre system, however, has provided evidence that advantages of cave colonization may exist. Firstly, the colonization of divergent habitat types was accompanied by a shift in resource use (Tobler 2008), indicating the exploitation of resources underused in ancestral habitats. Secondly, the presence of predators differs vastly among habitat types. Avian as well as piscine predators are absent in the cave habitats, but are common at least in non-sulfidic surface habitats (Tobler *et al.* 2006b, Tobler *et al.* 2007a, Riesch *et al.* 2009a), indicating that subterranean habitats may offer some protection against predation. Lastly, it was also hypothesized that cave habitats may offer an advantage in terms of reduced parasite exposure. Some parasites (e.g., the trematode *Uvulifer* sp.) that are highly abundant in non-sulfidic surface habitats have reduced prevalence in sulfidic surface habitats and are even completely absent in the sulfidic cave (Tobler *et al.* 2007a). Active colonization conferring benefits for cave colonizers thus seems a plausible scenario for the colonization of the caves in the Cueva del Azufre system.

Parapatric Divergence and Ecological Speciation

A predominant view in the literature is that obligate cave-dwellers evolved from facultative cave colonizers that became isolated through extinction of the surface ancestor (Barr 1968). Fluctuating environmental conditions during periods of climate change are thought to drive surface populations to extinction while cave populations sustain in suitable habitats (e.g., in terms of temperature or humidity); hence this hypothesis is known as the climate-relict model (Barr 1968, Wilkens 1973, Humphreys 1993, Trajano 1995, Allegrucci *et al.* 2005). Essentially, this is a two-step model of cave evolution, where caves were first colonized without restriction of gene flow with surface populations followed by allopatric genetic and phenotypic differentiation in geographically isolated caves after the local extinction of surface-dwelling ancestors.

The climate-relict model is inherently difficult to test in specific cases, since it fails to provide exclusive testable predictions. Even if phylogenetic and biogeographic patterns are congruent with the reconstruction of climatic changes in the past, it is impossible to exclude the possibility that troglobites diverged parapatrically from epigean ancestors before the respective surface populations went extinct. So, although many caves with their geological age and their environmental stability may indeed act as climatic refugia for phylogenetic lineages long extinct in adjacent surface habitats, the climate-relict hypothesis only offers limited insight to the ultimate mechanisms of the origin of troglobites.

An alternative hypothesis prevalent in the biospeleological literature is the 'adaptive shift' model, where epigean species invade caves to exploit novel resources and subsequently adapt to the divergent environmental conditions present (Howarth 1987). Divergence between epigean ancestors and cave inhabitants is usually assumed to arise parapatrically along the steep environmental gradient. Additional, sometimes very specific, models have been proposed to explain the origins of troglobites (see Danielopol and Rouch 2005 for an overview).

Generally, the investigation of the origins of subterranean organisms has been plagued by the lack of appropriate study systems. Many lineages of troglobites are phylogenetically old, making the elucidation of the mechanisms of initial colonization and divergence an improbable task. Future work will have to put more emphasis on phylogenetically young systems, where cave colonization occurred recently or is still in progress (e.g., the Cueva del Azufre system, see also Schilthuizen *et al.* 2005).

Molecular genetic work highlights that different populations of *P. mexicana* are still connected by low rates of gene flow, and, most importantly, all evidence points towards parapatric divergence of cave populations. However, the isolating mechanisms leading to genetic differentiation among populations of *P. mexicana* from different habitat types are only poorly understood so far. It is unlikely that populations in the divergent habitats are genetically incompatible (i.e. not inter-fertile), since the absence of intrinsic post-zygotic reproductive isolation has been documented even in more distantly related poeciliid species (Hubbs 1959, Schartl 1995, Ptacek 2002, Dries 2003, Rosenthal *et al.* 2003, Alexander and Breden 2004, Kittell *et al.* 2005). Likewise, a genetically-based preference for separate habitat types (Rice and Salt 1990, Johnson *et al.* 1996) is unlikely at least for the separation between surface- and cave-dwelling populations, as *P. mexicana* from the Cueva del Azufre – like surface-dwelling fish from El Azufre – exhibit photophilic behavior (Parzefall *et al.* 2007).

We propose that divergent selection caused by the abiotic environmental conditions gives rise to reproductive isolation. Several mechanisms, which may act in synchrony, seem possible in this system. (1) Selection could act directly on immigrants from divergent populations causing prezygotic isolations (Nosil *et al.* 2005). For example, *P. mexicana* from non-sulfidic habitats have been shown to have a high susceptibility to the toxic effects of H_2S (Peters *et al.* 1973, Tobler *et al.* 2008d). Likewise, fish from surface habitats have an impaired ability for intraspecific communication in darkness, which may lower their reproductive success in the cave environment (Plath *et al.* 2004c, 2006c). Selection against immigrants, however, should be unidirectional, unless adaptations to sulfidic and cave habitats come at a cost in non-sulfidic and surface habitats, respectively. For example, the reduced eye size could make cave *P. mexicana* more susceptible to predation or less efficient foragers in light compared to conspecifics from surface habitats. (2) *Poecilia mexicana* from different habitat types may be less attracted to conspecifics from divergent habitat types, which again may cause prezygotic isolation (Schluter 2000, Rundle and Nosil 2005). For example, in poeciliid fishes of the genus *Gambusia*, assortative mating for divergent body shapes caused by different predatory regimes mediates reproductive isolation among allopatric populations (Langerhans *et al.* 2007). Whether mating preferences for divergent phenotypes are present in the *P. mexicana* system remains to be studied. Besides intersexual selection, intrasexual selection by male-male competition may be operating. For example, all populations from cave and/or sulfidic habitats have reduced male sexual activity (Plath *et al.* 2003b, Plath 2008), and Cueva del Azufre fish are less aggressive (Parzefall 1974). Since reproductive success in poeciliids is determined by the number of successful copulations (Magurran and Seghers 1994), such males may be outcompeted by males from non-sulfidic surface fish due to higher aggression and higher rates of sexual behavior. (3) Divergent selection could act on hybrids of *P. mexicana* from different habitats (Hatfield and Schluter 1999, Schluter 2000), but to date no empirical evidence for this mechanism is available.

Future studies will need to pay careful attention to the mechanisms of reproductive isolation causing the observed small-scale population differentiation in the Cueva del Azufre system in order to test whether parapatric ecological speciation is indeed occurring. The current data, however, provide strong evidence for population differentiation and local adaptation along abiotic environmental gradients. Most other model systems studied to date found biotic ecological factors, i.e. interactions between species, such as resource use (Funk 1998, Ryan *et al.* 2007), predation (Langerhans *et al.* 2007), or parasitism (Blais *et al.* 2007), to be driving adaptive divergence among lineages.

Conservation Issues

Cave mollies are only known from the Cueva del Azufre system. The wider region is currently under rapid development, especially in terms of tourism. Deforestation, change in land use, and general population growth, which is accompanied by the spread of human settlements ever closer to the caves, must currently be considered as the main risk for the long term conservation of the caves and their inhabitants. During the last few years, a substantial proportion of the forests around the Cueva del Azufre were cut down, and signs of severe erosion can be seen. Furthermore, cattle farming as well as agriculture have increased in frequency. Like in other places in Mexico, the number of inhabitants is still increasing around the Cueva del Azufre. Settlements are now found directly above the Cueva del Azufre. Although no data are currently available, an increase of soil particle-, nutrient-, and potentially contaminant-influx (like human wastes and fertilizers) into the cave has to be expected, which may be a direct threat to the fragile cave ecosystems.

Also, the number and frequency of people visiting the caves may affect the organisms living in it. Nowadays, the Cueva del Azufre is a major source of attraction for tourists, and an ever-larger number of visitors come to the cave. Green tourisms clearly has developed as a major source of income for the local community, however, it needs to be assured that current and future development will be sustainable. Especially lamps of visitors that enter the deeper parts of the cave disturb bats and the cave fish. Also the waste left in the cave can attract alien species, which may be competitors to the cave adapted organisms, thereby threatening their existence.

FUTURE DIRECTIONS

Work on the cave molly, in comparison to other populations of *Poecilia mexicana*, has provided insight into the behavioral and evolutionary ecology of an extremophile cave fish. Moreover, molecular biological analyses so far indicated that extremophile poeciliids provide a unique opportunity to examine divergent local adaptation, a process that may ultimately cause (parapatric) small-scale speciation in cave and sulfidic habitats. By conducting similar analyses in other extremophile populations of *P. mexicana*, it may become possible to test a central hypothesis proposed here, namely that genetic differentiation and restricted gene-flow are more pronounced along harsh environmental gradients.

Another focus of future work should lie on the evolution of potential isolation mechanisms. For example, it seems promising to examine if assortative mating occurs at the interface of sulfidic and non-sulfidic, or surface and cave habitats, i.e. if females (and males) have evolved mating preferences for mates of their own, locally adapted population. Assortative mating might further promote speciation in sulfidic systems by preventing gene flow. Generally, research in the Cueva del Azufre system is one of the few examples where the ecology, behavior and evolution of several surface and cave populations were simultaneously considered. Applying multi-biological approaches to study other cave fish systems is clearly a promising challenge for future research.

Acknowledgements

We are deeply indebted to the people in the community of Tapijulapa, the Universidad Intercultural del Estado de Tabasco in Oxolotan as well as the Municipal de Tacotalpa for their continuous support and their hospitality during our visits. A number of people are acknowledged for contributing to this project: C.M. Franssen, F.J. García de León, O. Giere, K.E. Körner, A. and D. Möller, J. Parzefall, R. Riesch, L. Rosales Lagarde, and I. Schlupp. T. Schulz-Mirbach kindly provided information about otoliths. We thank J.K. Langhammer for providing access to an unpublished manuscript of R.R. Miller. Financial support came from the Deutsche Forschungsgemeinschaft (DFG) to M.P. and from the Swiss National Foundation (SNF PBZHA-121016) to M.T. The Mexican government and the local authorities provided permits to conduct our research (the most recent ones are: Permiso de pesca de fomento numbers: 291002-613-1577, DGOPA/5864/260704/-2408). The authors are indebted to C. Lamptey (M.P.) and C.M. Franssen (M.T.) for their support during the writing process. Finally, we wish to thank J. Parzefall for introducing us to the biology of the cave molly.

References

Abdel-Latif, H., E. Hassan and C. von Campenhausen. 1990. Sensory performance of blind Mexican cave fish after destruction of the canal neuromasts. *Naturwissenschaften* 77: 237-239.

Affonso, E. and F. Rantin. 2005. Respiratory responses of the air-breathing fish *Hoplosternum littorale* to hypoxia and hydrogen sulfide. *Comparative Biochemistry and Physiology C - Toxicology and Pharmacology* 141: 275-280.

Alexander, H.J. and F. Breden. 2004. Sexual isolation and extreme morphological divergence in the Cumana guppy: A possible case of incipient speciation. *Journal of Evolutionary Biology* 17: 1238-1254.

Allan, J. 1995. *Stream Ecology: Structure and Function of Running Waters.* Kluwer Academic Publishers, London.

Allegrucci, G., V. Todisco and V. Sbordoni. 2005. Molecular phylogeography of *Dolichopoda* cave crickets (Orthoptera, Rhaphidophoridae): A scenario suggested by mitochondrial DNA. *Molecular Phylogenetics and Evolution* 37: 153-164.

Andersson, M. 1994. *Sexual Selection.* Princeton University Press, New Jersey.

Bagarino, T. 1992. Sulfide as an environmental factor and toxicant: Tolerance and adaptations in aquatic organisms. *Aquatic Toxicology* 24: 21-62.

Bagarino, T. and R.D. Vetter. 1989. Sulfide tolerance and detoxification in shallow water marine fishes. *Marine Biology* 103: 291-302.

Bagarino, T. and I. Lantin-Olaguer. 1999. The sulfide tolerance of milkfish and tilapia in relation to fish kills in farms and natural waters in the Philippines. *Hydrobiologia* 382: 137-150.

Barr, T.C. 1968. Cave ecology and the evolution of troglobites. *Evolutionary Biology* 2: 35-102.

Bateman, A. 1948. Intra-sexual selection in *Drosophila. Heredity* 2: 349-368.

Begon, M.E., J.L. Harper and C.R. Townsend. 1996. *Ecology.* Blackwell Science, Oxford, 3rd edition.

Blais, J., C. Rico, C. Van Oosterhout, J. Cable, G.F. Turner and L. Bernatchez. 2007. MHC adaptive divergence between closely related and sympatric African cichlids. PLoS ONE 2:e734.

Brauner, C.J., C.L. Ballantyne, D.J. Randall and A.L. Val. 1995. Air-breathing in the armored catfish (*Hoplosternum littorale*) as an adaptation to hypoxic, acidic, and hydrogen sulfide-rich waters. *Canadian Journal of Zoology—Revue Canadienne de Zoologie* 73: 739-744.

Burt de Perera, T. 2004. Spatial parameters encoded in spatial map of the blind Mexican cave fish, *Astyanax fasciatus. Animal Behaviour* 68: 291-295.

Bush, A.O., J.C. Fernández, G.W. Esch and J.R. Seed. 2001. *Parasitism: The Diversity and Ecology of Animal Parasites.* Cambridge University Press, Cambridge.

Bussing, W.A. 1998. Peces de las aguas continentales de Costa Rica—Freshwater fishes of Costa Rica. Editorial de la Universidad de Costa Rica, San José.

Camassa, M.M. 2001. Responses to light in epigean and hypogean populations of *Gambusia affinis* (Cyprinodontiformes: Poeciliidae). *Environmental Biology of Fishes* 62: 115-118.

Carrico, R., W. Blumberg and J. Peisach. 1978. The reversible binding of oxygen to sulfhemoglobin. *Journal of Biological Chemistry* 253: 7212-7215.

Chapman, L.J. and C.A. Chapman. 1993. Desiccation, flooding, and the behavior of *Poecilia gillii* (Pisces: Poeciliidae). *Ichthyological Exploration of Freshwaters* 4: 279-287.

Chapman, L.J. and K.G. Hulen. 2001. Implications of hypoxia for the brain size and gill morphometry of mormyrid fishes. *Journal of Zoology* 254: 461-472.

Chapman, L., F. Galis and J. Shinn. 2000. Phenotypic plasticity and the possible role of genetic assimilation: Hypoxia-induced trade-offs in the morphological traits of an African cichlid. *Ecology Letters* 3: 387-393.

Chapman, L.J. C.A. Chapman, D.A. Brazeau, B. McLaughlin and M. Jordan. 1999. Papyrus swamps, hypoxia, and faunal diversification: variation among populations of *Barbus neumayeri*. *Journal of Fish Biology* 54: 310-327.

Chen, K. and J. Morris. 1972. Kinetics of oxidation of aqueous sulfide by O_2. *Environmental Science and Technology* 6: 529-537.

Cline, J. and F. Richards. 1969. Oxygenation of hydrogen sulfide in seawater at constant salinity, temerature, and pH. *Enviromental Science and Technology* 3: 838-843.

Colaco, A., F. Dehairs and Desbruyeres. 2002. Nutritional relations of deep-sea hydrothermal fields at the Mid-Atlantic ridge: a stable isotope approach. *Deep-Sea Research Part I—Oceanographic Research Papers* 49: 395-412.

Culver, D.C. 1982. *Cave Life—Evolution and Ecology*. Harvard Uiversity Press, Cambridge.

Curtis, C., T. Bartholomew, F. Rose and K. Dodgson. 1972. Detoxication of sodium ^{35}S-sulphide in the rat. *Biochemical Pharmacology* 21: 2313-2321.

Danielopol, D.L. and R. Rouch. 2005. Invasion, active versus passive. In: *Encyclopedia of Caves*, D.C. Culver and W. White (eds.) Elsevier Academic Press, Amsterdam.

Darnell, R.M. 1962. Fishes of the Rio Tamesi and related coastal lagoons in east-central Mexico. Publications of the Institute of Marine Sciences, University of Texas 8: 299-365.

Dries, L.A. 2003. Peering through the looking glass at a sexual parasite: Are Amazon mollies Red Queens? *Evolution* 57: 1387-1396.

Endler, J.A. 1987. Predation, light-intensity and courtship behavior in *Poecilia reticulata* (Pisces, Poeciliidae). *Animal Behaviour* 35: 1376-1385.

Fenolio, D., G. Graening, B. Collier and J. Stout. 2006. Coprophagy in a cave-adapted salamander: The importance of bat guano examined through nutritional and stable isotope analyses. *Proceedings of the Royal Society B-Biological Sciences* 273: 439-443.

Franssen, C., M. Tobler, R. Riesch, F.J. García de León, R. Tiedemann, I. Schlupp and M. Plath. 2008. Sperm production in an extremophile fish, the cave molly (*Poecilia mexicana* Poeciliidae, Teleostei). *Aquatic Ecology* 42: 685-692.

Funk, D.J. 1998. Isolating a role for natural selection in speciation: Host adaptation and sexual isolation in *Neochlamisius bebbianae* leaf beetles. *Evolution* 52: 1744-1759.

Gibert, J. and L. Deharveng. 2002. Subterranean ecosystems: A truncated functional biodiversity. *Bioscience* 52: 473-481.

Gordon, M.S. and D.E. Rosen. 1962. A cavernicolous form of the Poeciliid fish *Poecilia sphenops* from Tabasco, México. *Copeia* 1962: 360-368.

Graening, G. and A. Brown. 2003. Ecosystem dynamics and pollution effects in an Ozark cave stream. *Journal of the American Water Resources Association* 39: 1497-1507.

Graham, J. 2005. Aquatic and aerial respiration. In: *The Physiology of Fishes*, D. Evans and J. Claiborne (eds.). CRC Press, London.

Grether, G.F. 2000. Carotenoid limitation and mate preference evolution: a test of the indicator hypothesis in guppies (*Poecilia reticulata*). *Evolution* 54: 1712-1724.

Grieshaber, M.K. and S. Völkel. 1998. Animal adaptations for tolerance and exploitation of poisonous sulfide. *Annual Review of Physiology* 60: 33-53.

Harvey, P. and J. Bradbury. 1991. Sexual selection. In: *Behavioural Ecology: An Evolutionary Approach*, J. Krebs and N. Davies (eds.). Blackwell, Oxford, pp. 203-233.

Hatfield, T. and D. Schluter. 1999. Ecological speciation in sticklebacks: Environment dependent hybrid fitness. *Evolution* 53: 866-873.

Higham, T.E., C.D. Hulsey, O. Rican and A.M. Carroll. 2007. Feeding with speed: Prey capture evolution in cichlids. *Journal of Evolutionary Biology* 20: 70-78.

Hose, L. and J. Pisarowicz. 1999. Cueva de Villa Luz Tabasco, Mexico: reconnaissance study of an active sulfur spring cave and ecosystem. *Journal of Cave and Karst Studies* 61: 13-21.

Hose, L., A.N. Palmer, M. Palmer, D.E. Northup, P.J. Boston and H.R. DuChene. 2000. Microbiology and geochemistry in a hydrogen-sulphide-rich karst environment. *Chemical Geology* 169: 399-423.

Houde, A. 1997. *Sex, Color, and Mate Choice in Guppies*. Princeton University Press, Princeton.

Howarth, F.G. 1987. The evolution of non-relictual tropical troglobites. *International Journal of Speleology* 16: 1-16.

Hubbs, C. 1959. Population analysis of a hybrid swarm between *Gambusia affinis* and *G. heterochir*. *Evolution* 13: 236-246.

Hulsey, C.D. and F. Garcia de Leon. 2005. Cichlid jaw mechanics: Linking morphology to feeding specialization. *Functional Ecology* 19: 487-494.

Humphreys, W.F. 1993. Cave fauna in semi-arid tropical Western Australia: A diverse relict wet-forest litter fauna. *Mémoires de Biospéologie* 20: 105-110.

Hüppop, K. 2000. How do cave animals cope with the food scarcity in caves? In: *Ecosystems of the World 30: Subterranean Ecosystems*, H. Wilkens, D.C. Culver and W.F. Humphries (eds.). Elsevier Science, Amsterdam.

Ip, Y.K., S.S.L. Kuah and S.F. Chew. 2004. Strategies adopted by the mudskipper *Boleophthalmus boddaerti* to survive sulfide exposure in normoxia or hypoxia. *Physiological and Biochemical Zoology* 77: 824-837.

Jeyasingh, P.D. 2007. Plasticity in metabolic allometry: The role of dietary stoichiometry. *Ecology Letters* 10: 282-289.

Johnson, P., F. Hoppensteadt, J. Smith and G. Bush. 1996. Conditions for sympatric speciation: A diploid model incorporating habitat fidelity and non-habitat assortative mating. *Evolutionary Ecology* 10: 187-205.

Jorgensen, B.B. 1982. Ecology of the bacteria of the sulphur cycle with special reference to anoxic/oxic interface environments. *Philosophical Transactions of the Royal Society of London Series B-Biological Sciences* 298: 543-561.

Jorgensen, B.B. 1984. The microbial sulfur cycle. In: *Microbial Geochemistry*, W. Krumbein (ed.). Blackwell Science Publishers, Oxford.

Jorgensen, B.B. and T. Fenchel. 1974. The sulphur cycle of a marine sediment model system. *Marine Biology* 24: 189-201.

Kiester, R. 1979. Conspecifics as cues: a mechanism for habitat selection in the Panamanian grass anole (*Anolis auratus*). *Behavioral Ecology and Sociobiology* 5: 323-330.

Kirkpatrick, M. and M.J. Ryan. 1991. The evolution of mating preferences and the paradox of the lek. *Nature* 350: 33-38.

Kittell, M., M. Harvey, S. Contreras-Balderas and M.B. Ptacek. 2005. Wild-caught hybrids between sailfin and shortfin mollies (Poeciliidae, *Poecilia*): morphological and molecular verification. *Hidrobiologica* 15: 131-137.

Körner, K.E., I. Schlupp, M. Plath and E.R. Loew. 2006. Spectral sensitivity of mollies: comparing surface- and cave-dwelling Atlantic mollies, *Poecilia mexicana*. *Journal of Fish Biology* 69: 54-65.

Kramer, D. and M. Bryant. 1995. Intestine length in the fishes of a tropical stream: 2. Relationships to diet—the long and short of a convoulted issue. *Environmental Biology of Fishes* 42: 129-141.

Kramer, D.L. 1983. The evolutionary ecology of respiratory modes in fishes: An analysis based on the costs of breathing. *Environmental Biology of Fishes* 9: 145-158.

Krause, J. and G. Ruxton. 2002. *Living in Groups*. Oxford University Press, Oxford.

Krause, J., N. Hartmann and V. Pritchard. 1999. The influence of nutritional state on shoal choice in zebrafish, *Danio rerio*. *Animal Behaviour* 57: 771-775.

Langecker, T.G. 1989. Studies on the light reaction of epigean and cave populations of *Astyanax fasciatus* (Characidae, Pisces). *Mémoires de Biospéologie* 16: 169-176.

Langecker, T.G. 2000. The effect of continuous darkness on cave ecology and cavernicolous evolution. In. *Ecosystems of the World 30: Subterranean Ecosystems*, H. Wilkens, D.C. Culver and W.F. Humphreys (eds.). Elsevier Science, Amsterdam.

Langecker, T.G., H. Wilkens and J. Parzefall. 1996. Studies on the trophic structure of an energy-rich Mexican cave (Cueva de las Sardinas) containing sulfurous water. *Mémoires de Biospéologie* 23: 121-125.

Langerhans, R.B., M. Gifford and E. Joseph. 2007. Ecogical speciation in *Gambusia* fishes. *Evolution* 61: 2056-2074.

Lovatt Evans, C. 1967. The toxicity of hydrogen sulphide and other sulphides. *Quarterly Journal of Experimental Physiology* 52: 231-248.

Luther, G.W., S. Ma, R. Trouborst, B. Glazer, M. Blickley, R.W. Scarborough and M.G. Mensinger. 2004. The roles of anoxia, H$_2$S, and storm events in fish kills of dead-end canals of Delaware inland bays. *Estuaries* 27: 551-560.

Macalady, J.L., E.H. Lyon, B. Koffman, L.K. Albertson, K. Meyer, S. Galdenzi and S. Mariani. 2006. Dominant microbial populations in limestone-corroding stream biofilms, Frasassi cave system, Italy. *Applied and Environmental Microbiology* 72: 5596-5609.

MacArthur, R. and E. Pianka. 1966. On optimal use of patchy environment. *American Naturalist* 100: 603-609.

MacArthur, R., J. Diamond and J. Karr. 1972. Density compensation in island faunas. *Ecology* 53: 330-342.

Magurran, A.E. and B.H. Seghers. 1994. Sexual conflict as a consequence of ecology: Evidence from Guppy, *Poecilia reticulata*, populations in Trinidad. *Proceedings of the Royal Society of London Series B-Biological Sciences* 255: 31-36.

Meffe, G.K. and F.F. Snelson (eds.). 1989. *Ecology and Evolution of Lifebearing Fishes (Poeciliidae)*. Prentice Hall, New Jersey.

Menke, A. 1979. Family Belostomatidae—Giant water bugs. In: *The Semiaquatic and Aquatic Hemiptera of California (Heteroptera: Hemiptera)*, A. Menke (ed.). University of California Press, Berkeley.

Miller, R.R. 1983. Checklist and key to the mollies of Mexico (Pisces, Poeciliidae, *Poecilia*, Subgenus *Mollienesia*). *Copeia* 1983: 817-822.

Miller, R.R. 2005. *Freshwater Fishes of Mexico*. University of Chicago Press, Chicago.

Nakatsuru, K. and D.L. Kramer. 1982. Is sperm cheap? Limited male-fertility and female choice in the lemon tetra (Pisces, Characidae). *Science* 216: 753-755.

National Research Council. 1979. *Hydrogen Sulfide*. University Park Press, Baltimore.

Nicholls, P. 1975. The effect of sulphide on cytochrome aa3. Isosteric and allosteric shifts of the reduced alpha-peak. *Biochemistry Biophysics Acta* 396: 24-35.

Nosil, P., T.H. Vines and D.J. Funk. 2005. Perspective: Reproductive isolation caused by natural selection against immigrants from divergent habitats. *Evolution* 59: 705-719.

Nybakken, J. 2001. *Marine Biology, an Ecological Approach*. Benjamin Cummings, San Francisco.

Park, C., R. Nagel, W. Blumberg, J. Peisach and R. Maliozzo. 1986. Sulfhemoglobin: Properties of partially sulfurated tetramers. *Journal of Biological Chemistry* 261: 8805-8810.

Parzefall, J. 1969. Zur vergleichenden Ethologie verschiedener *Mollienesia*-Arten einschliesslich einer Höhlenform von *M. sphenops. Behaviour* 33: 1-37.

Parzefall, J. 1970. Morphologische Untersuchungen an einer Höhlenform von *Mollienesia sphenops* (Pisces, Poeciliidae). *Zeitschrift für Morphologie der Tiere* 68: 323-342.

Parzefall, J. 1973. Attraction and sexual cycle of poeciliids. In: *Genetics and Mutagenesis of Fish*, J. Schroeder (ed.). Springer Verlag, Berlin.

Parzefall, J. 1974. Rückbildung aggressiver Verhaltensweisen bei einer Höhlenform von *Poecilia sphenops* (Pisces, Poeciliidae). *Zeitschrift für Tierpsychologie* 35: 66-84.

Parzefall, J. 1979a. Genetics and biological significance of the aggressive behavior of *Poecilia sphenops* (Pisces, Poeciliidae): Studies on hybrids of epigeous and hypogeous living populations. *Zeitschrift für Tierpsychologie* 50: 399-422.

Parzefall, J. 1979b. Zur Genetik und biologischen Bedeutung des Aggressionsverhaltens von *Poecilia sphenops* (Pisces, Poeciliidae). *Zeitschrift für Tierpsychologie* 50: 399-422.

Parzefall, J. 1993. Behavioural ecology of cave-dwelling fishes. In: *Behaviour of Teleost Fishes*, T.J. Pitcher (ed.). Chapman and Hall, London, 2nd edition.

Parzefall, J. 2001. A review of morphological and behavioural changes in the cave molly, *Poecilia mexicana*, from Tabasco, Mexico. *Environmental Biology of Fishes* 62: 263-275.

Parzefall, J., C. Kraus, M. Tobler and M. Plath. 2007. Photophilic behaviour in surface- and cave-dwelling Atlantic mollies *Poecilia mexicana* (Poeciliidae). *Journal of Fish Biology* 71: 1225-1231.

Peek, A.S., R. Feldmann, R. Lutz and R.C. Vrijenhoek. 1998. Conspeciation of chemoautotrophic bacteria and deep-sea clams. *Proceedings of the National Academy of Sciences USA* 95: 9962-9966.

Peters, N. and G. Peters. 1968. Zur genetischen Interpretation morphologischer Gesetzmässigkeiten der degenerative Evolution. *Zeitschrift für Morphologie der Tiere* 62: 211-244.

Peters, N., G. Peters, J. Parzefall and H. Wilkens. 1973. Über degenerative und konstruktive Merkmale bei einer phylogenetisch jungen Höhlenform von *Poecilia sphenops* (Pisces, Poeciliidae). *Internationale Revue der Gesamten Hydrobiologie* 58: 417-436.

Petersen, L. 1977. The effect of inhibitors on the oxygen kinetics of cytochome c oxidase. *Biochimica et Biophysica Acta* 460: 299-307.

Plath, M. 2008. Male mating behavior and costs of sexual harassment for females in cavernicolous and extremophile populations of Atlantic mollies (*Poecilia mexicana*). *Behaviour* 145: 73-98.

Plath, M. and M. Tobler. 2007. Sex recognition in surface- and cave-dwelling Atlantic molly females (*Poecilia mexicana*, Poeciliidae, Teleostei): Influence of visual and non-visual cues. *Acta Ethologica* 10: 81-88.

Plath, M. and I. Schlupp. 2008. Parallel evolution leads to reduced shoaling behavior in two cave dwelling populations of Atlantic mollies (*Poecilia mexicana*, Poeciliidae, Teleostei). *Environmental Biology of Fishes* 82: 289-297.

Plath, M., K.E. Korner, I. Schlupp and J. Parzefall. 2001. Sex recognition and female preferences of Cave mollies *Poecilia mexicana* (Poeciliidae, Teleostei) in light in darkness. *Mémoires de Biospéologie* 28: 163-167.

Plath, M., K. Körner, J. Parzefall and I. Schlupp. 2003a. Persistence of a visually mediated mating preference in the cave molly, *Poecilia mexicana* (Poeciliidae, Teleostei). *Subterranean Biology* 1: 93-97.

Plath, M., J. Parzefall and I. Schlupp. 2003b. The role of sexual harassment in cave- and surface-dwelling populations of the Atlantic molly, *Poecilia mexicana* (Poeciliidae, Teleostei). *Behavioral Ecology and Sociobiology* 54: 303-309.

Plath, M., M. Arndt, J. Parzefall and I. Schlupp. 2004a. Size-dependent male mating behaviour in the Cave molly, *Poecilia mexicana* (Poeciliidae, Teleostei). *Subterranean Biology* 2: 59-64.

Plath, M., A. Brümmer and I. Schlupp. 2004b. Sexual harassment in a livebearing fish (*Poecilia mexicana*): Influence of population-specific male mating behaviour. *Acta Ethologica* 7: 65-72.

Plath, M., J. Parzefall, K. Körner and I. Schlupp. 2004c. Sexual selection in darkness? Female mating preferences in surface- and cave-dwelling Atlantic mollies, *Poecilia mexicana* (Poeciliidae, Teleostei). *Behavioral Ecology and Sociobiology* 55: 596-601.

Plath, M., M. Tobler and I. Schlupp. 2004d. Cave fish looking for mates: A visual mating preference in surface- and cave-dwelling Atlantic mollies (*Poecilia mexicana*, Poeciliidae). *Zeitschrift für Fischkunde* 7: 61-69.

Plath, M., K.U. Heubel, F.G. de Leon and I. Schlupp. 2005a. Cave molly females (*Poecilia mexicana*, Poeciliidae, Teleostei) like well-fed males. *Behavioral Ecology and Sociobiology* 58: 144-151.

Plath, M., K.U. Heubel and I. Schlupp. 2005b. Field observations on male mating behavior in surface- and cave-dwelling Atlantic mollies (*Poecilia mexicana*, Poeciliidae). *Zeitschrift für Fischkunde* 7: 113-119.

Plath, M., A. Brümmer, J. Parzefall and I. Schlupp. 2006a. Size-dependent male mating behaviour and sexual harassment in a population of Atlantic mollies (*Poecilia mexicana*) from a sulphur creek. *Acta Ethologica* 9: 15-21.

Plath, M., M. Rohde, T. Schroder, A. Taebel-Hellwig and I. Schlupp. 2006b. Female mating preferences in blind cave tetras *Astyanax fasciatus* (Characidae, Teleostei). *Behaviour* 143: 15-32.

Plath, M., U. Seggel, H. Burmeister, K.U. Heubel and I. Schlupp. 2006c. Choosy males from the underground: Male mating preferences in surface- and cave-dwelling Atlantic mollies (*Poecilia mexicana*). *Naturwissenschaften* 93: 103-109.

Plath, M., S. Hauswaldt, K. Moll, M. Tobler, F. Garcia de Leon, I. Schlupp and R. Tiedemann. 2007a. Local adaptation and pronounced genetic differentiation in an extremophile fish, *Poecilia mexicana*, inhabiting a Mexican cave with toxic hydrogen sulfide. *Molecular Ecology* 16: 967-976.

Plath, M., A. Makowicz, I. Schlupp and M. Tobler. 2007b. Sexual harassment in live-bearing fishes (Poeciliidae): Comparing courting and noncourting species. *Behavioral Ecology* 18: 680-688.

Plath, M., M. Tobler, F.J. Garcia de Leon, O. Giere and I. Schlupp. 2007c. Survival in an extreme habitat: the role of behaviour and energy limitation. Naturwissenschaften 94: 991-996.

Porter, M. and K. Crandall. 2003. Lost along the way: The significance of evolution in reverse. *Trends in Ecology & Evolution* 18: 541-547.

Poulson, T.L. 1963. Cave adaptation in amblyopsid fishes. *American Midland Naturalist* 70: 257-290.

Poulson, T.L. and W. B. White. 1969. The cave environment. *Science* 165: 971-981.

Poulson, T.L. and K.H. Lavoie. 2000. The trophic basis of subterranean ecosystems. In: *Ecosystems of the World 30: Subterranean Ecosystems*, H. Wilkens. D.C. Culver and W.F. Humphries (eds.). Elsevier Science, Amsterdam.

Proudlove, G. 2006. An account of the subterranean (hypogean) fishes described up to 2003 with a bibliography 1541-2004. International Society for Subterranean Biology, Moulis.

Ptacek, M.B. 2002. Patterns of inheritance of mating signals in interspecific hybrids between sailfin and shortfin mollies (Poeciliidae: *Poecilia*: *Mollienesia*). *Genetica* 116: 329-342.

Reiffenstein, R., W. Hulbert and S. Roth. 1992. Toxicology of hydrogen sulfide. *Annual Reviews of Pharmacology and Toxicology* 1992: 109-134.

Reis, R., E. Trajano and E. Hingst-Zaher. 2006. Shape variation in surface and cave populations of the armoured catfishes *Ancistrus* (Siluriformes: Loricariidae) from the Sao Domingos karst area, upper Tocantins River, Brazil. *Journal of Fish Biology* 68: 414-429.

Reznick, D. and J. Endler. 1982. The impact of predation on life history evolution in Trinidadian guppies (*Poecilia reticulata*). *Evolution* 36: 160-177.

Reznick, D., H. Bryga and J.A. Endler. 1990. Experimentally induced life-history evolution in a natural population. *Nature* 346: 357-359.

Reznick, D., M.J. Butler and H. Rodd. 2001. Life-history evolution in guppies. VII. The comparative ecology of high- and low-predation environments. *American Naturalist* 157: 126-140.

Rice, W. and G. Salt. 1990. The evolution of reproductive isolation as a correlated character under sympatric conditions: Experimental evidence. *Evolution* 44: 1140-1152.

Riesch, R., I. Schlupp and M. Plath. 2006a. Influence of male competition on male mating behaviour in the cave molly (*Poecilia mexicana*, Poeciliidae, Teleostei). *Journal of Ethology* 24: 27-31.

Riesch, R., M. Tobler, F.J. Garcia de Leon, I. Schlupp and M. Plath. 2006b. Reduction of the association preference for conspecifics in surface- and cave-dwelling Atlantic mollies, *Poecilia mexicana. Behavioral Ecology and Sociobiology* 60: 794-802.

Riesch, R., V. Duwe, N. Herrmann, L. Padur, A. Ramm, K. Scharnweber, M. Schulte, T. Schulz–Mirbach, M. Ziege and M. Plath. 2009a. Variation along the shy-bold continuum in extremophile fishes (*Poecilia mexicana, P. sulphuraria*). *Behavioral Ecology and Sociobiology* 63: 1515-1526.

Riesch, R., M. Tobler, I. Schlupp and M. Plath. 2009b. Reduced offspring number in the cave molly (*Poecilia mexicana*, Poeciliidae). *Environmental Biology of Fishes* 84: 89-94.

Riesch, R., M. Plath, F.J. García de León and I. Schlupp. In press-a. Convergent life–history shifts: toxic environments result in big babies in two clades of poeciliids. *Naturwissenschaften.*

Riesch, R., M. Plath and I. Schlupp. In press-b. Toxic hydrogen sulfide and dark caves: life history adaptations to extreme environments in a livebearing fish (*Poecilia mexicana*, Poeciliidae). *Ecology.*

Romero, A. and S.M. Green. 2005. The end of regressive evolution: Examining and interpreting the evidence from cave fishes. *Journal of Fish Biology* 67: 3-32.

Romero, A., A. Singh, A. McKie, M. Manna, R. Baker, K.M. Paulson and J.E. Creswell. 2002. Replacement of the troglomorphic population of *Rhamdia quelen* (Pisces: Pimelodidae) by an epigean population of the same species in the Cumaca Cave, Trinidad, West Indies. *Copeia* 2002: 938-942.

Rosen, D. and R. Bailey. 1963. The poeciliid fishes (Cyprinodontiformes), their structure, zoogeography and systematics. *Bulletin of the American Museum of Natural History* 126: 1-176.

Rosenthal, G. G., X. de la Rosa, S. Kazianis, M.J. Stephens, D.C. Morizot, M.J. Ryan and F.J. Garcia de Leon. 2003. Dissolution of sexual signal complexes in a hybrid zone between the swordtails *Xiphophorus birchmanni* and *Xiphophorus malinche* (Poeciliidae). *Copeia* 2003: 299-307.

Rundle, H.D. and P. Nosil. 2005. Ecological speciation. *Ecology Letters* 8: 336-352.

Ryan, M.J. and A. Keddy-Hector. 1992. Directional patterns of female mate choice and the role of sensory biases. *American Naturalist* 139: S4-S35.

Ryan, P., P. Bloomer, C. Moloney, T. Grant and W. Delport. 2007. Ecological speciation in south Atlantic island finches. *Science* 315: 1420-1422.

Sarbu, S.M., T.C. Kane and B.K. Kinkle. 1996. A chemoautotrophically based cave ecosystem. *Science* 272: 1953-1955.

Schartl, M. 1995. Platyfish and swordtails: A genetic system for the analysis of molecular mechanisms in tumor formation. *Trends in Genetics* 11: 185-189.

Schilthuizen, M., A. Cabanban and M. Haase. 2005. Possible speciation with gene flow on tropical cave snails. *Journal of Zoological Systematics and Evolutionary Research* 43: 133-138.

Schluter, D. 2000. *The Ecology of Adaptive Radiation*. Oxford University Press, Oxford.

Schluter, D. and J. McPhail. 1992. Ecological character displacement and speciation in sticklebacks. *American Naturalist* 140: 85-108.

Schulz-Mirbach, T., C. Stransky, J. Schlickeisen and B. Reichenbacher. 2008. Differences in otolith morphologies between surface- and cave-dwelling populations of *Poecilia mexicana* (Teleostei, Poeciliidae) reflect adaptations to life in an extreme habitat. *Evolutionary Ecology Research* 10: 1-22.

Schoener, T. 1971. Theory of feeding strategies. *Annual Review of Ecology and Systematics* 2: 369-404.

Spellman, S.J. and A.D. Johnson. 1987. *In vitro* encystment of the black spot trematode *Uvulifer ambloplitis* (Trematoda, Diplostomatidae). *International Journal of Parasitology* 17: 897-902.

Streever, W.J. 1996. Energy economy hypothesis and the troglobitic crayfish *Procambarus erythrops* in Sim's sink cave, Florida. *American Midland Naturalist* 135: 357-366.

Swart, C. and B. Felgenhauer. 2003. Structure and function of the mouthparts and salivary gland complex of the Giant Waterbug, *Belostoma lutarium* (Stal) (Hemiptera: Belostomatidae). *Annals of the Entomological Society of America* 96: 870-882.

Timmerman, C.M. and L.J. Chapman. 2004. Hypoxia and interdemic variation in *Poecilia latipinna*. *Journal of Fish Biology* 65: 635-650.

Tobler, M., H. Burmeister, I. Schlupp and M. Plath. 2006a. Regressive evolution of visually mediated preferences in the Cave molly (*Poecilia mexicana*, Poeciliidae, Teleostei). *Subterranean Biology* 4: 59-65.

Tobler, M., I. Schlupp, K. Heubel, R. Riesch, F.J. Garcia de Leon, O. Giere and M. Plath. 2006b. Life on the edge: Hydrogen sulfide and the fish communities of a Mexican cave and surrounding waters. *Extremophiles* 10: 577-585.

Tobler, M., I. Schlupp, F.J. Garcia de Leon, M. Glaubrecht and M. Plath. 2007a. Extreme habitats as refuge from parasite infections? Evidence from an extremophile fish. *Acta Oecologica* 31: 270-275.

Tobler, M., I. Schlupp and M. Plath. 2007b. Predation of a cave fish (*Poecilia mexicana*, Poeciliidae) by a giant water-bug (*Belostoma*, Belostomatidae) in a Mexican sulfur cave. *Ecological Entomology* 32: 492-495.

Tobler, M. 2008. Divergence in trophic ecology characterises colonisation of extreme habitats. *Biological Journal of the Linnean Society* 95: 517-528.

Tobler, M., I. Schlupp, R. Riesch, F.J. Garcia de Leon and M. Plath. 2008a. A new and morphologically distinct cavernicolous population of *Poecilia mexicana* (Poeciliidae, Teleostei). *Environmental Biology of Fishes* 82: 101-108.

Tobler, M., T.J. DeWitt, I. Schlupp, F.J. Garcia de Leon, R. Herrmann, P. Feulner, R. Tiedemann and M. Plath. 2008b. Toxic hydrogen sulfide and dark caves: phenotypic and genetic divergence across two abiotic environmental gradients in *Poecilia mexicana*. *Evolution* 62: 2643-2659.

Tobler, M., I. Schlupp and M. Plath. 2008c. Does divergence in female mate choice affect male size distributions in two cave fish populations? *Proceedings of the Royal Society London: Biology Letters* 4: 452-454.

Tobler, M., R. Riesch, F.J. Garcia de Leon, I. Schlupp and M. Plath. 2008d. Two endemic and endangered fishes, *Poecilia sulphuraria* (Alvarez, 1948) and *Gambusia eurystoma* Miller, 1975 (Poeciliidae, Teleostei), as only survivors in a small sulfidic habitat. *Journal of Fish Biology* 72: 523-533.

Tobler, M., C. Franssen and M. Plath. 2008e. Male-biased predation on a cave fish by a giant water bug. *Naturwissenschaften* 95: 775-779.

Tobler, M., R. Riesch, C.M. Tobler and M. Plath. 2009. Compensatory behavior in response to sulfide–induced hypoxia affects time budgets, feeding efficiency, and predation risk. *Evolutionary Ecology Research* 11: 935-948.

Torrans, E. and H. Clemens. 1982. Physiological and biochemical effects of acute exposure of fish to hydrogen sulfide. *Comparative Biochemistry and Physiology* 71C: 183-190.

Trajano, E. 1995. Evolution of tropical troglobites: Applicability of the model of Quaternary climatic fluctuations. *Mémoires de Biospéologie* 22: 203-209.

Trexler, J., R. Tempe and J. Travis. 1994. Size-selective predation of Sailfin mollies by two species of heron. *Oikos* 69: 250-259.

Trivers, R. 1972. Parental investment and sexual selection. In: *Sexual Selection and the Descent of Man*. B. Campbell (ed.). Aldine Publishing Company, Chicago.

Van Dover, C.L. 2000. *The Ecology of Deep-sea Hydrothermal Vents.* Princeton University Press, Princeton.

Wainwright, P.C. 1996. Ecological explanation through functional morphology: The feeding biology of sunfishes. *Ecology* 77: 1336-1343.

Walters, L. and V. Walters. 1965. Laboratory observations on a cavernicolous poeciliid from Tabasco, Mexico. *Copeia* 1965: 214-233.

Weber, J.M. and D.L. Kramer. 1983. Effects of hypoxia and surface access on growth, mortality, and behavior of juvenile Guppies, *Poecilia reticulata. Canadian Journal of Fisheries and Aquatic Sciences* 40: 1583-1588.

Wilkens, H. 1973. Über das phylogenetische Alter von Höhlentieren: Untersuchungen über die cavernicole Süsswasserfauna Yucatans. *Zeitschrift für zoologische Systematik und Evolutionsforschung* 11: 49-60.

Wilkens, H. 1979. Reduktionsgrad und phylogenetisches Alter: Ein Beitrag zur Besiedlungsgeschichte der Limnofauna Yukatans. *Zeitschrift fur zoologische Systematik und Evolutionsforschung* 17: 262-272.

Willis, L. and A. Brown. 1985. Distribution and habitat requirements of the Ozark cavefish, *Amblyopsis rosae. American Midland Naturalist* 114: 311-317.

Winemiller, K.O. 1993. Seasonality of reproduction by live-bearing fishes in tropical rainforest streams. *Oecologia* 95: 266-276.

Wolf, N. 1985. Odd fish abandon mixed-species groups when threatened. *Behavioral Ecology and Sociobiology* 17: 47-52.

Zeiske, E. 1968. Prädispositionen bei *Mollienesia sphenops* (Pisces, Poeciliidae) für einen Übergang zum Leben in subterranen Gewässer. *Zeitschrift für Vergleichende Physiologie* 58: 190-222.

CHAPTER 9

Subterranean Fishes of Brazil

Eleonora Trajano[1] and Maria Elina Bichuette[2]
[1] Departamento de Zoologia, Instituto de Biociências,
Universidade de São Paulo, São Paulo, SP, Brazil
E-mail: etrajano@usp.br
[2] Departamento de Ecologia e Biologia Evolutiva,
Universidade Federal de São Carlos, São Carlos, SP, Brazil
E-mail: bichuette@uol.com.br

INTRODUCTION

Brazilian subterranean ichthyofauna is remarkable worldwide not only due to the species richness, including troglomorphic and non-troglomorphic fishes, but also for the high ecological and evolutionary diversity, without a taxonomic correlation. More than 25 troglomorphic species showing different degrees of morphological specialization are currently known to occupy a large diversity of habitats, from epikarst to fast flowing base-level streams, in different karst regions. Such areas are submitted to distinct, sometimes contrasting climates, thus subterranean fishes must adapt to different ecological conditions, varying from accentuated food shortage, frequently seasonal, to a relative climatic stability and, in some cases, abundance of food resources.

The first known Brazilian troglobite was the heptapterid *Pimelodella kronei*, from the Upper Ribeira karst area, southern São Paulo State (SE Brazil), described as *Typhlobagrus kronei* Ribeira, 1907. The blind catfish from Iporanga was also the first South American subterranean organism studied in great detail, by Pavan (1945) and later on, by Trajano (1987) for their PhD theses, which included morphology, behavior and ecology analyzed within an evolutionary context. Therefore, this troglobitic fish represents an important historical reference in the development of Brazilian Subterranean Biology.

In the last two decades, studies on the taxonomy, natural history, population ecology and behavior of Brazilian troglobitic fishes have multiplied, encompassing several species living in different kinds of subterranean habitats and showing different degrees of reduction of eyes and melanic pigmentation (classical "troglomorphisms", Trajano 2005). However, this still represents a relatively small part of what there is to be investigated.

Herein, we present an overview on the state of art of the knowledge on Brazilian subterranean fishes, including both troglomorphic (morphologically specialized), probably troglobitic (exclusively subterranean) species, and non-troglomorphic, either troglophilic, trogloxenic or accidental fishes in caves.

I. BIODIVERSITY

An updated list of Brazilian subterranean troglomorphic fishes, with data on localities, predominant type of habitat (according to the classification in Trajano 2001a) and degree of reduction of eyes and melanic pigmentation, is presented in Table 1. As a consequence of the advancement of speleological activities throughout the country, information on "pale fish" is continuously arriving, and confirmed new records on subterranean species are added in a rate of 1-2 each year. Therefore, the present list and related inferences about patterns must be regarded as a current status of the knowledge on Brazilian subterranean fishes.

I.1 Taxonomic Diversity

I.1.1 Troglomorphic and Found Exclusively in Subterranean Habitats: Brazilian Troglobitic Fishes

The neotropical freshwater ichthyofauna is one of the most diversified in the world. Among the two dominant taxa, the Siluriformes (catfish) are mainly nocturnal and chemo-oriented, whereas most Characiformes are diurnal, oriented predominantly by vision. Therefore, it is not surprising the fact that the great majority of neotropical troglobitic fishes, including those found in Brazil, are siluriforms.

So far, 25 species of troglomorphic fishes, i.e., species with eyes and pigmentation reduced at least at some degree beyond that observed in their epigean (surface) congeners, indicating a troglobitic (exclusively subterranean) status, are known for Brazil (Table 1). All but two species are

Table 1 Brazilian troglobitic fishes, with localities (karst area/county, State), types of habitat (according to classification in Trajano 2001) and degree of troglomorphism. 1. slightly, but significantly reduced eyes/melanic pigmentation, with narrow intrapopulation variation; 2. wide variation, from slightly reduced to not visible externally eyes/pigmentation; 3. most individuals with deeply reduced eyes/pigmentation; 4. all individuals completely (externally) anophthalmic and/or pigmented; 4+. DOPA positive (see text); 4– DOPA negative; Polim. Polymorphic phenotypes

	Taxon	Locality	Habitat	Eyes	Pigm.
	CHARACIFORMES, *incertae sedis*				
1	Stygichthys typhlops	Jaíba, MG	Upper phreatic zone (fissures)	4	4+
	GYMNOTIFORMES				
2	Eigenmannia vicentespelaea	São Domingos, GO	Base-level stream	1	1
	SILURIFORMES				
	LORICARIIDAE				
3	Ancistrus cryptophthalmus	São Domingos, GO	Base-level streams	2	2
4	Ancistrus formoso	Bonito, MS	Flooded cave	4	4
	CALLICHTHYIDAE				
5	Aspidoras sp.	Posse, GO	Base level streams	2	1
	TRICHOMYCTERIDAE:				
6	Trichomycterus itacarambiensis	Itacarambi, MG	Base-level stream	2	Polim.
7	Trichomycterus sp.1*	Bonito, MS	Flooded caves	3	3
8	Trichomycterus sp. 2	Serra do Ramalho, BA	Vadose tributaries	3	3
9	Ituglanis passensis	São Domingos, GO	Base-level stream	3	3
10	Ituglanis bambui	idem	Vadose tributaries	3	3
11	Ituglanis epikarsticus	idem	Epikarst	3	3
12	Ituglanis ramiroi	dem	Vadose tributaries	3	3

Table 1 Contd.

Table 1 Contd.

13	*Ituglanis mamba***	Posse, GO	Base-level stream	1	1
14	*Ituglanis* sp.	Serra do Bodoquena, MS	Base-level + Vadose streams	2	2
15	*Glaphyropoma spinosum*	Igatu, BA	Vadose tributary	2	3
16	*Copionodon* sp.	Igatu, BA	Vadose tributary	2	2
	HEPTAPTERIDAE				
17	*Pimelodella kronei*	Alto Ribeira, SP	Base-level streams	3	2
18	*Pimelodella spelaea*	São Domingos, GO	Vadose tributary	1	1
19	*Rhamdia enfurnada*	Serra do Ramalho, BA	Base-level streams	2	2
20	*Rhamdia* sp.	Serra da Bodoquena, MS	Base-level streams	3	2
21	*Rhamdiopsis* sp. 1***	Chapada Diamantina, BA	Upper phreatic zone (caves)	4	4+
22	*Rhamdiopsis* sp. 2 (= *Taunayia* sp.)	Campo Formoso, BA	Upper phreatic zone (cave)	4	4–
23	*Rhamdiopsis* sp. 3	Cordisburgo, MG	Base-level stream	1	1
24	*Phreatobius cisternarum*	Amazonas delta (PA/AM)	Phreatic zone (alluvial fan)	3	3
25	*Phreatobius dracunculus*	Rio Pardo (RO)	Phreatic zone (alluvial fan)	4	4

* cited as *Trichomycterus* undesc. sp. 2 in Trajano (1997)

** cited as *Trichomycterus* sp. 3 in Trajano and Bichuette (2005)

*** cited as "new genus" in Trajano (1997, 2003)

siluriforms belonging to four families: the Trichomycteridae, Heptapteridae and Loricariidae (in this order) also include the great majority of troglobitic fishes found elsewhere in South and Central America, whereas the new Callichthyidae, genus *Aspidoras*, represents the first record of a troglobitic representative in the family. The only catfish family with troglobitic derivatives in South America (Ecuador and Peru – see Proudlove, this volume), which does not occur in Brazil, is Astroblepidae. Therefore, at least for the troglobitic species, the Brazilian subterranean ichthyofauna is representative of the general tendencies for the continental Neotropical region.

The Trichomycteridae, with 11 troglobitic species in Brazil (four genera – *Trichomycterus*, *Ituglanis*, *Glaphyropoma* and *Copionodon*), are well represented in other South American countries (*Trichomycterus* spp. in Venezuela, Colômbia, Bolívia and Argentina). Among the heptapterids (9 species in Brazil), the genus *Rhamdia*, with two troglomorphic species in Brazil and one in Venezuela, encompasses the highest number of troglobitic catfishes in Mexico. It is noteworthy that the Mexican species are related to the epigean *R. laticauda*, whereas the Brazilian species is closely related to *R. quelen*. The heptapterid genera *Rhamdiopsis* (3 species) and *Pimelodella* (2 species) do not have troglobitic derivatives in other neotropical countries; *Pimelodella* catfishes are widely distributed in South America, but the poorly known genus *Rhamdiopsis* would be restricted to Brazil.

The Amazonian genus *Phreatobius* encompasses two Brazilian subterranean species sampled in artificial wells, *Phreatobius cisternarum*, the first described troglomorphic species in the Neotropical region, and *P. dracunculus*; in addition, at least three undescribed species were found living in submerged litter banks at the margins of the Rio Negro. *P. cisternarum* was object of a detailed anatomical study by Reichel (1927), without any similar. This species is widely distributed around the Rio Amazonas delta, from northern State of Amapá to Belém, Pará State, also occurring in the island of Marajó. The recently described *P. dracunculus* was also collected in wells, in Rondônia, at the right margin of the Rio Amazonas. A third subterranean species accessible through wells, *P. sanguijuela*, was described from Bolívia, illustrating the wide distribution of this genus. The systematic position of *Phreatobius* within the Siluriformes is under study using molecular techniques; apparently it forms a trichotomy within the Pimelodoidea clade (*Conorhynchos* + Heptapteridae (*Phreatobius* + Pimelodidae + Pseudopimelodidae)) (Muriel-Cunha 2008).

The Loricariidae, by far the most speciose neotropical siluriform family, are relatively poorly represented among the troglobitic fishes, with three species (two in Brazil, one in Venezuela) belonging to the genus *Ancistrus*. The armored catfishes are grazers, generally feeding on aquatic plants and periphyton, therefore not expected to be frequent in subterranean habitats. Nevertheless, some species would be able to survive and form populations in caves rich in plant detritus, feeding on the thin layer of finely particulate organic covering rocky substrates.

The Neotropical electric-fishes, Order Gymnotiformes (knifefishes), are typically preadapted to the subterranean life, as they live in turbid waters and/or have nocturnal activity, orienting by electroreception. Hence, it is not surprising to find a troglomorphic species, *Eigenmannia vicentespelaea*, in a cave system in Central Brazil, and several non-troglomorphic in other caves (see below).

In addition to the troglomorphic catfishes and knifefish, there is one Brazilian characiform, *Stygichthys typhlops*, whose familial position is still uncertain. When described, with basis on a single specimen accidentally collected during a well drilling operation in the early 1960's and sent to the USA, this species has been hypothesized to be a tetra characin (Brittan and Bohlke 1965). No additional specimen was collected until 2003, when a research team visited the type-locality, Jaíba, and obtained fish from two artificial wells (there were reports on the past occurrence of "white fish" in other springs and wells, which are currently dry – see Bichuette and Trajano, this volume). Taxonomic studies on this material revealed that *S. typhlops* is not a characin, but it familial position within the characiforms is still unclear (Moreira *et al.*, in press).

Brazilian subterranean ichthyofauna is distinguished worldwide by its diversity at the familial (six families with troglobitic representatives) and generic (12 genera) levels. This is probably due, at least in part, to the high epigean fish diversity providing a great array of potential colonizers.

Nearly 2,600 freshwater fish species are estimated to occur in Brazil (M.T. Pizza, pers. comm.), corresponding to 20 percent of the freshwater species in the world (ca. 13,000 – Lévêque *et al.* 2008). Therefore, the number of known Brazilian troglobitic fishes is relatively modest. This is partly because the largest and richest region for the fish fauna, which is the Amazon basin, is relatively poor in subterranean habitats suitable for colonization and differentiation of hypogean populations.

The high diversity of Brazilian subterranean fishes when compared to other countries may be due to one or all the following factors: 1) general high ichthyofaunistic diversity in Brazil, providing a wide variety of potential colonizers of hypogean habitats; 2) high research efforts than in most other regions; and/or 3) the essentially comparative approach by Brazilian researchers, using phylogenetic criteria for the recognition of the troglobitic status, which is primarily a distributional attribute (the restriction to the subterranean environment). Because it is not scientifically possible to prove their absence from surface habitats, the second best criterion for recognition of troglobites is the occurrence of differentiation, expected to occur sometime after genetic isolation.

Therefore, in a preliminary classification, we consider troglobitic any subterranean fish showing eyes significantly smaller than in the closest known epigean relative (phenotypic plasticity renders pigmentation less suitable for this purpose). Likewise, wider intrapopulation variability than in epigean relatives indicates taxonomic identity. This approach, which may inflate the list of Brazilian troglobitic fishes compared to those from other countries, avoids the traps of subjectively establishing a minimum degree of eye and pigmentation reduction acceptable for a "true" troglobite – this is basically a gradistic approach, whose inadequacy for evolutionary studies has long been demonstrated. Or worst, of confusing a small eye with a reduced eye and, conversely, a relatively large eye with a non-reduced one.

I.1.2 Troglomorphic, but not Found in Typical Subterranean Habitats

As observed in other large rivers, such as those in the Congo basin (Thinès and Proudlove 1986), troglomorphic fishes have also been found in epigean habitats in the Amazonas basin. Such fish may not be classified as troglobites, which are, by definition, inhabitants of hypogean habitats. Amazonian troglomorphic fishes include psammophilous (sand-dwelling) and fossorial species, such as *Pygidianops* spp. and *Typhlobelus* spp. (Trichomycteridae), and deep/turbid water species, as the siluriforms *Bathycetopsis oliverai* (Cetopsidae*)*, *Micromyzon akamai* (Aspredinidae) and an undescribed pimelodid, and the gymnotiform *Orthosternarchus tamandua* (Trajano 1997a).

I.1.3 Non-troglomorphic, but Found in Subterranean Habitats (Caves)

Very little attention has been devoted to non-troglomorphic fishes found in subterranean habitats and few publications (e.g., Poly 2001) mention their occurrence. Therefore, it is unclear whether the relatively rich non-troglomorphic ichthyofauna recorded in several Brazilian caves is a particularity of such karst areas or simply an artifact due to the prevalent lack of information for other countries.

It may be difficult to distinguish between troglophilic, trogloxenic and accidental fish in caves, because this requires case-to-case population studies focusing on the distribution, permanence and reproduction in the subterranean environment. The long-term permanence of a considerable number of individuals at relatively large distances from the cave entrances indicates the presence of self-sustained source populations in the subterranean biotope, defining the troglophilic status (*sensu* Holsinger and Culver 1988). Based on these criteria, Table 2 contains the species hypothesized as forming troglophilic populations in at least one Brazilian cave. As expected, since such fish are also subterranean, they mostly belong to taxa considered as preadapted to the cave life, i.e., the siluriforms and gymnotiforms. An exception is the Erythrinidae characiforms, sit-and-wait predators rather frequent in some caves harboring large populations of potential prey, represented, for instance by the troglobitic *R. enfurnada* and *A. cryptophthalmus* catfishes. No troglomorphic erythrinid has been

Table 2 Updated list of Brazilian troglophilic fishes, with localities (karst area/ county, State) and types of habitat (Trajano 2001). Based on Mattox (2008) and unpublished data (marked with asterisk, locality including cave).

Taxon	Locality	Habitat
CHARACIFORMES: ERYTHRINIDAE		
1 *Erythrinus* sp.	Altamira-Itaituba, PA	Base-level stream
2 *Hoplerythrinus unitaeniatus*	São Domingos, GO	Base-level stream
3 *Hoplias* cf. *malabaricus**	Serra do Ramalho, BA[1]	Base-level stream
GYMNOTIFORMES		
4 *Gymnotus* cf. *carapo**	Cordisburgo, MG [2]	Base-level stream
SILURIFORMES: LORICARIIDAE		
5 *Isbrueckerichthys alipionis* *	Alto Ribeira, SP[3]	Vadose tributary
6 *Hypostomus* sp.*	São Domingos, GO[4]	Vadose tributary
7 *Parotocinclus* sp.	Presidente Olegário, MG	Vadose tributary
TRICHOMYCTERIDAE		
8 *Trichomycterus* sp. A*	Mambaí, GO[5]	Base-level stream
9 *Trichomycterus* sp. B*	Cordisburgo, MG[2]	Vadose tributary
10 *Trichomycterus* aff. *T. mimonha**	Montes Claros, MG[6]	Base-level stream
HEPTAPTERIDAE		
11 *Pimelodella transitoria*	Alto Ribeira, SP	Base-level stream
12 *Pimelodella* sp.*	Cordisburgo, MG[2]	
13 *Rhamdia* sp.*	Varzelândia, MG[7]	Base-level stream
14 *Imparfinis hollandi*	São Domingos, GO	Base-level stream

[1]Enfurnado Cave; [2]Morena Cave; [3]Santana Cave; [4]São Bernardo Cave; [5]Penhasco and Nova Esperança caves; [6]Lapa do Zu cave; [7]Zé Avelino Cave.
*unpubl. data

reported, but a few individuals without eye balls were found in a sandstone cave in Amazonia (Trajano and Moreira 1991).

Table 2 shows only species presenting good evidence to form troglophilic populations. Many other species with an unclear status were recorded in Brazilian caves (see, for instance, Trajano 1991, Bichuette and Trajano 2003). Once more, most are siluriforms belonging to the same families as the troglobitic species (Loricariidae, Callichthyidae, Trichomycteridae, and Heptapteridae), but the diversity of characiforms, especially characids, is greater. Tetra characins are relatively frequent, but usually as isolated, emaciated individuals, probably trapped by accident in caves, after floods for instance. Their frequency in Brazilian caves would just reflect their abundance in epigean water bodies.

I.2 Habitat Diversity

Troglomorphic fish have been found in practically all kinds of Brazilian subterranean habitats, from non-karst alluvial sediments to typical karst habitats, from dozens of meters deep into the water table to shallow streams. Apparently, type of habitat is not correlated with taxonomy. In fact, congeneric fish such as *Ituglanis* and *Trichomycterus* catfish may occupy habitats as diverse as streams, phreatic waters and the epikarst (Table 1). *I. epikarsticus* is the first fish species reported to live in an epikarst aquifer, and distribution evidence points to dispersion through the epikarst in *I. bambui* and *I. ramiroi*. For another genus of Trichomycteridae, *Glaphyropoma* sp., the type locality is a quartzitic cave. The local terrain is composed of arenites, siltites, argilites and conglomerates and represents the unique record of a Brazilian troglobitic fishes in this kind of rock. Caves in this area are formed by the erosion of soft rock components by rainwater penetrating through surface cracks and leaving spaces delimited by the harder components.

Streams are likely the ancestral habitat for the Brazilian troglobitic siluriforms, except perhaps for the large *Rhamdia* and *Pimelodella* species, which may also be frequent in large rivers (Trajano and Bockmann 1999). Therefore, the colonization of phreatic spaces is probably an ecological specialization (autapomorphy) of species such as *Rhamdiopsis* sp. 1 and sp. 2, *Ituglanis epikarsticus, I. bambui* and *I. ramiroi*. Wide spaces such as large subterranean lakes and cave pools would represent a secondary habitat for the above mentioned *Rhamdiopsis* species.

Another sequence of evolutionary events, without a necessary phreatic step, seems logical for *Ancistrus formoso* and *Trichomycterus* sp. 1 (Table 1), that live in the lower part of the Serra da Bodoquena karst area and have bulky bodies, adapted to fast-flowing waters in wide spaces. Probably, their ancestors lived in streams crossing the respective localities at Bonito County and progressively adapted to slow-moving water as the caves were flooded due to the subsidence of the Pantanal region (Bonito Co. is located at the border of this region).

All known Brazilian troglobitic fishes live in subterranean habitats accessible through caves, except for *S. typhlops* and *Phreatobius* spp., collected in artificial wells. For both, the colonization of ground waters was probably vertical, associated with droughts of epigean water bodies, streams in a karst area in the case of *S. typhlops* and submerged litter banks in the case of *Phreatobius* species.

I.3 Morphological Diversity

The habitat diversity promotes an enhancement of variation in body size and shape. Fish adapted to life in narrow subterranean spaces tend to be smaller and more elongated than their stream-dwelling relatives, both epigean and hypogean. *Rhamdiopsis spinosum* 1, *Ituglanis epikarsticus* and, to a lesser extent, also *I. bambui* and *I. ramiroi* would be instances of evolutionary reduction in body size allowing for life in phreatic waters, before (preadaptation) or after (apomorphy) the colonization of this habitat (Bichuette and Trajano 2004). An opposite tendency was observed for *Trichomycterus* sp. 1 found in flooded caves and possessing a bulky body with a well developed pre-dorsal fold. Life in small spaces is also probably the cause for loss of the lateral line in the body observed in *Stygichthys typhlops*, *Phreatobius* spp. and *Rhamdiopsis* sp. 1 (from Chapada Diamantina).

However, the most striking morphological characteristic of the Brazilian troglobitic ichthyofauna is the inter- and intrapopulation diversity in degree of reduction of eyes and melanic pigmentation (Table 1). As in the case with the habitat diversity, this variation has no clear a taxonomic correlation, but an apparent geographic correlation has been recognized (Trajano and Bichuette 2005).

Fish such as *Pimelodella spelaea* (Fig. 1), *Rhamdiopsis* sp. 3 (from Cordisburgo) and *Glaphyropoma spinosum* (Fig. 1) have slightly reduced eyes and pigmentation, and only the comparative analysis allowed for detection of significant differences in relation to epigean congeners. Eyes and pigmentation are also slightly reduced in *Eigenmannia vicentespelaea* (Fig. 1), but a few anophthalmic, very pale individuals have been found (Bichuette and Trajano 2006). On the other hand, *S. typhlops*, *A. formoso*, *Rhamdiopsis* sp. 1 (Fig. 1) and *Rhamdiopsis* sp. 2 (Fig. 1) populations are homogeneously unpigmented and eyeless (at least in the external view).

All degrees of variation are observed between these extremes, frequently with a mosaic distribution of character states revealing an independent differentiation of eyes and pigmentation. In *Rhamdia enfurnada*, both

Fig. 1 Contd..

Fig. 1 Contd..

Fig. 1 Brazilian troglobitic fishes, showing inter-specific differences in degree of reduction of eyes and melanic pigmentation: a. *Pimelodella spelaea* (Photo: Maria Elina Bichuette); b. *Eigenmannia vicentespelaea* (Photo: José Sabino); c. *Rhamdiopsis* sp. 1 (Photo: Adriano Gambarini); d. *Rhamdiopsis* sp. 2 (Photo: José Sabino); e. *Ancistrus formoso* (Photo: José Sabino); f. *Stygichthys typhlops* (Photo: Cristiano Moreira); g. *Glaphyropoma spinosum* (Photo: Adriano Gambarini).

Color image of this figure appears in the color plate section at the end of the book.

characters show a normal distribution within the only known population, but with independent states in each individual: paler specimens may have relatively large eyes, dark fish may exhibit deeply reduced eyes, and so on. In *Pimelodella kronei*, pigmentation has a normal distribution in the best known population, found in Areias de Cima cave, but 90 percent of the fish do not have externally visible eyes and the remaining 10 percent have variable vestigial eyes. In *R. enfurnada, Rhamdia* sp. and *P. kronei*, vestigial eyes are frequently assymetrical in size, which is in accordance with neutralistic theories of regressive evolution of eyes in cave fish.

All collected specimens of *P. cisternarum* have tiny vestigial eyes and traces of pigmentation, whereas *P. dracunculus* catfish are more advanced in degree of troglomorphism, having no pigment cells or eyes. Eyes are also absent in *P. sanguijuela*.

The *Ituganis* catfish species, represented by five formal species (*I. passensis I. bambui, I. epikarsticus, I. ramiroi* and *I. mambai*) (Fernandez and Bichuette 2002, Bichuette and Trajano 2004, 2008), present a mosaic of characters considering the eyes and body pigmentation. *I. bambui* and *I. mambai*, which present a lower degree of regression of body pigmentation and eyes, possibly have been isolated in the subterranean environment for a shorter time than *I. passensis, I. ramiroi* and *I. epikarsticus. I. bambui* and *I. mambai* have similar body pigmentation patterns and eye sizes with irregular light brown spots along the dorsum and flanks and the head with a dorsal portion darker than the rest of the body; the eyes are visible externally as round black spots, which are covered by skin and reduced in relation to epigean *Ituglanis* species. The more troglomorphic species also present a mosaic, but, apparently the variability is lower. *I. ramiroi* species has a pale yellowish color pattern with scattered chromatophores spots along the body; the eyes appear as very small round black spots covered by thin skin, in some cases no eyes are externally visible. *I. passensis* can be considered an intermediary case, showing coloration from yellowish to light gray, with scattered chromatophores. The eyes are reduced and, sometimes, not visible externally. *I. epikarsticus* is the most troglomorphic species of *Ituglanis*, with individuals totally depigmented, showing a translucid aspect. The eyes are small round black spots covered by thin skin, with a lower diameter compared to the other *Ituglanis* cave species (0.1 mm).

A most remarkable instance of mosaic distribution of characters states, including eyes, pigmentation and body shape, was reported for four populations of the armored cave catfish, *Ancistrus cryptophthalmus*, from São Domingos karst area, Central Brazil (Reis *et al.* 2006). The populations found in the caves Angélica, Bezerra, São Vicente I and Passa Três (part of São Vicente cave system) were compared to *Ancistrus* catfish found in the epigean stream reaches and ranked according to their degree of reduction of eyes and pigmentation. Different sequences were observed for eyes (from less to more reduced): epigean (eyes as in epigean congeners) → Bezerra → Angélica → São Vicente I → Passa Três (eyes not externally visible in adults), and for pigmentation: epigean (light black to brownish) → Bezerra (brownish to light brownish) → Passa Três (light brownish to dark yellow-greenish) → São Vicente I and Angélica (light brownish to light yellow). General body shape was compared through geometric analysis, and a third sequence was found when associating head shape and position of eyes and nostrils, and length of caudal region (Reis *et al.* 2006: Fig. 5): epigean + Angélica + Bezerra + Passa Três → São Vicente I. This mosaic distribution of character states among the four studied cave populations indicates a certain degree of isolation, with some independent differentiation in these populations.

An interesting case of polymorphism is represented by the only known population of *Trichomycterus itacarambiensis* (Fig. 2). Two distinct color morphotypes are present, one represented by completely depigmented individuals (one third of the population as verified in 1994) and another by variably pale, but clearly pigmented catfish (the remaining two thirds); 15 years later, this proportion was half to half, due to a population decrease affecting mostly the pigmented fish (Trajano *et al.* 2009). Variably reduced eyes are present in both phenotypes, showing a normal distribution in relation to eye size; externally visible eyes in depigmented fish are red colored (Trajano and Pinna 1996).

Such diversity in troglomorphic degree among related taxa provides excellent opportunities for the investigation of mechanisms of character regression. In fact, morphological and physiological studies indicate that the diversity in modes of pigmentation variation (continuous versus discontinuous, polymorphism etc.) results from a combination of different mechanisms of pigmentation reduction. The continuous variation in color, from darker to paler individuals, is associated with a decrease in density (and, in some cases, also size) of melanin cells in the skin. In populations with a discontinuous variation, as *T. itacarambiensis*, the depigmented phenotypes correspond to individuals that lose the ability to synthesize melanin. Physiological studies showed that highly specialized troglobitic species, with homogeneously depigmented populations, respond differently to the administration of L-DOPA *in vitro*: in most, classified as DOPA(+), melanin is then synthesized and melanin cells become visible (e.g., *Stygichthys typhlops, Ancistrus formoso, Rhamdiopsis* sp. 1), but in some, as *Rhamdiopsis* sp. 2 and the depigmented individuals of *T. itacarambiensis*, there is no response, and they are classified as DOPA (–). Apparently, a polygenic system is involved in the differentiation of melanophores, determining their density in the skin. On the other hand, monogenic systems would be involved in the ability to synthesize melanin, which

Fig. 2 Contd..

Fig. 2 Contd..

Fig. 2 Intra-specific variability in the degree of reduction of eyes and pigmentation in *Trichomycterus itacarambiensis*; a-b: pigmented fish with variably sized eyes; c-d: light pigmented *versus* albino specimens. (Photos: José Sabino).

Color image of this figure appears in the color plate section at the end of the book.

could be lost due to the blocking of different steps in this synthesis (Felice *et al.* 2008).

I.4 Regional Diversity and Speciation in Brazilian Troglobitic Fishes

Neither the troglobitic nor the troglophilic subterranean ichthyofaunas are evenly distributed throughout the Brazilian karst areas, with some regions conspicuously richer than others. So far, São Domingos karst area, in the Rio Tocantins drainage, Amazonas basin, is the richest in both troglomorphic (seven species) and non-troglomorphic species (Bichuette and Trajano 2003). However, the most specialized troglobites occur in semiarid karst areas of central and northern Bahia State (northeastern Brazil), where epigean drainages are mainly temporary and the blind fish are generally confined to phreatic water bodies.

The lack of a strict geographic correlation between total richness (troglobites + non-troglobites) and richness of troglobites alone indicate the existence of different factors: whereas general subterranean diversity seems to be mainly related to ecological factors, such as habitat diversity and availability of food resources, in addition to richness of potential colonizers (all very high in São Domingos caves), richness of troglobitic species and their degree of differentiation is better explained by historical factors, mainly vicariance leading to speciation (Trajano 2001b).

Ideally, different kinds of evidence, distributional, geological, geomorphological, paleoclimatic, morphological etc., should be put together in order to uncover the most likely processes leading to the diversification of the subterranean fauna and the origin of troglobites. It is generally accepted that allopatric speciation is the main cause of animal diversity, and there is no reason to suppose that subterranean organisms are an exception.

Disruption of connection between epigean (source of colonizers) and hypogean drainages is the most obvious, and probably the commonest cause of genetic isolation of aquatic organisms in subterranean habitats. In the case of cyclical events, such as the Quaternary glacial cycles, geographic isolation in dry phases would intercalate with the re-establishment of drainage continuity in wet phases, when surviving epigean populations may come back in contact with the subterranean ones. In this case, either hybridizing between these populations or secondary contact with maintenance of their taxonomic identities (if reproductive incompatibility was achieved by differentiation) may occur.

Isolation during past drier climates seems to be the case of the blind catfish, *Pimelodella kronei*, currently in secondary contact with its epigean relative, *P. transitoria*. The Alto Ribeira karst area is presently situated in a wet zone, but there is good evidence of dry, isolation phases in subtropical Brazil during the last 116,200 years (Cruz-Jr. *et al.* 2005). On the other hand, karst areas in Bahia State where troglobitic fish are found (Table 1) are located in currently semiarid regions, thus in a phase favoring differentiation. Among these species, *Rhamdiopsis* sp. 1 (Chapada Diamantina) and sp. 2 (Campo Formoso), *R. enfurnada* and *Trichomycterus* sp. 2 live in water bodies without direct epigean connections. For Campo Formoso, there is also evidence pointing to accentuated paleoclimatic fluctuations in the past 210,000 years (Wang *et al.* 2004), but dry phases were longer than in the Alto Ribeira. Longer isolation periods may account for the high degree of troglomorphism observed in *Rhamdiopsis* spp. from Bahia.

The localities with *Stygichthys typhlops* as well have no permanent, direct connections with epigean drainages. In this case, lowering of the water table due to extensive water pumping for irrigation projects in the last five decades is speeding up the disconnection process and threatening the species by habitat loss (Moreira *et al.*, in press).

Since no epigean congener was found in São Domingos karst area, in spite of intensive collecting efforts (Bichuette and Trajano 2003), extinction of epigean relatives may explain the origin of *Pimelodella spelaea* and the four troglobitic *Ituglanis* species found in the area. Topographical isolation may contribute for differentiation in these species, since they live basically in the upper karst level (vadose tributaries and epikearst – Bichuette and Trajano 2004). Topographic isolation may also be a cause for differentiation in *Rhamdiopsis* sp. 3 (from Cordisburgo), which live in a stream several meters above the base level in the area (unpubl. data).

More complicated is the case with the armored catfish, *Ancistrus cryptophthalmus*, from São Domingos. Topographic isolation may explain the morphological differentiation between the four cave populations (see above), because they inhabit stream reaches separated by waterfalls (Reis *et al.* 2006). However, there is a parapatric contact between epigean *Ancistrus* and *A. cryptophthalmus* at the large cave entrances of Angélica and Bezerra caves (sinkholes and resurgences). In the absence of strong evidence for extinction of epigean populations sometime in the past, with a secondary contact nowadays, parapatric speciation cannot be ruled out.

II. NATURAL HISTORY AND BIOLOGY

From about 1,700 references dealing strictly with natural history and ecology of subterranean fishes (G. Proudlove, pers. comm.), ca 35 are related with Brazilian cavefish, including habitat descriptions and comments on habitat use, population studies, reproduction and behavior. The first work, more extensive, was published by Pavan (1946) focusing the heptapterid catfish *Pimelodella kronei*. From 1990's on, a series of ecological studies were published focusing the Brazilian cavefishes (e.g., Trajano 1991, 1997a, b, Trajano and Bockmann 1999, Trajano 2001a,b, Trajano and Bichuette 2007). These works include information on population sizes and densities, fish movements and life stories aspects, such as feeding, reproduction and growth rates. Some of these aspects will be considered herein.

II.1 Diet and Feeding Behavior

Brazilian subterranean fishes are frequently generalist carnivores feeding on an opportunistic basis (trichomycterids, heptapterids and the electric fish species). Potentially these fishes can explore many kinds of resources, like crustaceans (amphipods and isopods), aquatics insects (juveniles to adults), mollusks, worms (annelids), besides several allochthonous items (terrestrial invertebrates); in the few cases where direct comparisons were made, it was shown that subterranean fishes belong to similar trophic guilds of their epigean relatives (Trajano 1989, Bichuette, unpubl. data). This suggests that epigean ancestors were already preadapted to feed in the hypogean environment.

Special cases must be considered herein, for example, cannibalism was recorded for *Stygichthys typhlops* (Moreira *et al.*, in press). There are records of cave catfish foraging on submerged bat guano (*Rhamdiopsis* sp. 1 – Mendes 1995, M.E. Bichuette, pers. obs.), but is not clear whether these fishes are feeding directly on the guano or on the invertebrates living on it, since most of the stomach items are invertebrates. The armoured catfishes *Ancistrus cryptophthalmus* are typically detritivorous (Trajano 2001a), probably feeding on the film of detritus covering rocks.

The high proportion of specimens captured with empty stomachs or few items inside is directly related with the effect of food scarcity of caves. For instance, *Trichomycterus itacarambiensis* and *Ituglanis* populations showed this pattern during the dry season (Trajano 1997b, Bichuette, unpubl. data). Furthermore, it was observed a decrease in the frequency of insects and an increase of oligochaete items throughout the dry period for *T. itacarambiensis*, probably related with the decrease in insect populations along the dry period.

In relation to feeding behavior, different foraging styles were observed for Brazilian cavefishes, related to the peculiar conditions of the subterranean realm. The heptapterids also feed at midwater and surface (Trajano 1989,

Fig. 3 Contd..

Fig. 3 Contd..

Fig. 3 Habitat and population study methods for Brazilian cavefishes (Photos: Adriano Gambarini): a-b. habitat of *Glaphyropoma spinosum* from Chapada Diamantina, northeastern Brazil; c. collecting procedure for mark-recapture studies and; d. measuring standard length for mark-recapture studies.

Color image of this figure appears in the color plate section at the end of the book.

M.E. Bichuette, pers. obs.), contrasting to the epigean relatives which are typical bottom-dwellers.

II.2   Life History

Troglobitic organisms usually show a precocial lifestyle, including life history traits such as large and yolky eggs, low fecundity, delayed and infrequent reproduction, slow growth rates and increase longevity (Culver 1982). For Brazilian cavefishes, some of these traits were verified, suggesting a precocial lifestyle: *T. itacarambiensis* and *A. cryptophthalmus* showed low proportions of reproductive females, producing relatively large eggs, slow individual growth rates and high longevity (Trajano 1997b, Trajano and Bichuette 2007). However, a precocial lifestyle may be a characteristic already present in epigean ancestors, favoring the colonization of the subterranean environment, as observed for *Pimelodella kronei* and *P. transitoria* (Trajano 1989).

In general, reproductive data for cavefishes are limited since several subterranean habitats are not reachable during the rainy periods and the population densities are frequently low. Seasonal reproduction was observed for some Brazilian cave catfishes: *Rhamdiopsis* sp. 1 and *T. itacarambiensis* showed females with developed gonads in the end of the rainy season, suggesting a reproductive peak in this period (Mendes 1995, Trajano 1997b). Trajano (1991) verified a very low number of *Pimelodella kronei* catfish showing mature gonads and no reproductive patterns were detected.

In relation to individual growth, very low rates (0.06 mm $\times$ month^{-1} in average) or negative growth, were recorded for *Trichomycterus itacarambiensis* (Trajano 1997b). This study was carried out along the dry season when the condition factor decreased, demonstrating the influence of food limitation in the life cycle of this species. Low rates of individual growth and negative growth was also observed for *Ancistrus cryptophthalmus* (Trajano and Bichuette 2007). For *Pimelodella kronei*, average growth rates were 1.0 mm $\times$ month^{-1}, and longevity around 10-15 years (Trajano 1991). These data are similar with that recorded for the putative sister-species *P. transitoria* (Gerhard 1999), showing a precocial lifestyle and possibly configurating a preadaptation to the cave life.

III.   POPULATION ECOLOGY

The population ecology of several Brazilian cavefishes was studied using either mark-recepture methods (MR) or visual census (VC). MR produces more reliable data on population sizes, and population densities may be calculated based on the one occupied by these populations. In the case of VC, densities are estimated directly from counts.

Trajano (2001a) proposed a classification based on mean population densities of cavefishes: (1) species with low population densities (< 0.1 ind $\times$ m^{-2}), for example, *Rhamdiopsis* sp. 1 from Chapada Diamantina, (six populations from 10 studied) *Rhamdiopsis* sp. 2 from Campo Formoso, and *Pimelodella kronei*; (2) species with median population densities (0.1-1.0 ind $\times$ m^{-2}), for example *Trichomycterus itacarambiensis, Ancistrus cryptophthalmus, Ituglanis bambui, I. passensis, I. ramiroi* and *Rhamdiopsis* sp. 1; there are no records for the third category proposed by Trajano (op. cit.), i.e., high population densities (> 1.0 ind $\times$ m^{-2}, on average).

Population sizes, based on MR studies, also varied from very small to comparable to relatively small populations of epigean species. The population size of the stream-dweller *Trichomycterus itacarambiensis* from Olhos D'água cave (5,000 m of habitat extension) was estimated to be 1,500 to 2,000 adults; *Pimelodella kronei* from Areias cave (5,000 m of habitat extension), 900-1,200 adults; *Ancistrus cryptophthalmus* from Angélica cave (8,000 m of habitat extension) reached 20,000 individuals but the population from Passa Três cave (2,000 m of habitat extension), had only 1,000 individuals; the trichomycterids from Goiás, *Ituglanis* spp., showed relatively small populations, with hundreds of individuals.

References

Bichuette, M.E. and E. Trajano. 2003. Epigean and subterranean ichthyofauna from São Domingos karst area, Upper Tocantins river basin, Central Brazil. *Journal of Fish Biology* 63: 1100-1121.

Bichuette, M.E. and E. Trajano. 2004. Three new subterranean species of *Ituglanis* from Central Brazil (Siluriformes: Trichomycteridae). *Ichthyological Explorations of Freshwaters* 15(3): 243-256.

Bichuette, M.E. and E. Trajano. 2006. Morphology and distribution of the cave knifefish *Eigenmannia vicentespelaea* Triques, 1996 (Gymnotiformes: Sternopygidae) from Central Brazil, with an expanded diagnosis and comments on subterranean evolution. *Neotropical Ichthyology* 4(1): 99-105.

Brittan, M.R. and J.E. Böhlke. 1965. A new blind characid fish from southeastern Brazil. *Notulae Naturae* 380: 1-4.

Cruz Jr., F.W., S.J. Burns, I. Karmann, W.D. Sharp, M. Vuille, A.O. Cardoso, J.A. Ferrari, P.L.S. Dias and O. Viana Jr. 2005: Insolation-driven changes in atmospheric circulation over the past 116,000 years in subtropical Brazil. *Nature* 434: 63-66

Culver, D.C. 1982. *Cave Life. Evolution and Ecology*. Harvard Press, Cambridge.

Felice, V., M.A. Visconti and E. Trajano. 2008. Mechanisms of pigmentation loss in subterranean fishes. *Neotropical Ichthyology* 6(4): 657-662.

Gerhard, P. 1999. Ecologia de populações e comportamento de quatro espécies de bagres Heptapterinae (Teleostei: Siluriformes) em riachos do Alto vale do rio Ribeira (Iporanga, São Paulo). MSc. Thesis – Instituto de Biociências, Universidade de São Paulo, São Paulo. 129 pp.

Holsinger, J.R. and D.C. Culver. 1988. The invertebrate cave fauna of Virginia and a part of Eastern Tennessee: Zoogeography and ecology. *Brimleyana* 14: 1-162.

Lèvêque, C., T. Oberdorff, D. Paugy, M.L.J. Stiassny and P.A. Tedesco. 2008. Global diversity of fish (Pisces) in freshwater. *Hydrobiologia* 595: 545-567.

Mattox, G.M.T., M.E. Bichuette, S. Secutti and E. Trajano. 2008. Surface and subterranean ichthyofauna in the Serra do Ramalho karst area, northeastern Brazil, with updated lists of Brazilian troglobitic and troglophilic fishes. *Biota Neotropica* 8(4): 145-152.

Mendes, L.F. 1995. Ecologia populacional e comportamento de uma nova espécie de bagres cavernícolas da Chapada Diamantina, BA (Siluriformes, Pimelodidae). MSc. Thesis-Instituto de Biociências, Universidade de São Paulo, São Paulo. 86 pp.

Moreira, C.R. M.E. Bichuette, O. Oyakawa, M.C.C. Pinna and E. Trajano. Rediscovery of *Stygichthys typhlops* Britton and Bohlke, 1965, an enigmatic subterranean characiform fish from Jaiba karst area, eastern Brazil. *Journal of Fish Biology*, in press.

Muriel-Cunha, J. 2008. Biodiversidade e sistemática molecular de Phreatobiidae (Ostariophysi, Siluriformes) - com uma proposta sobre sua posição filogenética em Siluriformes e uma discussão sobre a evolução do hábito subterrâneo, Unpubl. Ph.D. thesis, Universidade de São Paulo, Sao Paulo, 144 p.

Pavan, C. 1945. Os peixes cegos das cavernas de Iporanga e a evolução. *Boletim da Faculdade de Filosofia Ciências e Letras, Biologia Geral* 6: 79: 1-104.

Poly, W.J. 2001. Nontroglobitic fishes in Bruffey-Hills Creek Cave, West Virginia, and other caves worldwide. *Environmental Biology of Fishes* 62: 73-83.

Reichel, M. 1927. Étude anatomique du *Phreatobius cisternarum* Goeldi, silure aveugle du Brésil. *Revue Suisse de Zoologie* 34(16): 285-403 + 6 pl.

Reis, R.E., E. Trajano and E. Hingst-Zaher. 2006. Shape variation in surface and cave populations of the armoured catfish *Ancistrus* (Siluriformes: Loricariidae) from the São Domingos karst area, Upper Tocantins River, Brazil. *Journal of Fish Biology* 68: 414-429.

Thinès, G. and G. Proudlove. 1986. Pisces. In: *Stygofauna Mundi*, L. Botosaneanu (ed.). E.J. Brill, Leiden, pp. 709-733.

Trajano, E. 1987. Biologia do bagre cavernícola, Pimelodella kronei, e de seu provável ancestral, *Pimelodella transitoria* (Siluriformes, Pimelodidae). Ph.D. thesis, Instituto de Biociências, Universidade de São Paulo, São Paulo, 211 pp.

Trajano, E. 1989. Estudo do comportamento espontâneo e alimentar e da dieta do bagre cavernícola, *Pimelodella kronei*, e seu ancestral epigeo, *Pimelodella transitoria* (Siluriformes, Pimelodidae). *Revista Brasileira de Biologia* 49(3): 757-769.

Trajano, E. 1991. Populational ecology of *Pimelodella kronei*, troglobitic catfish from southeastern Brazil (Siluriformes, Pimelodidae). *Environmental Biology of Fishes* 30: 407-421.

Trajano, E. 1997a. Synopsis of Brazilian troglomorphic fishes. *Mémoires de Biospéologie* 24: 119-126.

Trajano, E. 1997b. Population ecology of *Trichomycterus itacarambiensis*, a cave catfish from Eastern Brazil (Siluriformes, Trichomycteridae). *Environmental Biology of Fishes* 50: 357-369.

Trajano, E. 2001a. Ecology of subterranean fishes: An overview. *Environmental Biology of Fishes* 62(1-3): 133-160.

Trajano, E. 2001b. Mapping subterranean biodiversity in Brazilian karst areas. *Karst Waters Institute Special Publication* 6: 67-70 (Proceedings of the Workshop "Mapping Subterranean Biodiversity", Moulis).

Trajano, E. 2005. Evolution of lineages. In: *The Encyclopedia of Caves*, D.C. Culver and W.B. White (eds.). Academic Press, San Diego, pp. 230-234.

Trajano, E. and J.R.A. Moreira. 1991. Estudo da fauna de cavernas da Província Espeleológica Arenítica Altamira-Itaituba, PA. *Revista Brasileira de Biologia* 51(1): 13-29.

Trajano, E. and F.A. Bockmann. 1999. Evolution of ecology and behaviour in Brazilian cave Heptapterinae catfishes, based on cladistic analysis (Teleostei: Siluriformes). *Mémoires de Biospéologie* 26: 123-129.

Trajano, E. and M.E. Bichuette. 2005. Diversity of subterranean fishes in Brazil. In: *World Subterranean Biodiversity*, J. Gibert (ed.). Proceedings of an International Symposium held on 8-10 December 2004 in Villeurbaine, pp. 161-163.

Trajano, E. and M.E. Bichuette. 2007. Population ecology of cave armoured catfish, *Ancistrus cryptophthalmus* Reis, 1987, from central Brazil (Siluriformes: Loricariidae). *Ecology of Freshwater Fish* 16: 105-115.

Trajano, E., S. Secutti and M.E. Bichuette. 2009. Population decline in a Brazilian cave catfish. *Trichomycterus itacarambiensis* Trajano and Pinna, 1996 (Siluriformes): reduced flashflood as a probable cause. *Speleobiology Notes* 1: 24-27.

Wang, X., A.S. Auler, R.L. Edwards, H. Cheng, P.S. Cristalli, P.L. Smart, D.A. Richards and C.-C. Shen. 2004. Wet periods in northeastern Brazil over the past 210 kyr linked to distant climate anomalies. *Nature* 432: 740-743.

CHAPTER 10

Subterranean Fishes of Africa

Roberto Berti[1] and Giuseppe Messana[2]
[1]Department of Evolutionary Biology,
University of Florence, Italy
[2]Institute for the study of Ecosystems, National
Research Council (CNR), Florence, Italy

INTRODUCTION

The extension of Africa comprising temperate to tropical latitudes, with a great quantity of different ecosystems, makes it an extremely variegated continent with high levels of biodiversity. A few zones have been explored from a biospeleological point of view, nonetheless the results obtained so far show a great subterranean diversity. Most of the biospeleological research begun in the first decades of the twentieth century and was achieved on Eastern (Ethiopia, Kenya and Somalia) Northern (especially Morocco) Central (Democratic Republic of the Congo) and Southern (Namibia and S. Africa) Africa and on Madagascar. Several researchers explored these territories and collected a large amount of zoological material.

It is in these years that the geographic and naturalistic exploration led to the discovery of the majority of the species of African stygobitic fishes, distributed only in the sub-Saharan hydrographical systems. The first cave dwelling fish was a little cyprinid discovered in the Democratic Republic of the Congo (Belgian Congo, at that time) and it is the first subterranean fish in the world that was officially protected.

Among the eight valid species of fishes known for the Afrotropical Region, six were described between 1921 and 1936. The beginning of the World War II, followed by the instability due to decolonization, was the cause of a long interruption of naturalistic research in Africa. This is the reason why for the description of the two new African subterranean fishes, we have to wait for the second half of the century.

It is also in this period that an extensive biological, physiological and behavioural research was conducted by Belgian and Italian researcher on the stygobitic fishes respectively from Central and Eastern Africa.

The present chapter enlightens how much is known about the distribution, biology and eco-physiology of each of the fish species inhabiting the subterranean waters of the African continent.

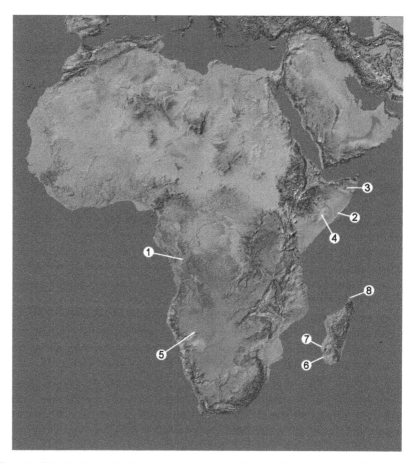

Fig. 1 Distribution of African subterranean fishes. The type localities of the eight species are indicated. 1: *Caecobarbus geertsii*; 2: *Phreatichthys andruzzii*; 3: *Barbopsis devecchii*; 4: *Uegitglanis zammaranoi*; 5: *Clarias cavernicola*; 6: *Typhleotris madagascariensis*; 7: *Typhleotris pauliani*; 8: *Glossogobius ankaranensis*.

Color image of this figure appears in the color plate section at the end of the book.

FAMILY CYPRINIDAE

Caecobarbus geertsii Boulenger, 1921

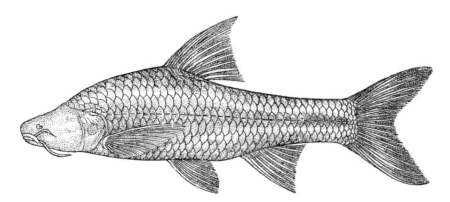

Fig. 2 *Caecobarbus geertsii* (after M. Poll; from Leleup 1956a, modified).

Historical Notes

The fish was discovered by some amateur cave explorers in 1917 in a cave near the town of Thysville, now Mbanga-Ngungu. One of them was Mr G. Geerts, at that time Managing Director of the Railway Company that exploited the railway from Matadi to Leopoldville via Thysville. Three years later Dr Henri Schouteden, a Belgian zoologist responsible for the zoological collections of the Congo Museum in Tervuren, went to Congo for a collecting trip. He met Mr Geerts, who told him about a blind depigmented fish living in some caves near Thysville and arranged for him the collection of three specimens (Thys van den Audenaerde 1999). Shipped to Belgium, the fish was described by Georges Albert Boulenger, the world's expert on African fishes, who established the new genus and species.

Caecobarbus is the first subterranean fish in the world to be officially protected, since the Belgian colonial authorities, in 1937, placed it on the list of fully protected animals, in the same category as gorilla and chimpanzee (Thys van den Audenaerde 1999). Moreover, this is the only hypogean fish whose trade is regulated by CITES, having been listed in CITES Appendix II in 1981 (Proudlove and Romero 2001, Proudlove 2006).

Derivatio Nominis

Caecobarbus: from *caecus* (blind, in Latin) + *barbus* (because of its close relationships with the genus *Barbus*);
geertsii: as a tribute to Mr G. Geerts.

Common Name

Congo blind barb (English), barbu aveugle (French).

Size

Up to 110 mm (Thinès 1955a, 1969)

Distribution Area and Ecological Notes

The species inhabits (Heuts 1952, Heuts and Leleup 1954, Leleup 1956a, 1956b) seven between caves and cave complexes (Table 1) located near and south of the town of Mbanga-Ngungu (formerly Thysville; 05°15'08"S, 14° 52'09"E), Western Democratic Republic of the Congo.

Table 1 Caves inhabited by *Caecobarbus geertsii* (from Heuts and Leleup 1954, modified).

Cave	Water temperature (°C)	Water hardness	pH	Altitude (m a.s.l.)	Tributary to
B 5 (Grotte des Gaz)	23	11.5	7.4	595	Fuma
B 6c (Grotte des Cascades)	-	-	-	528	Fuma
B 7 (Grande Grotte de Thysville)	22.5-23.4 °C	8-13.5	7.4-7.6	682.5	Fuma
B 11 (Grotte de Nenga)	22	15.5	7.3	540	Fuma
B 15a	22	10.75-12.75	7.5-7.6	515	Fuma
B 15b	21-22	13-13.75	-	-	Fuma
B 15e	-	-	-	-	Fuma
B 15f	-	-	-	500	Fuma
B 16a	21.5	15.5	7.8	640	Kokosi
B 16b	-	-	-	-	Kokosi
B 20	21.5	12	-	-	Kokosi

The caves colonized by *Caecobarbus* belong hydrographically to two streams (Fuma and Kokosi, respectively), both tributary to the Kwilu River (Lower Congo basin). The distribution area is quite small, since all collection sites are placed within a radius of 7/8 km at the most. The area is included in limestone formations, laid down during the Pre-Cambrian or at least during the Cambrian; the genesis of the caves is thought to take place between late Tertiary and mid-Holocene (Heuts 1952).

The *Caecobarbus* habitats show pronounced seasonal cycles of activity. During the rainy season they receive heavy inflows of water carrying nutrients from the surface and eventually leading to complete submergence; during the dry season the subterranean water flow either is not perceivable or it is weak and of endogenous origin. Therefore, periodic changes occur in food resources availability, fish density and water chemical features, the latter ones showing cyclic variations (1 to 3 times wide) in $Ca(HCO_3)_2$ concentration and in free CO_2 contents. No subterranean continuity is thought to exist between cave complexes, thus each complex seems to be an isolated subterranean unit; this is confirmed by both chemical features and population analyses (Heuts 1952, Heuts and Leleup 1954).

The aquatic associated fauna is quite rich, and almost exclusively composed of invertebrates. These include both potential preys as culicids larvae, planarians, oligochaetes, crustaceans (ostracods, copepods, isopods), the dryopid beetle *Troglelmis leleupi* Jeannel, and potential predators as the crab *Potamonautes biballensis* or dytiscids larvae (Leleup 1956a, 1957). The only available data about vertebrates concern (Heuts and Leleup 1954, Leleup 1956a) the presence of few epigean fishes, as some siluriforms – most likely *Clarias* sp. (B15 and B6 cave complexes) and *Labeo* sp. specimens (B6 cave complex).

Systematics and Phylogenetic Relationships

The species is generally thought to be phylogenetically related to the genus *Barbus* Cuvier (Boulenger 1921, Pellegrin 1928, 1929a, 1929b, Gianferrari 1932, 1934, Petit 1938a, Bruun and Kaiser 1943, Heuts 1953, Matthes 1963, Vandel 1965, Thinès 1969). In *Caecobarbus* the pharyngeal teeth are arranged over two instead of three rows as it occurs in *Barbus*; on the ground of this feature some authors (Pellegrin 1932, Gianferrari 1937a) suggested its close relationships with a primitive *Barbus* type.

More recently Howes (1991) proposed the species to belong to a barbin lineage of the subfamily Cyprininae.

Morpho-physiological Notes

When adult, *Caecobarbus* is a blind, apparently anophthalmic, cavefish. In small individuals (under 10 mm in length) the eye is still visible by transparency through the skin of the orbital region as a dark spot (Heuts 1952); almost all histological components of the eye of epigean fish larvae are recognizable, moreover both optic nerve and chiasma are still present (Quaghebeur 1955). Then a degenerative process occurs, and the adult eye is vestigial, deeply receded inside the skull and consisting only of histological remnants, disorderly arranged and covered by connective tissue and skin (Gérard 1936, Quaghebeur 1955). Moreover, eye degeneration is accompanied by the reduction of optic lobes (Quaghebeur 1957) as well as by a rearrangement of the circumorbital dermal bones (Petit 1941).

The metabolic rate, as measured by oxygen consumption per unit weight, is three times lower than in the epigean *Barbus conchonius*, and the existence of a low growth rate was established by scale readings (Heuts 1952, 1953).

The scales are still present, but very thin and soft (Boulenger 1921, Petit 1938a). The skin colour is reddish, due to the loss of all external melanine pigmentation, while black pigment cells are present on the peritoneum. A spot of guanine pigment on the operculum is shown by the populations of two cave systems (B16 and B20, respectively); it is absent in the remnant ones (Heuts 1952). Other differences exist between different populations; this is the case of a) the highest ray of the dorsal fin, which is not serrated in six out of the seven studied populations, and it is in the seventh (from cave B5) population, b) the age distribution, deduced from scale readings, which appears to be typical for each population, and c) the growth rate curve (Heuts 1952). The variability existing between populations for these morphological and physiological features strongly supports the hypothesis that the species could be structured in distinct local populations (Heuts 1952, Heuts et Leleup 1954).

Histological surveys showed that a) the thyroid is little developed, with a lowered functional activity (Olivereau and Francotte-Henry 1955), b) the hypophysis lacks thyrotropic cells, and this could explain the reduction of the thyroid gland (Olivereau and Herlant 1954). The aforementioned findings appear to be in complete accordance with both physiological features and morphological traits, as lowered growth rate, reduced metabolic rate and loss of external melanine pigmentation (Olivereau and Francotte-Henry 1955).

Besides these degenerative features, one compensative trait is shown. Thus, whereas the trunk lateral line system is normally developed (Boulenger 1921), an increased amount of sensory buds – neuromasts or taste buds – is present on the head (Heuts 1952, 1953).

Ethological Notes

The first behavioural observations on *Caecobarbus* were few remarks on its spontaneous behaviour in captivity describing the swimming characteristics and the feeding behaviour (Petit and Besnard 1937).

The first quantitative studies are those by Thinès (1953, 1955b) on the light sensitivity: tested in an alternative chamber apparatus *Caecobarbus* shows an evident photophobic behaviour, which doesn't vary in intensity as a function of the light wavelength. In the field, fishes observed in a light gradient at the cave entrance exhibited different behaviours as a function of their age, the young fishes being apparently indifferent to light and the old ones becoming photophobic for light intensities higher than 8 lux (Thinès 1958a).

Feeding behaviour studies highlighted how *Caecobarbus* react to food administration with an immediate polarization of its swimming activity towards the bottom where the fish starts an active food search (Thinès and Wissoq 1972). The stereotype of polarisation on the substrate has an important role in the behavioural repertoire of *Caecobarbus*, since it appears both after mechanical stimulation (Thinès and Wissoq 1972) and as response to alarm substance, which is thought to elicit a real feeding behaviour (Thinès and Legrain 1973).

Chemical stimulations not only trigger feeding behaviour of *Caecobarbus* (Thinès and Wissoq 1972), they are also effective in influencing its topographic orientation. In fact, when placed in a choice-apparatus, adult fishes show a significant tendency to orient towards the section in which water from a tank occupied by familiar conspecifics is introduced (Berti and Thinès 1980). In despite of this apparent tendency to aggregation, neither schooling nor shoaling behaviour is shown, as evidenced by Jankowska and Thinès (1982).

FAMILY CYPRINIDAE

Phreatichthys andruzzii Vinciguerra, 1924

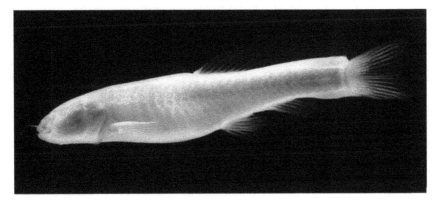

Fig. 3 *Phreatichthys andruzzii* (photo by R. Innocenti).
Color image of this figure appears in the color plate section at the end of the book.

Historical Notes

The first six specimens were collected at an unspecified date by lieutenant Giovanni Zaccarini, a geographer who played an active role in the ichthyologic exploration of the inland waters of Italian Somaliland. Dr Alcibiade Andruzzi, Navy medical major and at that time director of the Colonial Health Service of Somalia, took care of some naturalistic investigations on behalf of the Museo Civico di Storia Naturale di Genova. Thus, when he received the material collected by Zaccarini, shipped the fishes to Prof. Decio Vinciguerra, ichthyologist of that museum, who examined and described them as belonging to a new genus and species.

Derivatio Nominis

Phreatichthys: from φρέαρ *phrear* (well, in Greek) + ἰχθύς *ichthys* (fish, in Greek);
andruzzii: in homage to A. Andruzzi.

Common Name

None

Size

Up to 106 mm in total length and to 87 mm in standard length (personal observations).

Distribution Area and Ecological Notes

The species inhabits the subterranean waters of the Central Somalia. From a geological point of view, the hydrographic system where *Phreatichthys* occurs is included in evaporative formations of the middle-lower Eocene Age, connected in the West with biogenic limestone massive formations of the lower Eocene.

The distribution area of *Phreatichthys* is rather small, the distance between the two farthest collection sites being of about 30 km. The habitat is probably a continuous hydrographic system, but still the species seems to be structured in distinct local populations, each probably occupying its own area (Berti and Zorn 2001), and among which gene flow seems to be either restricted or absent (Cobolli Sbordoni *et al.* 1996). The water table is very superficial, lying at a depth varying from some meters to few tens of centimeters under the ground surface. Besides the type locality, six collection sites are known (Table 2).

Table 2 Known collection sites of *Phreatichthys andruzzii*; type locality: Bud Bud (from Ercolini *et al.* 1982, modified).

Locality	Latitude	Longitude	Altitude (m)	T(°C)
Bud Bud (3 sites)	04°11'19"N	46°28'27"E	137	29
Gal Ef (2 sites)	04°11'53"N	46°28'16"E	139	-
El Dirri	04°20'12"N	46°36'24"E	136	29
Diptias	04°07'02"N	46°27'16"E	137	31
Damairor	04°08'15"N	46°27'52"E	137	30
Gheriale	04°08'22"N	46°29'05"E	133	29
El Dauao (2 sites)	04°10'58"N	46°29'18"E	136	31

The collection stations are either hand dug wells (Bud Bud 1, Gal Ef 2, Damairor) or springs (El Dirri, Diptias, El Dauao 1) or sites where the water table become superficial after the *caliche* had collapsed (Gal Ef 1, Gheriale, El Dauao 2, Bud Bud 2 and 3).

In the wells of Bud Bud, together with *Phreatichthys*, were collected the isopod *Acanthastenasellus forficuloides* new species and genus (Chelazzi and Messana 1985) and a new amphipod of the genus *Afridiella*, *A. pectinicauda* (Ruffo 1982).

Systematics and Phylogenetic Relationships

The species was suggested to be phylogenetically related to the genus *Barbus* on the ground of a purely morphomeristic examination (Vinciguerra 1924). The descriptor's statement was in principle shared by other authors (Pellegrin 1932, Gianferrari 1932, 1934, 1937a, Bruun and Kaiser 1943, Matthes 1963, Vandel 1965, Thinès 1969, Ercolini *et al.* 1982, Howes 1991) and was never rejected until Banister (1984) pointed out evident osteological similarities to the genus *Garra*.

The hypothesis of a close phylogenetic relationship between *Phreatichthys* and the genus *Garra* is supported by a recent molecular phylogenetic study (Colli *et al.* 2008). The study, conducted on mitochondrial sequences corresponding to the whole cytochrome *b* gene and to a fragment of the 12s rRNA gene, shows that *Phreatichthys* is firmly placed within a well-supported clade composed exclusively by Asiatic and Middle Eastern species of the genus *Garra*.

Finally, a rather close relationship with the other Somalian cave cyprinid, *Barbopsis devecchii*, is shown by the results of an electrophoretic study (Cobolli-Sbordoni *et al.* 1996). Allozymes analyses suggest that both species could originate from a common epigean ancestor, widely spread in the area in the past.

Morpho-physiological Notes

Phreatichthys andruzzii is one of the best adapted forms to the subterranean habitat, as shown by the extent of its degenerative features. Scales and pigmentation are fully absent (Vinciguerra 1924). The maximal eye differentiation is reached 36 hours after egg laying, followed by a degeneration so rapid and hard that one month later a rudimentary cyst is all that remains (Berti *et al.* 2001). In the adult the anophthalmia is complete, accompanied by the loss of the optic nerves; a strong reduction of the entire encephalon, and in particular of the optic lobes, is also present (Ercolini and Berti 1975).

No compensative improvement is revealed by any other sensory system. A comparative survey of structural and ultrastructural features of the olfactory epithelium revealed no significant qualitative nor quantitative compensations in relation to the degeneration of the visual organ (Delfino *et al.* 1981). While the cephalic lateral line system is perfectly developed, according to the typical teleostean pattern, the trunk system appears heavily degenerated. The canal is not continuous, wide interruptions are present and only few and short tubes persist along its course (Berti, unpublished data).

Finally, an histological and histochemical survey carried out on the epidermis cells (Bianchi 1978a) showed, besides the characteristics of the muciparous cells, the presence of club cells, the alarm substance cells.

The high degree of *Phreatichthys* adaptation to the subterranean life is also confirmed by the presence of an extremely reduced metabolic rate. During the tests carried out with a respirometer *Phreatichthys* shows an oxygen consumption 2 to 2.5 times lower than *Puntius lateristriga*, while under anoxia the time survival results 3 times higher (Ercolini *et al.* 1987). Such a high capability to survive under anoxia is probably to be related to a very peculiar mechanism of respiratory compensation. In all Osteichthyes so far studied, the spleen stores and releases erythrocytes according to the animal's respiratory needs. *Phreatichthys*, on the contrary, uses its liver rather than spleen as site of accumulation in the respiratory compensation processes, like it occurs in the amphibian *Rana esculenta* (Frangioni *et al.* 1997).

Phreatichthys seems to tolerate large variations of water salinity, as indicated by the 1:3 ratio of the conductivity values recorded at Bud Bud and, respectively, at El Dirri (Ercolini *et al.* 1982)

The species is very easy to breed; both spontaneous and hormone induced reproductions were repeatedly obtained in captivity. *Phreatichthys* is a long-lived fish; at present wild individuals collected in 1977 and 1980 are still alive (personal observations).

Ethological Notes

In spite of the complete anophthalmia, the adult is clearly photosensitive. When the fish had to choose between a dark and a light sector (monochromatic equi-energetic lights of different wavelength), a clear photophobic behaviour is shown, particularly under blue light (Ercolini and Berti 1975).

Compared with *Barbus filamentosus* tested in absence of any visual sensory stimulation, *Phreatichthys* shows a more efficient feeding behaviour. The difference seems to be related to both the tendency to swim closer to the bottom and the behavioural stereotypes of food localization and intake. The food search is clearly guided by chemical stimulation, but some data indicate that also mechanical stimulation could play some role (Berti and Masciarelli 1993).

The chemical information is effective in orienting the fish locomotory activity also. *Phreatichthys andruzzii* can discriminate both between the odour of familiar and unfamiliar conspecifics (Berti *et al.* 1982) and between the odour of unfamiliar conspecifics and heterospecifics (Berti *et al.* 1989), showing different behavioural responses to the three chemical cues: indifference to the odour of familiar conspecifics, low avoidance for unfamiliar conspecifics, higher avoidance for heterospecifics (Berti and Zorn 2001). It was hypothesised that avoidance of areas in which odours of unfamiliar fishes are detected indirectly assures that each specimen remains within its habitual living area. The response by *P. andruzzii* to extraneous fish odours could be part of a behavioural mechanism of spatial orientation based on chemical information.

Recently, it has been shown that fish swimming can be topographically polarised by self-odour perception. When an unfamiliar area is experimentally scented with fish self-odour, *Phreatichthys* behaves as if the area was previously explored. The fish prefers an odour-free area to a self-odour-scented one, and when offered the choice between a familiar and an unfamiliar area, it prefers the unexplored environment. Avoidance of self-odour-scented areas would allow effective exploration of the subterranean environment, thus minimizing the risks of repeatedly exploring the same water volumes (Paglianti *et al.* 2006).

FAMILY CYPRINIDAE

Barbopsis devecchii Di Caporiacco, 1926

Fig. 4 *Barbopsis devecchii* (photo by R. Innocenti; from Ercolini & Berti 1978).

Historical Notes

During the spring of 1924 (Di Caporiacco 1927) the Florentine professors G. Stefanini and N. Puccioni made several naturalistic investigations in the northern part of the Italian Somalia. Both were whole naturalists, a paleontologist the former and an anthropologist the latter. When they visited the village of Talèh they didn't fail to collect the cyprinids living in those wells. Back home, they delivered the fishes to Di Caporiacco who described them as a new genus and species.

Derivatio Nominis

Barbopsis: from *Barbus* (because of its likeness to the genus *Barbus*) + ὄψις *opsis* (semblance, likeness, in Greek)

devecchii: as a tribute to Cesare Maria De Vecchi, governor of Italian Somalia in those years.

Synonyms

Barbopsis devecchi (Di Caporiacco 1926)
Eilichthys microphthalmus (Pellegrin 1929a, 1929b)
Barbopsis stefaninii (Gianferrari 1930)
Zaccarinia stefaninii (Gianferrari 1934)

Common name

None

Size

Up to 103.4 mm in total length and to 83.4 mm in standard length (Poll 1961).

Distribution Area and Ecological Notes

The distribution area of *Barbopsis* is quite large, stretching over a wide part of the Wadi Nogal Valley, in the North-Eastern Somalia. According to the collection sites so far known (Table 3), it extends along a NW-SE axis from Talèh to Eil, for at least 200 km in length. Eil is the site where the two specimens of *Barbopsis* described by Pellegrin (1929a, 1929b) as *Eilichthys microphthalmus* were collected. A survey of the region, made in 1980, greatly improved the knowledge of the species distribution, but was unable to confirm the presence of the fish in that site (Ercolini *et al.* 1982).

Table 3 Known collection sites of *Barbopsis devecchii*; type locality: Talèh (from Ercolini *et al.* 1982, modified).

Locality	Latitude	Longitude	Altitude (m)	T(°C)
Talèh	09°09′21″N	48°24′22″E	645	28
Gibaganle	08°30′51″N	48°39′14″E	395	29
El Goddomei	08°31′52″N	48°44′57″E	380	26
Bug Der	08°35′51″N	48°47′34″E	376	30
Callis	08°23′15″N	49°05′09″E	312	30.5
Eil	07°58′44″N	49°49′00″E	25	-

The geological nature of the territory, which includes the collection wells, is rather uniform: it is formed by evaporitic formations of the lower and middle Eocene (Talèh evaporites: anhydrites, gypsum, dolomites and clays), except in the zone of Eil where an oligocenic formation prevails, with marls and biological limestones.

The habitat of *Barbopsis* is probably constituted by a continuous water layer, with fissures and small intercommunicating cavities. In Talèh, the type locality, and surroundings, are present caves sometimes conspicuous (Messana *et al.* 1985). *Barbopsis devecchii* is able to tolerate wide variations of some environmental characteristics showing, as total salinity and hardness, 1:2 ratios among different sites (Ercolini *et al.* 1982).

Some interesting stygobitic Crustacea have been found in the waters where *Barbopsis* lives. The wells of Bug Der and Gibaganle are inhabited also by the microphthalmic decapod *Caridina lanzana* (Ercolini *et al.* 1982), while in Gibaganle, Callis and Eil the cirolanid isopod *Haptolana somala* was found (Messana and Chelazzi 1984). Finally the springs of Eil host, together with *H. Somala*, also the isopod *Stenasellus migiurtinicus* (Messana and Chelazzi 1984). Probably, these crustaceans are the main trophic resource of *Barbopsis*, which has a carnivorous attitude in captivity (personal observations). In our opinion should be rejected the hypothesis of a mainly phytophagous habit, based on the study of the feeding mechanism and of the gut content of a single specimen (Matthes 1963). Moreover such a kind of diet doesn't fit (Proudlove 2006) to a species so much adapted to cave life.

Systematics and Phylogenetic Relationships

The phylogenetic affinity between *Barbopsis* and the genus *Barbus* was stated since the original description of *B. devecchii* (Di Caporiacco 1926) and was also suggested for both *Eilichthys microphthalmus* (Pellegrin 1929a, 1929b, 1932) and *Barbopsis stefaninii* (Gianferrari 1930, 1932, 1934, 1937a). Both latter species were considered invalid by Poll (1961) and synonimised with *B. devecchii*. However the close relationship of *Barbopsis* with the genus *Barbus* was widely validated (Bruun and Kaiser 1943, Vandel 1965, Thinès 1969, Ercolini *et al.* 1982, Howes 1991).

In the years some doubts arose about the value of such phylogenetic affinities. Matthes (1963) stated that *Barbopsis* derived from a "primitive *Barbus* type", thus sharing the opinion Gianferrari (1937a) had formed on the ground of the arrangement of the pharyngeal teeth, while Banister (1984), in his paper on subterranean populations of *Garra barreimiae*, stated that the relationships of *Barbopsis* are unclear, and that it does not have any close relatives among the epigean fishes of north-eastern Africa. Some indication on the topic can be given by a genetic study carried out with allozyme molecular markers (Cobolli-Sbordoni *et al.* 1996) on *Barbopsis* and *Phreatichthys*, the other cave cyprinid from Somalia. The paper revealed a rather close relationship between the two species suggesting a common origin from an extinct surface-dwelling ancestor.

We consider it worth, in this chapter, to take the opportunity to clarify a small taxonomic mystery. Gianferrari in a paper of 1934, where she lists all the African cavefishes, cites the cyprinids from Callis – previously described as *Barbopsis stefaninii* (Gianferrari 1930) – as *Zaccarinia stefaninii*, thus creating a new taxon in a completely anomalous manner as no description has ever appeared in any journal so far. The fact should have been brought to the author's attention as there is no trace of *Zaccarinia stefaninii* in her successive articles.

Morpho-physiological Notes

Barbopsis devecchii shows less degenerative features than other African cave cyprinids. The depigmentation is incomplete with different degrees of reduction, the scales are thin but never lacking, the eyes are variously reduced in size, sometimes sunken under the skin but still discernible. Optic nerves and chiasma are present, but the optic lobes are widely reduced and the retina is more or less degenerated according to the degree of eye reduction. In cases of marked microphthalmia the retina is strongly disrupted and neither retinal layers nor cones and rods are always recognizable; on the contrary, slightly microphthalmic fish show the typical succession of cellular layers and both types of visual cells are still noticeable (Ercolini and Berti 1978).

In teleosts the pseudobranch is an organ directly involved in the eye's functioning. In *Barbopsis* the pseudobranch tends to degenerate and disappears sooner than the eye; among the 64 specimens examined by Bianchi and Ercolini (1984), 59 had both eyes, 4 had one eye and 1 had no externally visible eyes, while 14 had both pseudobranchs, 9 had one pseudobranch and 41 had no pseudobranchs. Furthermore, no correlation existed between number of acidophil cells, pseudobranch volume and external diameter of the ipsilateral eye; thus pseudobranch degeneration and eye reduction seem to occur independently. Similarly, eye's reduction in size doesn't seem to be related to the degree of depigmentation (Della Croce 1963).

Finally, a comparative survey of structural and ultrastructural features of the olfactory epithelium (Delfino *et al.* 1981) revealed no significant degenerative nor compensative traits.

As well as the other Somalian cave cyprinid *Phreatichthys andruzzii* does, *Barbopsis* exhibits a reduced metabolic rate. When tested in a respirometer chamber specimens of the latter species showed an oxygen consumption about 2 times lower than *Puntius lateristriga*, while under anoxia their time survival resulted 1.5 to 2 times higher (Ercolini *et al.* 1987).

Ethological Notes

At present, the behaviour of *Barbopsis* is almost unexplored. As far as we know only two studies, on the response to light and, respectively, on the tendency to shoaling, were performed as yet. The former (Ercolini and Berti 1978) dealt with fish responses to equi-energetic, achromatic and monochromatic light stimuli; the results suggest that *Barbopsis,* although it exhibits an evident photophobia, can still adapt itself to light even in cases of marked eye reduction; this capacity determines a substantially photo-indifferent response in animals previously kept under a normal light-darkness cycle. The latter one (Jankowska and Thinès 1982) showed the absence of any tendency to aggregate, since the mean density of fish group fell within the theoretical limits of random distribution values.

FAMILY CLARIIDAE

Uegitglanis zammaranoi Gianferrari, 1923

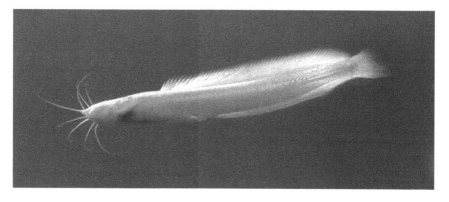

Fig. 5 *Uegitglanis zammaranoi* (photo by R. Berti; from Ercolini *et al.* 1982, modified).

Color image of this figure appears in the color plate section at the end of the book.

Historical Notes

The species was discovered by an Italian army officer, the major Vittorio Tedesco Zammarano, who collected the first specimens from a well of the village of Uegit, Southern Somalia, at an unknown date. Sent to Italy, the fish was described by Dr Luisa Gianferrari, curator of zoology at the *Museo di Storia Naturale di Milano.*

Derivatio Nominis

Uegitglanis: from Uegit (the type locality) + γλάνις *glanis* (catfish, in Greek);
zammaranoi: in honour of the collector.

Synonyms

Uegitglanis zammaranoi subsp. *baidoaensis* (Parenzan 1938)

Common Name

None

Size

Up to 250 mm in total length (Thinès 1958b).

Distribution Area and Ecological Notes

The species inhabits the subterranean waters of a vast plateau of Southern Somalia, between the course of Juba and Webi Shebeli, the two main rivers of the region. At present four collection sites (Table 4) are known; the presence of *Uegitglanis* in stations quite distant from one another (about 130 km between Manas and Siggia) suggests that the fish habitat is a wide and continuous underground hydrographic system, with tunnels, fissures and cavities, even of large dimensions such as in Manas (Thinès 1958b, Messana *et al.* 1985).

Table 4 Known collection sites of *Uegitglanis zammaranoi*; type locality: Uegit (from Ercolini *et al.* 1982, modified).

Locality	Latitude	Longitude	Altitude (m)	T(°C)
Uegit (2 sites)	03°48′26″ N	43°14′49″E	371	30.5
Baidoa (2 sites)	03°06′58″ N	43°38′48″E	433	28.5
Manas	02°53′28″ N	43°27′42″E	428	29.5
Siggia	03°59′18″ N	43°24′48″E	416	27

The distribution area is in Upper Jurassic formations, with organogenetic and oolitic limestones, marls and gypsum (Merla *et al.* 1979). In spite of the uniform geological structure, however, the physico-chemical features of the waters inhabited by *Uegitglanis* show a wide variability. The species clearly tolerates remarkable variations in water composition, at least as regards salinity and hardness, and in some cases even highly foul waters, with high contents of H_2S (Thinès 1958b, Ercolini *et al.* 1982). Such wide tolerance towards some chemical environment factors is confirmed by Authors' experience in breeding the species, kept in laboratory for about 10 years in very good conditions (unpublished data).

Reproduction in captivity has never been observed, in the wild reproduction is presumed to occur between January and March, a period characterized by a dry, hot climate; females with developed ovaries were collected only in March, from the well of Manas.

Together with *Uegitglanis* cohabits *Stenasellus pardii*, a stygobitic isopod found both at Baidoa and Uegit (Ercolini *et al.* 1982), which is probably the principal if not the exclusive trophic resource of the fish.

Systematics and Phylogenetic Relationships

Systematic position and phylogenetic affinities of *Uegitglanis* are quite uncertain. The original description (Gianferrari 1923) stated a direct derivation from the genus *Clarias*, on the basis, among other things, of the presence of a suprabranchial organ and a bony capsule enveloping the swim bladder. The presence of both structures was yet denied by other studies, and thus the close relationship with *Clarias* was questioned. Besides this, the characteristics also of both the skull and the Weberian apparatus induced David (1936) to suggest that *Uegitglanis* occupies an intermediate position between clariids and bagriid, and Chardon (1968) to create the new family Uegitglanidae. Contrarywise, a quite different relationship was proposed by Trewavas, when she stated that the closest genus was to be identify with *Gymnallabes* (Trewavas 1936).

Molecular studies could clarify the phylogenetic relationships of this species, at the moment considered a primitive Clariidae (Teugels and Adriaens 2003)

Finally, the validity of the subspecies *baidoaensis*, erected by Parenzan (1938) on the ground of both differences in size and a supposed habitat fragmentation, has to be excluded (Ercolini *et al.* 1982).

Morpho-physiological Notes

The adaptation degree to the hypogean environment shown by *Uegitglanis* is one of the highest among all cavefish. The eye disappeared and the depigmentation is almost complete, light traces of pigment being shown by both the meninx and the tegument between the nostrils. Besides the absence of any ocular residue, anatomical and histological studies showed the total disappearance of optic nerves and chiasma. The whole encephalon is reduced in size, and the reduction is particularly enhanced at the level of the optic lobes (Ercolini and Berti 1977).

The thyroid follicles (Olivereau 1960a, 1960b) are reduced in number and not fully differentiated; moreover, they are characterized by some lowering of activity levels.

On the contrary, at macroscopic, structural and ultrastructural level the olfactory organ (Bianchi *et al.* 1978, Bianchi 1978b) does not exhibit any degenerative nor compensative traits as regards the olfactory rosette and the cytological components of the olfactory neuroepithelium. Likewise, histological and histochemical studies performed about skin cells (Bianchi 1975) as well as about encephalic neurosecretory cells (Bianchi 1978c) do not show any significant divergences from epigean species.

Many other aspects of the fish morphology and anatomy have been taken into consideration, such as dentition (Gianferrari 1925), swim bladder (Gianferrari 1937b), cranial osteology (David 1936) and Weberian apparatus (Chardon 1968). The suprabranchial organ is fully absent (David 1936, Chardon 1968); the tendency to lose this air-breathing structure is common to few other clariid genera and, with regard to *Uegitglanis*, is probably to be related to its cave-dwelling lifestyle. Tortonese (1952) suggested that the resultant inability to perform aerial respiration should be compensated by the improved vascularisation of both anal and dorsal fin.

Ethological Notes

Behavioural aspects so far investigated concern photic sensitivity and aggressive behaviour.

Notwithstanding its complete anophthalmia *Uegitglanis*, as well as other cave fish do, revealed to be photosensitive. Kept for long time under 12/12 light-dark cycle, *Uegitglanis* showed slightly photophobic responses to achromatic light during the lighted phase, whereas during the dark phase it didn't exhibit any definite reaction (Ercolini and Berti 1977). Some field observations, made in the wells of Uegit, Iscia Baidoa, and Manas in the course of both day- and night-time, confirmed the laboratory data (Berti, personal observations).

Agonistic interactions have been noticed in the field also (Berti and Ercolini 1979), and laboratory observations allowed to define an exhaustive description of the aggressive behaviour. Specimens raised in isolation and tested in pairs immediately exhibit distinct agonistic behaviour, setting up clear dominant-subordinate relationships. The immediate effect of dominance seems to be the possession of the bottom by the dominant fish which banishes the subordinate to midwater (Ercolini *et al.* 1981). Morphological and quantitative analysis of the various aggressive patterns show that the behavioural repertoire is very rich and highly articulate, similar to that of epigean siluriforms (Berti *et al.* 1983).

As expected for a fish characterized by a such territorial behaviour no tendency to school or shoal is exhibited, as evidenced by Jankowska and Thinès (1982).

FAMILY CLARIIDAE

Clarias cavernicola Trewavas, 1936

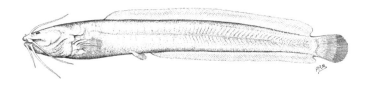

Fig. 6 *Clarias cavernicola* (from Skelton & Teugels 1991; reproduced by kind permission of the South African Institute for Aquatic Biodiversity).

Historical Notes

The first report about this species is probably that of the geologists Jaeger and Waibel (1921). It is however only in 1933 that the first specimens were collected, during the Dr Karl Jordan's expedition. Later on the species was described by Dr Ethelwyn Trewavas, ichthyologist at the British Museum of Natural History of London.

Derivatio Nominis

cavernicola: from *caverna* (cave, in Latin) + *colere* (to dwell, in Latin).

Common Name

Cave catfish (English), spelonkbaber (Afrikaans), blinde Höhlenfisch (German)

Size

Up to 161 mm in standard length (Bruton 1995).

Distribution Area and Ecological Notes

Clarias cavernicola inhabits the crystal-clear waters of a subterranean small lake, lying at the bottom of the Aigamas Cave (19°26'30"S, 17°17'26"E; Skelton and Teugels 1991), the type locality, about 25 km north of Otavi, Namibia. The distribution area is very restricted, although there are reports of similar fishes from other caves of the same dolomitic limestone region, such as Ajgab Cave, about 50 km west of Aigamas Cave, or Gamarab Cave, near Outjo (Jubb 1969), the type locality is at present the one site where the presence of the species was ascertained.

The cave has two entrances, about 100 m distant; the first one is a 4 m chimney opening into a large chamber, the second one is a narrow opening giving access to the cave whose floor descends steeply before dropping almost vertically for about 20 m to the water surface (Skelton 1987, 1990a, 1990b). When the geologists Jaeger and Waibel (1921) visited the cave, the water surface was about 18 m in length and 2.5 m wide, the water depth was about 70 m, the temperature 24.5°C. Since then, some important alterations probably occurred in the subterranean habitat, as suggested by the fact that the water level has fallen by about 20 m (Skelton 1987, 1989), and the water temperature was considerably increased, reaching in 1987 the value of 27°C (Bruton 1995).

The fishes seem to spent most part of their activity in the upper water layers, seldom descending deeper than 15 m. They congregate and feed exclusively over a shelf that slopes from just below the water surface to several meters depth, and where trophic materials falling from the cave tend to collect (Skelton 1990a, 1990b). The food supply seems to be principally of exogenous nature, and consists of the carcasses of dead bats and wild animals trapped in the cave, their parasites and these insects feeding on such remains, of insects contained in bat and baboon feces, of coprophagous invertebrates (Trewavas 1936, Skelton 1987, 1990a, 1990b, Bruton 1995). Moreover, it seems very likely that fish feed also on the small shrimps present in the lake as well as that the larger fish cannibalize smaller individuals (Skelton 1989), an event not unusual in the cave environment (Berti and Zorn 2001).

Both the aforementioned presence of small shrimps (Skelton 1989) and the finding inside the fish stomach of some invertebrates such as a curved white grub and a flatworm (Trewavas 1936), get us to hypothesize that an associated fauna could inhabit the waters of the cave. In our opinion it should be of great interest to investigate these aspects of the Agamas Cave ecology.

So far no studies appeared on the ecology of Aigamas Cave, and the population size is not exactly known. According to different assessments the presence of 150-200 (Sefton *et al.* 1986) or 200-400 individual (Bruton 1995) was hypothesized.

Systematics and Phylogenetic Relationships

Although some species had been proposed as closely related to *C. cavernicola* the identification of its epigean ancestor, as well as it occurs for many other cave fish, is still largely doubtful. The original description (Trewavas 1936) hypothesized a relationship of *C. cavernicola* with either *C. allaudi* or *C. submarginatus* or *C. dumerilii*. Jubb (1958, 1969) substantially agreed on the hypothesis of a phylogenetic relationships with *C. dumerilii*, because of both the morphological resemblance between the two fishes and the distribution area of the epigean species, inhabiting the Cunene and Ocavango rivers, to the north of the Otavi district. On the contrary, in his systematic outline of the African species of the genus *Clarias*, Teugels (1982a, 1982b, 1986) rejected the relationships with *C. dumerilii*, confirming that one with *C. allaudi* and *C. submarginatus*, i.e. the two other fishes mentioned by Trewavas (1936). According to Teugels (1982, 1986), very closely related species could be *C. nigromarmoratus* and yet more *C. theodorae*. The latter one was proposed as the closest species because of its extraordinary similarity in shape and size of both head and body, and in the position of the anal fin origin (Skelton and Teugels 1991). Moreover the distribution area of *C. theodorae* is partially situated in two basins (Cunene and Ocavango) adjacent to the Aigamas Cave.

Morpho-physiological Notes

Although *C. cavernicola* shows some degenerative traits clearly related to cave life, the extent of its adaptation to the underground habitat is not so high as in *Uegitglanis zammaranoi*, the other African cave clariid.

In some specimens the eyes are absent (Teugels 1986) but in most individuals they are variously developed, often sunk in the socket, with the surface skin more or less opaque (Trewavas 1936). The great variability that seems characterize the eye degeneration is confirmed by the results of a morphological study about brain and sense organs in *C. cavernicola* and in the epigean species *C. gariepinus* and *C. ngamensis* (Bok 1968). While the brain doesn't show the striking reductions in size characterizing other cave fishes, the cerebellum being its only reduced region, and while the olfactory apparatus shows a slight decrease in development, not so marked as to suggest that the cavefish has poorly developed olfactory powers, the eyes and optic nerves of *C. cavernicola* have degenerated to a large degree. Furthermore, there was found to be considerable variation in the degree of degeneration of these two structures both among different fish and also in the same individual, and the large variation in diameter of the optic nerve was found to correlate with that one of the eye. Histological sections of the most degenerated eyes show that choroid and retinal layers are completely disorganized, and a few large vacuolated cells are all that remains of the lens tissue; no rods or cones were identified (Bok 1968).

C. cavernicola as well as most cave-dwelling fishes, is devoid of pigment; in life the skin colour is creamish flushed with pink or orange (Skelton and Teugels 1991).

Unlike *Uegitglanis* the suprabranchial organ is still present but, in common with many clariid genera (Teugels and Adriaens 2003), is greatly reduced; therefore the suprabranchial chamber is only partly filled (Teugels

1986, Skelton and Teugels 1991). The reduction of the suprabranchial organ has probably to be related to a loss of its functional significance, due to either a lowered oxygen consumption or an high oxygen content of the water or both, since *C. cavernicola* very seldom even comes to the surface to gulp air (Hennig 1977).

In spite of the great attention this species deserves, no real scientific studies were made on its biology and life history (Skelton 1987, 1990a, 1990b). No data are available on breeding habits, and several attempts to breed captive individuals artificially have failed (Skelton 1990a, 1990b, Bruton 1995).

Ethological Notes

At present, the behaviour of *C. cavernicola* is completely unexplored. The only available records concern a few extemporary observations about its spontaneous behaviour in both the field and aquarium (Scheide 1977, Hennig 1977, Skelton 1987, 1990).

FAMILY ELEOTRIDAE

Typhleotris madagascariensis Petit, 1933

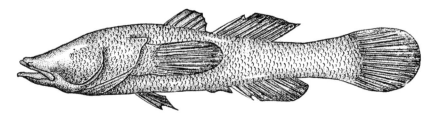

Fig. 7 *Typhleotris madagascariensis* (from Arnoult 1959a).

Historical Notes

The species was discovered about 1930 (the collection date is unknown) by Henry Perrier de la Bathie and Mr Ursch, controller of the *Eaux et Forêts*. Georges Petit described the fish as new genus and species three years later.

Derivatio Nominis

Typhleotris: from τυφλός *typhlos* (blind, in Greek) + *Eleotris* (because of its close relationship to that genus);

madagascariensis: from Madagascar.

Common Name

None

Size

Up to 67 mm in standard length (Arnoult 1959a) and to 100 mm in total length (Maugé 1986).

Distribution Area and Ecological Notes

The species inhabits the subterranean waters of a karst, included in the Eocene or upper Cretaceous limestone formations (Arnoult 1959b, Kiener 1964, Decary and Kiener 1970) of the Mahafaly Plateau, southwestern Madagascar. The type locality is the Mitoho Cave (24°03′S, 43°45′E, 50 m a.s.l.; Goodman *et al.* 2005), Toliara (formerly Tuléar) Province. Besides the type locality, six other sites are known as inhabited by the fish. From north to south, they are the well of Ambilanilalika (about 40 km south of the Onilahy River mouth), and the caves of Malazamanga, Lavaboro, Lalia, Andramanoetse and Nikotsy, just north of Itampolo. The size of the distribution area is rather uncertain; the eastern boundaries are still unknown but, however, the fish habitat probably lengthen over a large area south of the Onihaly River, since the distance between the two farthest sites is about 90 km (Kiener 1964, Dekary and Kiener 1971).

The water table lies at a depth varying from 15 (Mitoho Cave) to 27 meters (well of Ambilanilalika) under the ground surface (Petit *et al.* 1935, Poisson and Besairie 1947, Angel 1949).

Mitoho Cave, the type locality, was an underground chamber whose ceiling weakened and partially collapsed, exposing a small lake that represents the only accessible part of the subterranean hydrographic system. Observed in the field, the fishes didn't show any photophobic behaviour, searching for food even in the lighted portion of the cave (Petit 1933, Petit *et al.* 1935). The diet is suggested to consist of either an aquatic micro-fauna of epigean origin (Petit *et al.* 1935) or plankton (Kiener 1964) or, more likely, cave crustacean and insects (Arnoult 1959a, 1959b). In Mitoho Cave, the associated aquatic fauna comprehends both crustaceans and insects: among the former, the cave decapod *Typhlopatsa pauliani*, the isopod *Anopsilana poissoni*, some copepods (*Tropocyclops confinis* and *Diaptomus* sp.) and some eyed tanaidaceans, among the latter the hemipteran *Microvelia mitohoi* (Decary and Kiener 1970, Remillet 1973).

Systematics and Phylogenetic Relationships

The validity of the genus and its close phylogenetic relationships with the epigean genus *Eleotris*, stated by the original description (Petit 1933), has always been shared by all authors dealing with this matter. On the contrary, its inclusion in the family Eleotridae (Petit 1933, Petit *et al.* 1935, Petit 1941, Thinès 1955a, Arnoult 1959a, Maugé 1986, Bauchot *et al.* 1991, Stiassny and Raminosoa 1994, Lévêque 1997, Eschmeyer 1998, Romero and Paulson 2001) has been repeatedly called into question by some author assigning the genus to the family Gobiidae (Thinès 1969, Thinès and Proudlove 1986). Finally, Hoese and Gill (1993), on the strength of a study of the phylogenetic relationships of eleotridid fishes, suggested that *Typhleotris* together with other eleotrid genera was to be considered belonging to the family Gobiidae, subfamily Butinae. The matter is quite controversial (Proudlove 2006), therefore we prefer to follow the traditional systematic status and to consider *Typhleotris* as belonging to the family Eleotridae.

Morpho-physiological Notes

Typhleotris madagascariensis appears to be highly adapted to cave life. The adult results fully blind: eyes are neither externally visible (Petit 1933, Petit *et al.* 1935, Arnoult 1959a) nor found, sunken in the head soft tissues, under dissection (Petit 1941). Unfortunately no histological studies were performed to verify the degeneration degree, and the presence of eye remnants can't be excluded at all. As it occurs in the cavefishes of the genus *Astyanax* (Breder 1944, Yamamoto *et al.* 2003) and in *Caecobarbus geertsii* (Petit 1941) the eye degeneration induces a deep rearrangement of the craniofacial skeleton, that in *T. madagascariensis* causes a reduced mandibular mobility (Petit 1941). Degenerative features are shown by the tegumentary system also. The scales are still present but very thin and soft; they are both ctenoid and, predominantly, cycloid (Petit 1933, Petit *et al.* 1935, Angel 1949), with the latter ones being degenerated ctenoid scales (Petit 1938b). The fishes are variously depigmented, a high intraspecific variability occurring; the skin colour (light yellow, in alcohol) varies in life from white, with a blood-red spot on the operculum (Petit 1933, Petit *et al.* 1935), to brown-pinkish to brown (Angel 1949). On the contrary, no variability is shown by the amount of vertebrae, without exception in number of 25 (Decary and Kiener 1970).

Up to now, no physiological studies were conducted on *T. madagascariensis*.

Ethological Notes

In the occasion of the fish discovery, Perrier de la Bathie observed for few days its spontaneous behaviour in the field (Petit 1933, Petit *et al.* 1935). Those observations, together with the few ones made in the same cave some forty years later (Kiener 1964), constitute the only behavioural records till now known.

FAMILY ELEOTRIDAE

Typhleotris pauliani Arnoult, 1959

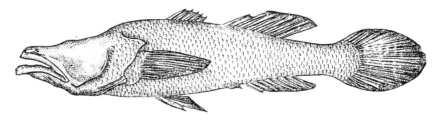

Fig. 8 *Typhleotris pauliani* (from Arnoult 1959a).

Historical Notes

The first specimens were collected in July 1956 by Renaud Paulian, Director of the I.R.S.M. (*Institut de Recherche Scientifique de Madagascar*), in a cave near the *Baie des Assassins* (Murderers' Bay), between the towns of Morombe and Toliara (formerly Tuléar), south-western Madagascar. The new species was described three years later by Jacques Arnoult, of the zoological laboratory of the *Muséum National d'Histoire Naturelle de Paris*.

Derivatio Nominis

pauliani: in homage to R. Paulian, collector of the first specimens.

Common Name

None

Size

Up to 56 mm in standard length (Arnoult 1959a, 1959b) and to 65 mm in total length (Maugé 1986).

Distribution Area and Ecological Notes

The type locality is the Andranomaly Cave, near the *Baie des Assassins* (about 22°12′30″S–43°17′30″E) and south-east of the village of Andolombezo. The fish habitat is constituted by some pools, 150 m in diameter, 1,5 m in depth; the water temperature is about 22°C. The fish diet probably consists of small aquatic animals and insects, and bat guano (Arnoult 1959a, 1959b, Remillet 1973).

On the outskirts of Andolombezo two other caves named Safora and Ankilivona (Decary and Kiener 1971) are known as inhabited by the fish. At present no other sites are known as housing the species, it is therefore impossible to define or even hypothesize the extent of the fish distribution area. Although *Typhleotris pauliani* and *T. madagascariensis* both inhabit the subterranean waters of the same karst, in the Mahafaly Plateau, their habitats seem to be fully distinct. Not only the nearest collection sites of the two species are more than 300 km apart but also, and above all, any hypothetical connection should be broken off by the valleys of Onilahy and Fiherenana rivers (Arnoult 1959b, Paulian 1961, Kiener 1964).

Systematics and Phylogenetic Relationships

On the ground of the high similarity and the evident close relationships with *Typhleotris madagascariensis*, Arnoult didn't erect any new genus for the cavefish he described and only established a new species.

The validity of the genus and its close phylogenetic relationships with the epigean genus *Eleotris* stated by Petit (1933), has always been shared by all authors dealing with this matter. On the contrary, its inclusion in the family Eleotridae (Petit 1933, Petit *et al.* 1935, Petit 1941, Thinès 1955a, Arnoult 1959a, Maugé 1986, Bauchot *et al.* 1991, Stiassny and Raminosoa 1994, Lévêque 1997, Eschmeyer 1998, Romero and Paulson 2001) has been repeatedly called into question by some author assigning the genus to the family Gobiidae (Thinès 1969, Thinès and Proudlove 1986). Finally, Hoese and Gill (1993), on the strength of a study of the phylogenetic relationships of eleotridid fishes, suggested that *Typhleotris* together with other eleotrid genera was to be considered belonging to the family Gobiidae, subfamily Butinae. The matter is quite controversial (Proudlove 2006), therefore we prefer to follow the traditional systematic status and to consider *Typhleotris* as belonging to the family Eleotridae.

Morpho-physiological Notes

Typhleotris pauliani appears to be well adapted to the hypogean environment. No traces of eyes are externally visible and the pigmentation is fully absent, thus the skin colour is white with a light pink spot under the *operculum* in life whereas the fish colour turns to "old ivory" in alcohol (Arnoult 1959a, 1959b). The scales are almost exclusively cycloid and just a few ctenoid scales are present (Arnoult 1959b).

Neither morpho-anatomical nor histological nor physiological studies were done.

Ethological Notes

With the only exception of a few field observations on the swimming modality (Arnoult 1959b), the behaviour of *T. madagascariensis* is still fully unknown.

FAMILY GOBIIDAE

Glossogobius ankaranensis Banister, 1994

Fig. 9 *Glossogobius ankaranensis* (photo by Jane Wilson).
Color image of this figure appears in the color plate section at the end of the book.

Historical Notes

The fish was firstly seen and collected by members of the 1986 Crocodile Caves Expedition to Madagascar (Wilson 1995, 1996). Deposited in the collections of the Natural History Museum of London, was described by Keith E. Banister who established the new species *Glossogobius ankaranensis*.

Derivatio Nominis

ankaranensis: from the Ankarana karst, where the species lives.

Common Name

None

Size

Up to about 70 mm in total length (Wilson *et al.* 1988, Wilson 1996).

Distribution Area and Ecological Notes

The species inhabits the subterranean waters of the Ankarana massif, an outcrop of Middle Jurassic limestones in the northern Madagascar. At present the fish is known only from the type locality, the Second River Cave (about 12°50'S, 49°10'E). This is a cavity about 150 m long, and the river flows slowly through about 100 m of low passage. The river is about 2 m deep and 5 m wide, in September the water temperature is 21°C. The Second River Cave is the only accessible part of the fish habitat so far known, but the species is thought to people also other sites of the hydrographic subterranean system (Wilson 1996). Despite the small dimensions of the massif (about 28 km in length and 8 km in width), the cave complex is very long, with 98 km of cave passages already surveyed (Wilson *et al.* 1988) and more still to be found (Wilson 1985, 1987a): it seems very probable that the fish habitat extends over other parts, either still unknown or inaccessible, of the karst.

G. *ankaranensis* occurs in great numbers in the watercourse, calmly swimming even in a lighted portion of the cave, and lives in sympatry with G. *callidus*, a related species normally eyed and pigmented (Wilson 1996). The epigean gobiid, smaller in size than G. *ankaranensis*, does not seem to compete with this one; plausibly arrived in the cave with the rainy season flood waters (Wilson 1987b, Wilson *et al.* 1988) it appears to be well adapted to cave life, since it is present with very large populations (Wilson 1996). Finally an undetermined eel, probably belonging to a new species, lives in most caves; up to now no specimens have been caught (Wilson 1987b).

The Ankarana underground waters house also a rich invertebrate fauna. The copepod *Mesocyclops leuckarti*, the epigean decapod *Macrobrachium moorei*, an undetermined depigmented crab and eleven decapods (eight belonging to the genus *Caridina* and three to the genus *Parisia*) are at present known (Remillet 1973, Wilson 1987b). Some decapods with various degrees of cave life adaptation are endemic of the Ankarana streams (Gurney 1984), while some new species belonging to the genera *Caridina* and *Parisia* are waiting for their description (Wilson 1987b). Although no fish had been observed feeding on these crustaceans, such a rich invertebrate fauna probably constitutes the main part of the diet of *G. ankaranensis*, together with organic matter washed in by floods, dead insects, bat guano, swift droppings and feces of forest mammals coming in to drink (Banister 1994, Wilson 1996).

Systematics and Phylogenetic Relationships

In the original description (Banister 1994) the author pointed out that the new species fitted within the definition of the genus *Glossogobius* as outlined by Koumans (1953), hence there were no good reasons for erecting a new genus. With regard to its phylogenetic relationships the descriptor suggested that the new species is closely related to *G. giuris*, a gobiid very similar in many characters to *G. ankaranensis* and widespread over the Indo-Pacific area.

Morpho-physiological Notes

Glossogobius ankaranensis shows a high degree of adaptation to cave life. The eye degeneration is almost complete, the only traces of eyes are two minute black spots deeply embedded in the tissues. The fish are devoid of pigment and are pink in colour when alive, ivory when preserved in alcohol. The scales, both ctenoid and cycloid, are thin and weak; no perforated lateral line scales are present (Banister 1994, Wilson 1996). At present no histological data are available.

Ethological Notes

The ethology of *Glossogobius ankaranensis* is up to now completely unexplored. The only known behavioural features are some field observations on the swimming modalities. They are slow-moving in a browsing swimming pattern, gently oscillating in the horizontal plane (Wilson 1996).

Acknowledgements

We would like to thank the following colleagues and friends who kindly provided help in the preparation of this chapter: Jane Wilson, who provided the *Glossogobius ankaranenensis* picture; Gie Robeyns, who supplied precious information about *Caecobarbus geertsii*; Alessandro Cianfanelli, for his valuable help in both text revision and figures arrangement.

References

Angel, F. 1949. Contribution à l'étude du *Typhleotris madagascariensis*, poisson aveugle, cavernicole, du Sud-Ouest de Madagascar. *Bulletin du Muséum national d'histoire naturelle, 2ᵉ série* 21: 56-59.

Arnoult, J. 1959a. *Poissons des eaux douces*. Faune de Madagascar: Publications de l'Institut de Recherche Scientifique, Tanarive-Tsimbazaza. Volume 10.

Arnoult, J. 1959b. Une nouvelle espèce de poisson aveugle de Madagascar: *Typhleotris pauliani* n. sp. *Memoires de l'Institut Scientifique de Madagascar. Série A* 13: 133-138.

Banister, K.E. 1984. A subterranean population of *Garra barreimiae* (Teleostei: Cypinidae) from Oman, with comments on the concept of regressive evolution. *Journal of Natural History* 18: 927-938.

Banister, K.E. 1994. *Glossogobius ankaranensis*, a new species of blind cave goby from Madagascar (Pisces: Gobioidei: Gobiidae). *Aqua* 1: 25-28.

Bauchot, M.L., M. Desoutter, D.F. Hoese and H.K. Larson. 1991. Catalogue critique des types de Poissons du Muséum national d'Histoire naturelle - (Suite) - Sous ordre des Gobioidei. *Bulletin du Muséum national d'histoire naturelle, 4ᵉ série, Section A* 13 (1-2, Supplément): 1-82.

Berti, R. and A. Ercolini. 1979. Aggressive behaviour in the anophthalmic phreatic fish *Uegitglanis zammaranoi* Gianferrari (Claridae: Siluriformes). *Monitore zoologico italiano (N.S.) Supplemento* 13: 197.

Berti, R. and G. Thinès. 1980. Influence of chemical signals on the topographic orientation of the cave fish *Caecobarbus geertsi* Boulenger (Pisces, Cyprinidae). *Experientia* 36: 1384-1385.

Berti, R. and L. Masciarelli. 1993. Comparative performances of non-visual food search in the hypogean cyprinid *Phreatichthys andruzzii* and in the epigean relative *Barbus filamentosus*. *International Journal of Speleology* 22: 121-130.

Berti, R. and L. Zorn. 2001. Locomotory responses of the cave cyprinid *Phreatichthys andruzzii* to chemical signals from conspecifics and related species: New findings. *Environmental Biology of Fishes* 62: 107-114.

Berti, R., G. Thinès and B. Lefèvre. 1982. Effets des informations chimiques provenant d' un milieu habité par des congénères sur l' orientation topographique du poisson cavernicole *Phreatichthys andruzzii* Vinciguerra (Pisces, Cyprinidae). *International Journal of Speleology* 12: 103-117.

Berti, R., A. Ercolini and A. Cianfanelli. 1983. The aggressive behavior repertoire of an anophthalmic phreatic fish from Somalia. *Experientia* 39: 217-219.

Berti, R., R. Vezzosi and A. Ercolini. 1989. Locomotory responses of *Phreatichthys andruzzii* Vinciguerra (Pisces: Cyprinidae) to chemical signals of conspecifics and of closely related species. *Experientia* 45: 205-207.

Berti, R., J.P. Durand, S. Becchi, R. Brizzi, N. Keller and G. Ruffat. 2001. Eye degeneration in the blind cave-dwelling fish *Phreatichthys andruzzii*. *Canadian Journal of Zoology* 79: 1278-1285.

Bianchi, S. 1975. Preliminary observations on the epidermis of *Uegitglanis zammaranoi* Gianferrari (Claridae: Siluriformes.) *Monitore zoologico italiano (N. S.) Supplemento* 6: 91-101.

Bianchi, S. 1978a. Osservazioni preliminari sulle cellule mucose di *Phreatichthys andruzzii* Vinciguerra, pesce freatico anoftalmo della Somalia. *Rivista di Istochimica normale e patologica* 22: 15-18.

Bianchi, S. 1978b. Le cellule mucose e l'istochimica dei mucopolisaccaridi nell'epitelio olfattivo di *Uegitglanis zammaranoi* Gianferrari (Clariidae, Siluriformes), pesce freatico anoftalmo della Somalia. *Rivista di Istochimica normale e patologica* 22: 101-107.

Bianchi, S. 1978c. Indagine preliminare sulla topografia delle cellule neurosecernenti a secreto fucsinoparaldeide positivo nel S.N.C. di *Uegitglanis zammaranoi* Gianferrari, pesce freatico anoftalmo della Somalia. *Boll. Zool.* 45 (Supplement): 8.

Bianchi, S. and A. Ercolini. 1984. Observations on the morphology and regression of the pseudobranchs in *Barbopsis devecchii* Di Caporiacco microphthalmic phreatobic cyprinid from Somalia. *Monitore zoologico italiano (N.S.) Supplemento* 18: 223-230.

Bianchi, S., G. Delfino and A. Ercolini. 1978. Morphology and fine structure of the olfactory organ in *Uegitglanis zammaranoi* Gianferrari (Claridae Siluriformes), anophthalmic phreatic fish from Somalia. *Monitore zoologico italiano (N.S.) Supplemento* 10: 157-171.

Bok, A. 1968. *A Morphological Study of the Brain and Sense Organs of the Cave Fish, Clarias cavernicola Trewavas.* Zoology Honours Project, Rhodes University, Grahamstown.

Boulenger, G.-A. 1921. Description d'un poisson aveugle découvert par M. G. Geerts dans la grotte de Thysville (Bas-Congo). *Revue Zoologique Africaine* 9: 252-253.

Breder, C.M. Jr. 1944. Ocular anatomy and light sensitivity studies on the blind fish from Cueva de los Sabinos, Mexico. *Zoologica* 29: 131-144.

Bruton, M.N. 1995. Threatened fishes of the world: *Clarias cavernicola* Trewavas,1936 (Clariidae). *Environmental Biology of Fishes* 43: 162.

Bruun, A.F. and E.W. Kaiser. 1943. *Iranocypris typhlops* n. gen., n. sp., the first true cave-fish from Asia. *Danish Scientific Investigations in Iran* 4: 1-8.

Chardon, M. 1968. Anatomie comparée de l'appareil de Weber et des structures connexes chez les Siluriformes. *Annales du Musée royal de l'Afrique centrale. Série in 8vo (Sciences zoologiques)* 169: 1-277.

Chelazzi, L. and G. Messana. 1985. *Acanthastenasellus forficuloides* n. gen. n. sp., a stenasellid isopod (Asellota) from Somalian phreatic layer. *Monitore zoologico italiano (N.S.) Supplemento* 20: 43-54.

Cobolli Sbordoni, M., E. De Matthaeis, M. Mattoccia, R. Berti and V. Sbordoni. 1996. Genetic variability and differentiation of hypogean cyprinid fishes from Somalia. *J. Zool. Syst. Evol. Research* 34: 75-84.

Colli, L., A. Paglianti, R. Berti, G. Gandolfi and J. Tagliavini. 2008. Molecular phylogeny of the blind cavefish *Phreatichthys andruzzii* and *Garra barreimiae* within the family Cyprinidae. *Environmental Biology of Fishes* doi: 10.1007/s10641-008-9393-z.

David, L. 1936. *Uegitglanis,* Silure aveugle de la Somalie Italienne - Chainon entre Bagrides et Clariides. *Revue de Zoologie et de Botanique Africaines* 28: 369-388.

Decary, R. and A. Kiener. 1970. Les cavités souterraines de Madagascar. *Annales de Spéléologie* 25: 409-440.

Decary, R. and A. Kiener. 1971. Inventaire schématique des cavités de Madagascar. *Annales de Spéléologie* 26: 31-46.

Delfino, G., S. Bianchi and A. Ercolini. 1981. On the olfactory epithelium in cyprinids: A comparison between hypogean and epigean species. *Monitore zoologico italiano (N.S.) Supplemento* 14: 153-180.

Della Croce, N. 1963. III Spedizione del Prof. G. Scortecci in Somalia. Osservazioni su *Barbopsis devecchii* Di Caporiacco (Cyprinidae). *Bollettino dei musei e degli istituti biologici dell'Università di Genova* 32 (1962-1963): 15-28.

Di Caporiacco, L. 1926. Un nuovo genere di Ciprinide somalo delle acque di pozzo. *Monitore zoologico italiano* 37: 23-25.

Di Caporiacco, L. 1927. Pesci raccolti in Somalia nel 1924, dalla spedizione Stefanini e Puccioni. *Monitore zoologico italiano* 38: 84-89.

Ercolini, A. and R. Berti. 1975. Light sensitivity experiments and morphology studies of the blind phreatic fish *Phreatichthys andruzzii* Vinciguerra from Somalia. *Monitore zoologico italiano (N.S.) Supplemento* 6: 29-43.

Ercolini, A. and R. Berti. 1977. Morphology and response to light of *Uegitglanis zammaranoi* Gianferrari, anophthalmic phreatic fish from Somalia. *Monitore zoologico italiano (N.S.) Supplemento* 9: 183-199.

Ercolini, A. and R. Berti. 1978. Morphology and response to light of *Barbopsis devecchii* Di Caporiacco (Cyprinidae) microphthalmic phreatic fish from Somalia. *Monitore zoologico italiano (N.S.) Supplemento* 10: 299-314.

Ercolini, A., R. Berti and A. Cianfanelli. 1981. Aggressive behaviour in *Uegitglanis zammaranoi* Gianferrari (Clariidae, Siluriformes), anophthalmic phreatic fish from Somalia. *Monitore zoologico italiano (N.S.) Supplemento* 14: 39-56.

Ercolini, A., R. Berti, L. Chelazzi and G. Messana. 1982. Researches on the phreatobic fishes of Somalia: Achievements and prospects. *Monitore zoologico italiano (N.S.) Supplemento* 17: 219-241.

Ercolini, A., R. Berti, L. Chelazzi and G. Messana. 1987. Oxygen consumption in hypogean and epigean cyprinids (Pisces). *Monitore zoologico italiano (N.S.) Supplemento* 22: 23-30.

Eschmeyer, W.N. (ed.). 1998. *Catalog of Fishes.* California Academy of Sciences, San Francisco, 3 vols.

Frangioni, G., R. Berti and G. Borgioli. 1997. Hepatic respiratory compensation and haematological changes in the cave cyprinid, *Phreatichthys andruzzii. Journal of Comparative Physiology* B 167: 461-467.

Gérard, P. 1936. Sur l'existence de vestiges oculaires chez *Coecobarbus geertsi. Mémoires du Musée royal d'histoire naturelle de Belgique*, 2e série 3: 549-552.

Gianferrari, L. 1923. *Uegitglanis zammaranoi* un nuovo Siluride cieco africano. *Atti della Società italiana di Scienze naturali* 62; 1-3.

Gianferrari, L. 1925. Su la dentatura di *Uegitglanis zammaranoi* Gianf. *Atti della Società italiana di Scienze naturali* 63: 326-334.

Gianferrari, L. 1930. Un nuovo ciprinide somalo (*Barbopsis stefaninii* Gianf.). *Atti della Società italiana di scienze naturali e del Museo civico di storia naturale* 69: 106-111.

Gianferrari, L. 1932. Elenco dei pesci cavernicoli e loro distribuzione geografica. *Atti della Società italiana di scienze naturali e del Museo civico di storia naturale* 71: 217-223.

Gianferrari, L. 1934. I pesci cavernicoli ciechi africani. *Rassegna faunistica* 1: 33-36.

Gianferrari, L. 1937a. Nuove osservazioni sui Ciprinidi di Callis (Somalia Italiana). *Atti della Società italiana di scienze naturali e del Museo civico di storia naturale* 76: 198-204.

Gianferrari, L. 1937b. Su la vescica natatoria di *Uegitglanis zammaranoi* Gian. *Atti della Società italiana di Scienze naturali* 76: 195-197.

Goodman, S.M., D. Andriafidison, R. Andrianaivoarivelo, S.G. Cardiff, E. Ifticene, R.K.B. Jenkins, A. Kofoky, T. Mbohoahy, D. Rakotondravony, J. Ranivo, F. Ratrimomanarivo, J. Razafimanahaka and P.A. Racey. 2005. The distribution and conservation of bats in the dry regions of Madagascar. *Animal Conservation* 8: 153-165.

Gurney, A.R. 1984. Freshwater shrimp genera Caridina and Parisia (Decapoda: Caridea: Atyidae) of Madagascar, with descriptions of four new species. *Journal of Natural History* 18: 567-590.

Hennig, H. 1977. Note on the blind catfish *Clarias cavernicola*. *S.W.A. Wissenshaftliche Gesellschaft. Verein für Höhlenforschung, Arbeitsberichte* 10: 11-12.

Heuts, M.J. 1952. Ecology, variation and adaptation of the blind African cave fish *Caecobarbus geertsii* Blgr. *Annales de la Société royale zoologique de Belgique* 82 (1951): 155-230.

Heuts, M.J. 1953. Regressive evolution in cave animals. *Symposia of the Society for Experimental Biology* 7: 290-309.

Heuts, M.J. and N. Leleup. 1954. La Géographie et l'Ecologie des Grottes du Bas-Congo. Les Habitats de *Caecobarbus geertsi* Blgr. *Annales du Musée royal du Congo Belge. Série in 8vo (Sciences Zoologiques)* 35: 1-71.

Hoese, D.F. and A.C. Gill. 1993. Phylogenetic relationships of eleotridid fishes (Perciformes: Gobioidei). *Bulletin of Marine Science* 52: 415-440.

Howes, G.J. 1991. Systematics and biogeography: An overview. In: *Cyprinid Fishes — Systematics, Biology and Exploitation*, I.J. Winfield and J.S. Nelson (eds). Chapman & Hall, London, pp. 1-33.

Jaeger, F. and L. Waibel. 1921. *Beiträge zur Landeskunde von Südwestafrika — Band II. Landschaften des nördlichen Südwestafrika*. E.S. Mittler und Sohn, Berlin (quoted by Skelton 1990a).

Jankowska, M. and G. Thinès. 1982. Etude comparative de la densité de groupes de poissons cavernicoles et épigés (Characidae, Cyprinidae, Clariidae). *Behavioural Processes* 7: 281-294.

Jubb, R.A. 1958. A cave-dwelling cat-fish *Clarias cavernicola* Trewavas 1936, in South West Africa. *Piscator* 43: 56-57.

Jubb, R.A. 1969. Fishes of the dolomitic limestone caves and sink holes of southern Africa. *Piscator* 75: 15-20.

Kiener, A. 1964. De la présence de certaines populations ichthyologiques dans les eaux souterraines des formations karstiques de la côte Ouest du Madagascar. *Bulletin du Madagascar* 219: 573-592.

Koumans, F.P. 1953. Gobioidea. In: *The Fishes of the Indo-Australian Archipelago*, M. Weber and L.F. De Beaufort (eds). E.J. Brill Ltd, Leiden. Volume 10.

Leleup, N. 1956a. La Faune cavernicole du Congo Belge et Considérations sur les Coléoptères reliques d'Afrique intertropicale. *Annales du Musée royal du Congo Belge. Série in 8vo (Sciences Zoologiques)* 46: 1-171.

Leleup, N. 1956b. Description de deux grottes du Bas-Congo. *Notes Biospéologiques* 11: 73-80.

Leleup, N. 1957. La faune cavernicole du Congo Belge. *Folia scientifica Africae centralis* 3: 31-33.

Lévêque C. 1997. *Biodiversity Dynamics and Conservation: The Freshwater Fish of Tropical Africa.* Cambridge University Press, Cambridge.

Matthes, H. 1963. A comparative study of the feeding mechanisms of some African Cyprinidae (Pisces, Cypriniformes). *Bijdragen tot de dierkunde* 33: 3-35.

Maugé, A.L. 1986. Eleotridae. In: *Checklist of the Freshwater Fishes of Africa (CLOFFA)*, J. Daget, J.-P. Gosse and D.F.E. Thys van den Audenaerde (eds). ISNB, Bruxelles, MRAC, Tervuren, ORSTOM, Paris, Volume 2, pp. 389-398.

Merla, G., E. Abbate, A. Azzaroli, P. Bruni, P. Canuti, M. Fazzuoli, M. Sagri and P. Tacconi. 1979. *A Geological Map of Ethiopia and Somalia (1973) 1: 2.000.000 and Comment, with a Map of Major Landforms.* Consiglio Nazionale delle Ricerche, Firenze.

Messana, G. and L. Chelazzi. 1984. *Haptolana somala* n. sp., a phreatobic cirolanid isopod (Crustacea) from the Nogal Valley (Northern Somalia). *Monitore zoologico italiano (N.S.) Supplemento* 19: 291- 298.

Messana, G., L. Chelazzi and N. Baccetti. 1985. Biospeleology of Somalia. Mugdile and Showli Berdi caves. *Monitore zoologico italiano (N.S.) Supplemento* 20: 325-340.

Olivereau, M. 1960a. Quelques aspects anatomiques et physiologiques de la glande thyroïde des Poissons. *Annales de la Société royale zoologique de Belgique* 90(2): 83-98.

Olivereau, M. 1960b. Etude anatomique et histologique de la glande thyroïde d'*Uegitglanis zammaranoi* Gianferrari, poisson aveugle et cavernicole, comparaison avec un Clariidae voisin, *Clarias buthupogon* A. Dum. *Annales de la Société royale zoologique de Belgique* 90(2): 99-116.

Olivereau, M. and M. Herlant. 1954. Étude histologique de l'hypophyse de *Caecobarbus geertsii* Blgr. *Bulletin de l'Académie royale de Belgique. Classe des sciences* 40: 50-57.

Olivereau, M. and M. Francotte-Henry. 1955. Étude histologique et biométrique de la glande thyroïde de *Caecobarbus geertsi* Blgr. *Annales de la Société royale zoologique de Belgique* 86(1): 129-150.

Paglianti, A., G. Messana, A. Cianfanelli and R. Berti. 2006. Is the perception of their own odour effective in orienting the exploratory activity of cave fishes? *Canadian Journal of Zoology* 84: 871-876.

Parenzan, P. 1938. Pesci delle acque sotterranee della Somalia. *Bollettino di Pesca, di Piscicoltura e di Idrobiologia* 14: 755-768.

Paulian, R. 1961. *La zoogéographie de Madagascar et des îles voisines.* Faune de Madagascar: Publications de l'Institut de Recherche Scientifique, Tanarive-Tsimbazaza. Volume 13.

Pellegrin, J. 1928. Les Poissons cavernicoles aveugles d'Afrique. *Compte rendu de l'Association française pour l'avancement des sciences*: 409-412.

Pellegrin, J. 1929a. *L'Eilichthys microphthalmus* Pellegrin, Poisson cavernicole de la Somalie italienne. *Bulletin du Muséum national d'histoire naturelle, 2ᵉ série* 1: 363-364.

Pellegrin, J. 1929b. Sur un poisson cavernicole africain microphthalme. *Compte rendu hebdomadaire des séances de l'Académie des sciences* 189: 204-205.

Pellegrin, J. 1932. Les Cyprinidés cavernicoles d'Afrique. *Archivio zoologico italiano* 16: 622-629.

Penney, A.J. 1988. Notes on the collection of blind cave catfish from Aigamas Cave during Swex '87. *The Bulletin of the South African Spelaeological Association* 29: 14-15.

Petit, G. 1933. Un poisson cavernicole aveugle des eaux douces de Madagascar: *Typhleotris madagascariensis* gen. et. sp. nov. *Compte rendu hebdomadaire des séances de l'Académie des sciences* 197: 347-348.

Petit, G. 1938a. Au sujet de *Caecorbarbus geertsi* Blgr. *Bulletin de la Société zoologique de France* 63: 135-141.

Petit, G. 1938b. Sur *Typhleotris madagascariensis* G. Petit. *Bulletin du Muséum national d'histoire naturelle, 2ᵉ série* 10: 491-495.

Petit, G. 1941. Sur la morphologie cranienne de deux poissons cavernicoles aveugles: *Typhleotris madagascariensis* G. Petit et *Coecobarbus geertsi* Blgr. *Compte rendu des travaux de la Faculté des sciences de Marseille* 1: 36-40.

Petit, G. and W. Besnard. 1937. Sur le comportement en aquarium du *Caecobarbus geertsii* Boulenger. *Bulletin du Muséum national d'histoire naturelle, 2ᵉ série* 9: 50-53.

Petit, G., L. Germain, Th. Monod and F. Angel. 1935. Contribution à l'étude faunistique de la réserve naturelle du Manampetsa (Madagascar). *Annales des Sciences Naturelles, Zoologie, 10ᵉ série* 18: 421-481.

Poisson, H. and H. Besairie. 1947. Note sur une nouvelle station d'un poisson cavernicole aveugle du plateau calcaire Mahafaly. *Bulletin de l'Academie Malgache N.S.* 26: 167-168.

Poll, M. 1961. Contribution a l'étude des poissons d'eau douce de Somalie appartenant aux genres *Gobius* et *Barbopsis. Bollettino dei musei e degli istituti biologici dell'Università di Genova* 31: 15-35.

Proudlove, G.S. 2006. *Subterranean Fishes of the World — An Account of the Subterranean (Hypogean) Fishes Described up to 2003 with a Bibliography 1541-2004.* International Society for Subterranean Biology, Moulis.

Proudlove, G.S. and A. Romero. 2001. Threatened fishes of the world: *Caecobarbus geertsii* Boulenger, 1921 (Cyprinidae). *Environmental Biology of Fishes* 62: 238.

Quaghebeur, M. 1955. *Onderzoekingen over blinde grotvissen in verband met de reductie in de ontwikkeling van hun ogen*. Thesis, University of Louvain (quoted by Thinès, 1969).

Quaghebeur, M. 1957. *Onderzoek over de reductie van de ogen, van de oogkas en van het tectum opticum bij blinde grotvissen*. Thesis, University of Louvain (quoted by Thinès, 1960).

Remillet, M. 1973. Aperçu de la faune souterraine à Madagascar. In: *Livre du cinquantenaire de l'Institut de Spéologie "Émile Racovitza"*, T. Ordigan (ed.). Editura Academiei Republicii Socialiste România, Bucureşti, pp. 135-160.

Romero, A. and K.M. Paulson. 2001. It's a wonderful hypogean life: A guide to the troglomorphic fishes of the world. *Environmental Biology of Fishes* 62: 13-41.

Ruffo, S. 1982. Studi sui Crostacei Anfipodi. 92. Nuovi Anfipodi di acque sotterranee della Somalia. *Monitore zoologico italiano (N.S.) Supplemento* 17: 97-113.

Sefton, M., J. Martini and R. Ellis. 1986. Cave description—Aigamas Cave. *The Bulletin of the South African Spelaeological Association* 27: 91-92 (quoted by Penney, 1988).

Skelton, P.H. 1987. *South African Red Data Book—Fishes*. South African National Scientific Programmes Report No. 137

Skelton, P.H. 1989. Gems of a different kind. *Rössing* (Winhoek, Namibia) October 1989: 6-11.

Skelton, P.H. 1990a. The status of fishes from sinkholes and caves in Namibia. *Namibia Scientific Society, Journal* 42: 75-83.

Skelton, P.H. 1990b. The status of fishes from sinkholes and caves in Namibia. *Bulletin of the South African Speleological Association* 31: 77-81.

Skelton, P.H. and G.G. Teugels. 1991. A review of the clariid catfishes (Siluroidei, Clariidae) occurring in southern Africa. *Revue d'Hydrobiologie Tropicale* 24: 241-260.

Stiassny, M.L.J. and N. Raminosoa. 1994. The fishes of the inland waters of Madagascar. In: *Biological Diversity of African Fresh- and Brackish Water Fishes*. G.G. Teugels, J.-F. Guégan and J.-J. Albaret (eds). *Annales du Musée royal de l'Afrique centrale (Sciences zoologiques)* 275: 133-148.

Teugels, G.G. 1982a. A systematic outline of the African species of the the genus *Clarias* (Pisces: Clariidae), with an annotated bibliography. *Annales du Musée royal de l'Afrique centrale. Série in 8vo (Sciences zoologiques)* 236: 1-249.

Teugels, G.G. 1982b. Preliminary data of a systematic outline of the African species of the genus *Clarias* (Pisces: Clariidae). *Revue de Zoologie africaine* 96: 731-748.

Teugels, G.G. 1986. A systematic revision of the African species of the genus *Clarias* (Pisces: Clariidae). *Annales du Musée royal de l'Afrique centrale (Sciences zoologiques)* 247: 1-199.

Teugels, G.G. and D. Adriaens. 2003. Taxonomy and phylogeny of Clariidae — An overview. In: *Catfishes*, G. Arratia, B.G. Kapoor, M. Chardon and R. Diogo (eds), Science Publishers, Inc., Enfield (NH), USA, pp. 465-487.

Thinès, G. 1953. Recherches expérimentales sur la photosensibilité du poisson aveugle *Caecobarbus geertsii* Blgr. *Annales de la Société royale zoologique de Belgique* 84: 231-265.

Thinès, G. 1955a. Les poissons aveugles (I)—Origine, taxonomie, répartition géographique, comportement. *Annales de la Société royale zoologique de Belgique* 86(1): 5-128.

Thines, G. 1955b. Étude comparative de la photosensibilité des poissons aveugles *Caecobarbus geertsii* Blgr et *Anoptichthys jordani* Hubbs et Innes. *Annales de la Société royale zoologique de Belgique* 85 (1954): 35-58.

Thinès, G. 1958a. Beobachtungen über die Phototaxis und die Thermotaxis des blinden Höhlenfisches, *Caecobarbus geertsi* Blgr (Cyprinidae). *Experientia* 14: 381-382.

Thinès, G. 1958b. Observations sur les habitats de l'*Uegitglanis zammaranoi* Gianferrari 1923, Clariidae aveugle de la Somalie Italienne. *Revue de Zoologie et de Botanique africaines* 52: 117-124.

Thinès, G. 1960. Sensory degeneration and survival in cave fishes. *Symposia of the Zoological Society of London* 3: 39-52.

Thinès, G. 1969. *L'évolution régressive des Poissons cavernicoles et abyssaux*. Masson et Cie., Paris.

Thinès, G. and N. Wissocq. 1972. Etude comparée du comportement alimentaire de deux poissons cavernicoles (*Anoptichthys jordani* Hubbs et Innes et *Caecobarbus geertsi* Blgr). *International Journal of Speleology* 4: 139-169.

Thinès, G. and J.-M. Legrain. 1973. Effets de la substance d'alarme sur le comportement des poissons cavernicoles *Anoptichthys jordani* (Characidae) et *Caecobarbus geertsi* (Cyprinidae). *Annales de Spéléologie* 28: 291-297.

Thinès, G. and G. Proudlove. 1986. Pisces. In: *Stygofauna mundi—A Faunistic, Distributional, and Ecological Synthesis of the World Fauna Inhabiting Subterranean Waters (including the Marine Interstitial)*. L. Botosaneanu (ed.), E.J. Brill/Dr. W. Backhuys, Leiden, pp. 709-733.

Thys van den Audenaerde, D.F.E. 1999. A peacock, a blind barbel, a railway wagon and a disappointed ichthyologist. *Ichthos* 62: 8-11.

Tortonese, E. 1952. Intorno al Siluroide cieco *Uegitglanis zammaranoi* (Gianf.) con particolare riguardo ai dispositivi respiratorii. *Bollettino dell'Instituto e Museo di Zoologia della Università di Torino* 3 (1951-1952): 163-170.

Trewavas, E. 1936. Dr. Karl Jordan's expedition to South-West Africa and Angola: The fresh-water fishes. *Novitates Zoologicae* 40: 63-74.

Vandel, A. 1965. *Biospeleology—The Biology of Cavernicolous Animals*. Pergamon Press Ltd., Oxford.

Vinciguerra, D. 1924. Descrizione di un ciprinide cieco proveniente dalla Somalia Italiana. *Annali del Museo civico di storia naturale Giacomo Doria* 51 (1923): 239-243.

Wilson, J.M. 1985. Ecology of the crocodile caves of Ankarana. *Cave Science* 12: 135-138.

Wilson, J.M. 1987a. The crocodile caves of Ankarana, Madagascar. *Oryx* 21: 43-47.

Wilson, J. 1987b. The crocodile caves of Ankarana: expedition to northern Madagascar, 1986. Other cave fauna. *Cave Science* 14: 115-116.

Wilson, J. 1995. *Lemurs of the Lost World — Exploring the Forests and Crocodile Caves of Madagascar.* Impact Books, London. 2nd edition.

Wilson, J.M. 1996. Conservation and ecology of a new blind fish *Glossogobius ankaranensis* from the Ankárana Caves, Madagascar. *Oryx* 30: 218-221.

Wilson, J.M., P.D. Stewart and S.V. Fowler. 1988. Ankarana—a rediscovered nature reserve in northern Madagascar. *Oryx* 22: 163-171.

Yamamoto, Y., L. Espinasa, D.W. Stock and W.R. Jeffery. 2003. Development and evolution of craniofacial patterning is mediated by eye-dependent and-independent processes in the cavefish *Astyanax. Evolution & Development* 5: 435-446.

Further Reading

Botosaneanu, L. (ed.). 1986. *Stygofauna Mundi—A Faunistic, Distributional, and Ecological Synthesis of the World Fauna Inhabiting Subterranean Waters (Including the Marine Interstitial).* E.J. Brill, Leyden.

Culver, D.C. and W.B. White (eds.). 2005. *Encyclopedia of Caves.* Elsevier/Academic Press, Amsterdam/Boston.

Gunn, J. (ed.). 2005. *Encyclopedia of Caves and Karst Science.* Fitzroy Dearbone, New York.

Jeannel, R. and E.-G. Racovitza. 1914. Énumération des Grottes visitées, 1911-1913 (5e série) Biospeologica XXXIII. *Archives de Zoologie Expérimentale et Générale* 53: 325-558.

Juberthie, C. and V. Decou (eds.). 1994-2001. *Encyclopaedia Biospeologica.* Société Internationale de Biospéologie, Moulis. Volume 1 (1994), Volume 2 (1998), Volume 3 (2001).

Wilkens, H., D.C. Culver and W.F. Humphreys (eds.). 2000. *Subterranean Ecosystems, Ecosystems of the World.* Elsevier, Amsterdam. Volume 30.

CHAPTER 11

Subterranean Fishes of China

Chen Zi-Ming[1], Luo Jing[1], Xiao Heng[1] and Yang Jun-Xing[2]
[1] School of Life Science, Yunnan University, Kunming,
Yunnan, 650091, P. R. China. E-mail: cziming@hotmail.com
[2] Kunming Institute of Zoology, Chinese Academy of Sciences, Kunming,
Yunnan, 650223, P. R. China

INTRODUCTION

Subterranean or hypogean fishes are herein defined as those living in [an occluded, semi-occluded] a cave or subterranean river system during their whole life. Among these, species restricted to subterranean habitats (troglobites) usually show a series of autapomorphies (troglomorphisms), including morphological, physiological, behavioral and ecological features, which can be directly or indirectly related to the hypogean life, being either adaptive or not (Trajano 2001). The most common morphological traits characterizing cavefishes and other troglobites are the reduction of eyes and pigmentation when compared to their surface relatives. Eyes are vestigial or absent and dark pigmentation may reduce to the point that the body appears to be transparent. In addition, the numbers and sizes of mechano- and chemo-reception organs are often increased (Greenwood 1967, Chen *et al.* 1997, Chen *et al.* 1998, Chen *et al.* 2001).

To a certain degree, it can be said that cavefishes are living fossils in the process of evolution (Chen *et al.* 2001). Because of their unusual appearance, local people also call them "transparent fishes", "glass fishes", "blind fishes" or "eyeless fishes". Hypogean fishes have always been a source of amazement and amusement throughout the history of ichthyology (Romero and Paulson 2001).

The authors contributed equally to this work

There are 3.44 million km² of karst landform areas in China, the largest part, about 0.55 million km², concentrated in southwest China in [namely] Yunnan, Guizhou, Guangxi, Chongqing, Sichuan, Hubei and Hunan Provinces (Li *et al.* 2002). In these regions, well developed karst landforms, abundant precipitation and multitudinous river systems provide good conditions for the formation of caves and evolution of troglobitic fishes (Chen *et al.* 2001).

I. A BRIEF HISTORY OF DISCOVERY AND DESCRIPTION

The earliest record of the hypogean fishes in the world is *Sinocyclocheilus hyalinus* Chen and Yang, 1993 (Chen *et al.* 2001). This fish species occurs in Alugu Cave, Luxi County, Yunnan Province and was called 'transparent fish' on a stone epigraph located at the entrance of Alugu Cave and dated to China's Ming Dynasty in 1,541 (Xie 1541). It was not formally described until 1993. After 1993, many Chinese cavefishes have been discovered and described.

On 23 April 1976, a little fish (*Typhlobarbus nudiventris* Chu and Chen 1982) was found 100 meters deep in Jianshui County, Yunnan Province, where the local people were digging to find a new water source. Following that, nine blind loaches, *Triplophysa gejiuensis* (Chu and Chen 1979), were discovered in an underground river about 400 m deep in Gejiu County, Yunnan Province, in 1978. After these two discoveries, the interest in Chinese hypogean fishes has been aroused among ichthyologists in China and abroad resulting in an upsurge of exploration and study. The number of troglomorphic fishes described for China increased rapidly, from eight species in 1990 (Chen 1990), to 11 in 1993 (Chen and Yang 1993), 21 in 2001 (Chen *et al.* 2001), 31 in 2006 (Zhao and Zhang 2006) and 51 in 2008 (Fig. 1).

II. DIVERSITY OF HYPOGEAN FISHES OCCURRING IN CHINA

II.1 Features of Hypogean Fishes

So far, 51 species of troglomorphic fishes have been discovered and formally described (Table 1). The composition of the Chinese hypogean fish fauna has the following features:

 (1) The richness of troglobitic fish species is great compared to the total number (151) recorded worldwide (Proudlove, this volume);

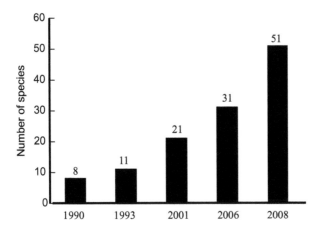

Fig. 1 The increased number of described hypogean fish species occurring in China through time.

(2) All the troglobitic fishes occurring in China belong to the Order Cypriniformes;

(3) Most of these troglobitic fishes are in two genera in two families. The genus *Sinocyclocheilus* (Cyprinidae) has the largest number of troglobitic species (28), accounting for 54.9% of the total. The genus *Triplophysa* (Balitoridae) has 10 troglobitic species, accounting for 19.6% of the total;

(4) Many species, especially in Cyprinidae, have more developed barbels than their epigean congeners. As one of the important sense organs, barbels may function importantly in the evolutionary transition from the earth's epigean water to the dark hypogean water environments (Chen *et al.* 2001);

(5) All the known troglobitic species seem to be endemic to very small areas and to have small populations. Some species, such as *T. nudiventris* Chu and Chen, 1982, have been collected only from wells. Several species, such as *T. nudiventris* Chu and Chen, 1982, *Longanalus macrochirous* Li, Ran and Chen, 2006, *T. rosa* Chen and Yang, 2005, are only known from a single specimen. Only a few species are represented by several small populations, such as *S. rhinocerous* Li and Tao, 1994 (Li *et al.* 2000);

(6) The number of specimens by which new species have been described is usually low: 27 species (52.9% of the total) were described from five specimens or less, and eight (15.7%) were described from a single specimen (Table 1); in other cases, the type material is in different collections (at least 13 up to now), hampering comparisons and further research (Zhao and Zhang 2006a).

Table 1. The taxa, distribution localities, habitats, degree of troglomorphism and specimens of the hypogean fishes occurring in China (the species were grouped by family following Nelson's (1994) systematic order), types of habitat (according to classification in Trajano 2001) and degree of troglomorphism. Abs.= not visible externally; Var. = variable in the population; Red. = slightly reduced; Polim. = population with distinct morphs. 1. slightly, but significantly reduced eyes/melanic pigmentation, with narrow intrapopulation variation; 2. wide variation, from slightly reduced to not visible externally eyes/pigmentation; 3. most individuals with deeply reduced eyes/pigmentation; 4. all individuals completely (externally) anophthalmic and/or depigmented; 4+. DOPA positive (melanin is then synthesized and melanin cells become visible when in light); 4– DOPA negative (there is no response to light); Polim. Polymorphic phenotypes (according to classification in Trajano and Bichuette, this volume)

Taxon	Locality	Habitat	Eyes	Pigm.	Specimens
Order Cypriniformes					
Family Cyprinidae					
Subfamily Barbinae					
1 *Sinocyclocheilus albeoguttatus* (Zhou and Li 1998)	Leye, Guangxi	Upper phreatic zone (cave)	4	4	1
2 *Sinocyclocheilus altishoulderus* (Li and Lan 1992)	Taiping township, Donglan, Guangxi	Base-level stream	3	3	13
#3 *Sinocyclocheilus anatirostris* (Lin and Luo 1986)	Lingyun and Leye County, Guangxi	Base-level stream	4	4	6
#4 *Sinocyclocheilus angularis* (Zheng and Wang 1990)	Panxian, Guizhou	Flooded caves	3	3	3
#5 *Sinocyclocheilus anophthalmus* (Chen and Chu 1988)	Jiuxiang township, Yiliang, Yunnan	Base-level stream	4	4	8
6 *Sinocyclocheilus aquihornes* (Li and Yang 2007)	Qiubei, Yunnan	Base-level stream	4	4	1
7 *Sinocyclocheilus bicornutus* (Wang and Liao 1997)	Xingren and Zhenfeng County, Guizhou	Base-level stream	3	4+	9
8 *Sinocyclocheilus brevibarbatus* (Zhao et al. 2008)	Gaoling Town, Du'an County, Guangxi	Base-level stream	1	1	7

Table 1 Contd...

Table 1 Contd.

No.	Species	Locality	Habitat			
9	Sinocyclocheilus broadihornes (Li and Mao 2007)	Shilin, Yunnan	Base-level stream	3	3	9
#10	Sinocyclocheilus cyphotergous (Dai 1988)	Luodian, Guizhou	Base-level stream	3	4	4
11	Sinocyclocheilus donglanensis (Zhao et al. 2006)	Donglan, Guangxi	Flooded cave	Abs	Abs	5
12	Sinocyclocheilus furcodorsalis (Chen et al. 1997)	Tian'e, Guangxi	Upper phreatic zone (cave)	4	4	4
13	Sinocyclocheilus guangxiensis (Zhou and Li 1998)	Longhuang cave, Tongle township, Leye, Guangxi	Base-level stream	4	4	3
14	Sinocyclocheilus guilinensis (Ji 1985)	Guilin, Guangxi	Upper phreatic zone (cave)	1	1	7
15	Sinocyclocheilus halfibindus (Li and Lan 1992)	Fengchen township, Fengshan, Guangxi	Base-level stream	4	4	25
16	Sinocyclocheilus huaningensis (Li 1998)	Dalongtan, Huaning, Yunnan	Upper phreatic zone	1	1	26
17	Sinocyclocheilus hugeibarbus (Li and Ran 2003)	Libo, Guizhou	Base-level stream	3	3	5
#18	Sinocyclocheilus hyalinus (Chen and Yang 1993)	Alugu cave, Luxi, Yunnan	Base-level stream	4	4+	2
19	Sinocyclocheilus jiuxuensis (Li and Lan 2003)	Jiuxu township, Hechi, Guangxi	Base-level stream	3	4	6
20	Sinocyclocheilus liboensis (Li et al. 2004)	Libo, Guizhou	Upper phreatic zone (cave)	1	1	2
21	Sinocyclocheilus lingyunensis (Li et al. 2000)	Lingyun, Guangxi	Base-level stream	3	4	13
22	Sinocyclocheilus longifinus (Li 1998)	Dalongtan, Huaning, Yunnan	Upper phreatic zone	1	1	2
#23	Sinocyclocheilus microphthalmus (Li 1989)	Luolou township, Lingyun, Guangxi	Base-level stream	3	4	15
#24	Sinocyclocheilus rhinocerous (Li and Tao 1994)	Xinzhai, Luoping, Yunnan	Flooded caves	2	3	2
25	Sinocyclocheilus tianeensis (Li et al. 2003)	Tian'e, Guangxi	Upper phreatic zone (cave)	4	4	11
26	Sinocyclocheilus tianlinensis (Zhou et al. 2004)	Pingshan township, Tianlin, Guangxi	Upper phreatic zone (cave)	4	4	1

Table 1 Contd.

Table 1 Contd.

27	*Sinocyclocheilus tileihornes* (Mao et al. 2003)	A'gan, Luoping, Yunnan	Base-level stream	3	4	7
28	*Sinocyclocheilus xunleensis* (Lan et al. 2004)	Xunle township, Huangjiang, Guangxi	Upper phreatic zone (cave)	4	4	5
#29	*Typhlobarbus nudiventris* (Chu and Chen 1982)	Yangjieba, Jianshui, Yunnan	Upper phreatic zone (cave)	4	4	3
	Subfamily Labeoninae					
30	*Longanalus macrochirous* (Li et al. 2006)	Libo, Guizhou	Upper phreatic zone (cave)	Abs	Abs	1
	Family Balitoridae Subfamily Nemacheilinae					
31	*Heminoemacheilus hylimus* (Lan et al. 1996)	Bao'an Township, Du'an County, Guangxi	Base-level stream	4	4	6
#32	*Oreonectes anophthalmus* (Zheng 1981) Wuming, Guangxi	Qifeng mountaion, zone (cave)	Upper phreatic	4	4	7
33	*Oreonectes furcocaudalis* (Zhu and Cao 1987)	Rongshui County, Guangxi	Base-level stream	4	4	3
34	*Oreonectes microphthalmus* (Du et al. 2008)	Du'an County, Guangxi	Upper phreatic zone	4	4	6
#35	*Oreonectes retrodorsalis* (Lan et al. 1995)	Liuzhai township, Nandan, Guangxi	Base-level stream	4	4	7
36	*Oreonectes translucens* (Zhang et al. 2006)	Xia'ao Village, Du'an County, Guangxi	Upper phreatic zone (cave)	4	4	3
#37	*Paracobitis longibarbatus* (Chen et al. 1998)	Libo, Guizhou	Flooded caves	4	4	3
38	*Paracobitis maolanensis* (Li et al. 2006)	Maolan, Guizhou	Upper phreatic zone (cave)	4	4	1
39	*Paracobitis posterodarsalus* (Li et al. 2006)	Nandan, Guangxi	Upper phreatic zone (cave)	4	4	1
40	*Triplophysa aluensis* (Li and Zhu 2000)	A'lugu cave, Luxi, Yunnan	Base-level stream	3	3	1

Table 1 Contd..

Table 1 Contd.

#41	*Triplophysa gejiuensis* (Chu and Chen 1979)	Kafan, Gejiu, Yunnan	Base-level stream	4	4	9
42	*Triplophysa nandanensis* (Lan et al. 1995)	Nandan, Guangxi	Upper phreatic zone (cave)	4	4	11
43	*Triplophysa qiubeiensis* (Li et al. 2008)	Nijiao, Qiubei, Yunnan	Upper phreatic zone (cave)	4	4	11
44	*Triplophysa rosa* (Chen and Yang 2005)	Wulong, Chongqing	Upper phreatic zone (cave)	4	4	1
*45	*Triplophysa* sp. (Chen et al. 2010)	Leye, Guangxi	Upper phreatic zone (cave)	3	3	10
#46	*Triplophysa shilinensis* (Chen and Yang 1992)	Weiboyi, Shilin, Yunnan	Upper phreatic zone (cave)	4	4	2
47	*Triplophysa tianeensis* (Chen et al. 2004)	Tian'e, Guangxi	Base-level stream	4	4	6
48	*Triplophysa xiangxiensis* (Yang et al. 1986)	Longshan, Xiangxi, Hunan	Base-level stream	4	4	3
#49	*Triplophysa yunnanensis* (Yang 1990) Yiliang, Yunnan Family Cobitidae Subfamily Cobitinae	Jiuxiang township,	Base-level stream	1	1	4
50	*Protocobitis polylepis* (Zhu et al. 2008)	Wuming, Guangxi	Upper phreatic zone		4	2
51	*Protocobitis typhlops* (Yang and Chen 1994)	Xia'ao, Du'an, Guangxi	Upper phreatic zone (cave)	4	4	7

Total: 1 order 3 families 4 subfamilies 8 genera 51 species.

*unpublished
#Being listed in 'China Species Red List' (Wang and Xie 2004).

II.2 Geographic Distribution Patterns

All the 45 described Chinese troglobitic fishes occur in south and southwest China with 27 species in Guangxi, 14 in Yunnan, 8 in Guizhou, one in Hunan and one in Chongqing (Fig. 2). There are 0.55 million km² of karst in southwest China, namely Yunnan, Guizhou, Guangxi, Chongqing, Sichuan, Hubei and Hunan Provinces, where factors such as the well developed karst landforms, abundant precipitation and multitudinous river systems provide good conditions for the evolution of troglobitic populations, and abundant potential for the discovery of more troglobitic species. By 1987, there were more than 70 cave species being described in China. Besides these, many other new species have been discovered. These include snails, shrimps, spiders, crabs, and diplopods. There were two new families of diplopod (Paracortinidae, Sinocallipodidae), five new genera of diplopod and one new genus of crabs (Ran and Chen 1998). Just in Guizhou, there were 85 cave animal species reported by 2007 (Internet resource, http://gzsb.gog.com.cn/system/2007/10/27/010151816. shtml).

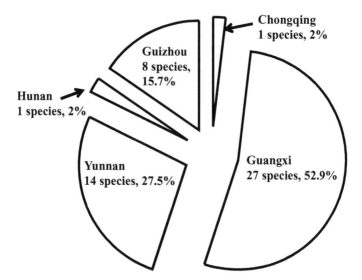

Fig. 2 The distribution pattern of the number of described hypogean fish species in different provinces (or regions) in China.

Although so many cave species have been discovered, still there are likely that many species remain to be discovered because: (1) the area of karst is so vast that not all caves potentially harbouring troglobites have been explored [or been explored] to the extent needed to discover them; and (2) in other cases, cavefishes and other animals have been reported but no specimens were sent to ichthyologists or other experts. There is no doubt that more and more troglobitic species (including fishes) will be discovered through time.

II.3 The Questionable Status of Some Cavefish Species

The status and validity of some cavefish species is questionable:

(1) *Sinocyclocheilus guilinensis* Ji, 1985. This species was first mentioned in Zhou's (1985) paper about cavefishes occurring in Guangxi and has never been mentioned in later literature. From its simple description it is similar to *S. jii* Zhang and Dai, 1992. However Zhao *et al.* (2008a) compared the specimens to those of *S. jii* and found that *Sinocyclocheilus guilinensis* is a valid species.

(2) *Sinocyclocheilus yimenensis* Li and Xiao, 2005. Although all the *Sinocyclocheilus* species are associated with cave or other underground habitats, this does not mean that they are all troglobites. Only those with typical troglobite characteristics, such as degeneration of eyes and greatly reduced pigmentation etc., are treated as such. Although *S. yimenensis* was collected from an underground river (Li *et al.* 2005), it has no characteristics of troglobites. Hence, we treat it as an epigean species found accidentally in caves.

(3) *Paralepidocephalus yui* Tchang, 1935. This species was first reported from Yilong Lake in Yunnan (Tchang 1935). It could be also collected from cave environments, such as Jiuxiang Limestone Cave (Yang and Chen 1993b), but it has no typical troglobite characteristics, and is treated as an epigean species herein.

III. CURRENT RESEARCH STATUS ON CAVEFISHES IN CHINA

Currently, most studies in China are focused on discovering and describing new species. Detailed studies in depth on other subjects, such as physiology, behavior, ecology, etc., are scarce or lacking. The limited information on some of these subjects is herein documented:

III.1 Biology and Ecology

When kept in light in the laboratory, individuals of species such as *S. anophthalmus*, may become dark due to the synthesis of melanin or development of more pigmented melanophores (Chen *et al.* 2001). Li *et al.* (2000) described the conditions of three collecting localities of *Sinocyclocheilus rhinocerous*. One locality is a seasonal lake, about 7000 m^2 in summer and dries in winter. The second locality is a cave about 80 m^2 and 15 m below ground; the water level is consistent all [around the] year. The third locality is an outlet of a spring in summer, flowing only in the rainy season, from May to October. *S. malacopterus, Discogobio brachyphysallidos* and *Rhodeus sinensis* are also found in this spring outlet. Li and Tao (2005) have bred *S. rhinocerous* in captivity. In the lab, these fish are omnivores, eating steamed rice, raw noodles, paste and earthworms. The optimum water temperature is 12-20°C; the high lethal temperature is 28°C and when the temperature is lower than 4°C and higher than 0°C the fish stop feeding and moving.

III.2 Morphology, Anatomy and Troglomorphic [Characteristic] Evolution

There are 59 species in genus *Sinocyclocheilus*, but only 28 species are hypogean. This genus provides an ideal basis for research on the evolution of troglomorphic characters. By comparing epigean and hypogean species in *Sinocyclocheilus*, an evolutionary trend from surface dwellers to underground dwellers has been observed for characteristics such as size of the eyes, pigmentation, scales, barbels, lateral line and frontal organ (Chen *et al.* 1994). Eyes tend to become vestigial in subterranean species. The barbels, which the fish use to probe for food, are longer and thicker than in surface water dwellers and so on (Chen *et al.* 1994, Wang and Chen 2000, Chen *et al.* 2001).

Li *et al.* (1997) observed the tissues of the horn-like frontal organ of *S. rhinocerous* and found no nerves or glands. They concluded that this strange derived organ had no relationship with feeling; it may act as an organ to protect the soft head if the fish collides with the stony wall in the dark.

Chen *et al.* (2001) pointed out that cavefish, as well as some troglobitic invertebrates, are often distantly related phylogenetically, but have evolved similar characteristics. This is a clear example of the convergent evolution of adaptation to the dark cave environment.

III.3 Morphological and Genetic Evolution

The genus *Sinocyclocheilus*, which is endemic to China, occurs in karst cave waters and surface rivers or lakes in Yunnan, Guizhou and Guangxi. It includes at least 60 species of which 28 are hypogean. It provides a good model system, in which closely related species specialized to the life in subterranean habitats possess similar phenotypes. This allows us to study evolutionary developmental mechanisms and the role of natural selection and adaptation in cave animals. Based on osteological and external characteristics, Shan and Yue (1994) revealed that all *Sinocyclocheilus* species in their study were grouped into two monophyletic clusters, one including all surface species and the other all cave species except *S. anophthalmus* that groups with the first cluster. The distinctive morphological divergences between these two groups were variations in the frontal organ, number of lateral line scales and presence of radii in the scales. Species in the group containing all but one of the cave species shared the following character states: lower lateral line scale numbers (less than 45 or body naked resulting from scale regression), radii present only in the scales apical region, and an acutely upheaved frontal–parietal area. They regarded these characters as having a phylogenetic significance. Thus they divided the genus *Sinocyclocheilus* [was subdivided] into two subgenera, *Sinocyclocheilus* (for the surface species plus *S. anophthalmus*) and *Gibbibarbus* (for the remaining troglobitic ones).

Wang and Chen (1999) re-examined the distinguishing characters between the two subgenera and found out that scale numbers within *Sinocyclocheilus* varied in a continuous way and the divergence in scale radius between the two groups was probably due to convergence. Consequently, they did not support the view that this genus could be subdivided into those two subgenera, *Sinocyclocheilus* and *Gibbibarbus*. However, they also treated the states of frontal–parietal area as one of phylogenetically informative characters. Thus, all cave species in their study clustered as a subclade and nested within the surface species.

Xiao *et al.* (2005) determined nucleotide sequences of complete mitochondrial cytochrome *b* gene (1140 bp) and partial *ND4* gene (1032 bp) of 31 recognized in group *Sinocyclocheilus* species and one outgroup species, *Barbodes laticeps*. Phylogenetic trees were reconstructed using maximum parsimony, Bayesian, and maximum likelihood analyses. Their results showed that cave species were polyphyletic and occurred in all the five major clades, indicating that adaptation to subterranean environments had occurred multiple times during the evolutionary history of *Sinocyclocheilus*. In addition, Li *et al.* (2008) added five more species, but their results did not change the number of clades within the genus. They also estimated the divergence time of the genus *Sinocyclocheilus* as the late Miocene, about 11 million years ago (Ma). In addition, except for *S. rhinocerous* and *S. hyalinus*, all species distributed in Yunnan formed a strongly supported monophyletic group. Their results supported that the diversification of *Sinocyclocheilus* in Yunnan may correlate with the uplifting of the Yunnan Plateau, and Yunnan might be the diversification centre of *Sinocyclocheilus*. Obviously, based on the genetic results, it is not acceptable to subdivide the genus *Sinocyclocheilus* into *Sinocyclocheilus* and *Gibbibarbus* based on convergent troglobitic traits, such as reduction of eyes and pigmentation.

IV. CONSERVATION

Apparently, all the hypogean fish species occurring in China are endemic to small areas and have small populations. For some species, such as *Typhlobarbus nudiventris*, the only access to them has been limited to occasional collections from wells. As already mentioned, several species, such as *Typhlobarbus nudiventris*, *Longanalus macrochirous*, *Triplophysa rosa*, are known from a single specimen each; only a few species are represented by several small populations, such as *Sinocyclocheilus rhinocerous* (Li *et al.* 2000). Proudlove (2001) concluded that the hypogean fishes are susceptible to five main threats: habitat degradation, hydrological manipulations, environmental pollution, overexploitation, and introduced exotic species. Any threat could have serious consequences for cave species survival.

In 1998, four hypogean fish species (*Sinocyclocheilus anophthalmus*, *Typhlobarbus nudiventris*, *Triplophysa gejiuensis*, *Oreonectes anophthalmus*) were included in the "China Red Data Book of Endangered Animals, Pisces" (Yue and Chen 1998). In 2004, this number had increased to 14 (Wang and Xie 2004) (Table 1). Hypogean fish species are receiving more and more attention for their conservation. But, in view of the large number of troglobitic species and their vulnerability, conservation actions are not sufficient.

Cavefish have evolved in relatively stable cave environments for long periods of time so when caves are altered by human activities the hypogean fishes living there may not survive. For example, after Alugu Cave was developed as a scenic spot, *S. hyalinus* almost became extinct because of environmental deterioration (Yang and Chen 1994). With the fast development of China's economy and fast increase in population, such problems may increase in the future. So, for the hypogean fish survival, a wider range of consequences resulting from human activities—such as irrigation, hydropower construction, tourism, and underground water use—should be circumspectly considered. And much more positive conservation measures—including habitat protection, population recovery and scientific research—should be supported and implemented.

The distributional locality of hypogean fishes is usually a point-like area and most of them are located in remote rural mountains. So far, there is no report that introduced exotic species pose threat for their survival.

V. DISCUSSION

The unique morphology and rarity of hypogean fishes make them very precious. Their discovery in China not only adds a new feature to the Chinese fresh-water fish fauna, but also enriches the content of Chinese speleology (Chen *et al.* 2001). Considering that 150 hypogean fish species have been listed for the world (Proudlove, this volume), the diversity of 45 hypogean fish species in China is high compared with the total number. Additionally, there are 3.44 million km² of karst landform area in China, and most of these karst landform areas are poorly explored by experts on hypogean creatures. It is undoubtedly true that the number of hypogean fish species will increase with continued exploration and study. The increasing number of new species discovered and described enrich Chinese fish fauna and fish fauna of the world, and provide many opportunities for research. On the other hand, many hypogean fish species may disappear due to habitat disturbance (irrigation, hydropower construction, tourism exploitation, and underground water use) if conservation measures are not taken.

Although many hypogean fish species have been described, there are still numerous problems with the published papers. In the case of *Sinocyclocheilus*, for instance, some species descriptions were too simple without proper comparisons with other related species. In addition, papers were not reviewed by peers. Also, type specimens are scattered in different research institutions or private collections and this poses obstacles for the examination of specimens for new species description. Finally, experts in the same field do not openly communicate.

In spite of the rich Chinese hypogean fish fauna, studies on evolution, ecology, physiology, conservation and behaviour are just beginning or have not started. We need not only more study but also more finanacial support.

Acknowledgments

We thank Dr. Eleonora Trajano and Dr. Thomas Poulson for improving the manuscript and helping with the language. This study was supported by the National Science Foundation of China (30970326, U0936602, 30870291), the Cooperation Project between the Chinese Academy of Sciences and Guizhou Province (2005-2008), The National Basic Research Program (also called 973 Program) (2003CB415105) and the Scientific Research Foundation of Yunnan University (2008YB004).

References

Chen, X.Y. and J.X. Yang. 2005. *Triplophysa rosa* sp. nov.: A new blind loach from China. *Journal of Fish Biology* 66: 599-608.

Chen, X.Y., G.H. Cui and J.X. Yang. 2004. A new cave-dwelling fish species of genus *Triplophysa* (Balitoridae) from Guangxi, China. *Zoological Research* 25: 227-231. (In Chinese with English abstract.)

Chen, Y.R. 1990. Research status on Chinese cavefishes. *Biology Information* 2(3): 117-119. (In Chinese)

Chen, Y.R. and J.X. Yang. 1993. A synopsis of cavefishes from China. *Proceedings of XI International Congress of Speleology*: 121-122.

Chen, Y.R. and J.X. Yang. 1994. Exploitation and conservation of cave fish in tourist resources. In: *Proceedings of International Symposium on the Exploitation and Protection of Karst and Cave Scenic Tourist Resources*, L.H. Song (Ed.). Earthquake Press, Beijing, pp. 149-153.

Chen, Y.R., J.X. Yang and G.C. Xu. 1992. A new blind loach of *Triplophysa* from Yunnan stone forest with comments on its phylogenetic relationship. *Zoological Research* 13(1): 17-23. (In Chinese with English abstract.)

Chen, Y.R., J.X. Yang and Z.G. Zhu. 1994. A new fish of the genus *Sinocyclocheilus* from Yunnan with comments of its characters adaptation. *Acta Zootaxonomica Sinica* 19(2): 246-253. (In Chinese with English abstract.)

Chen, Y.R., J.X. Yang and J.H. Lan. 1997. One new species of blind cavefish from Guangxi with comments on its phylogenetic status (Cypriniformes: Cyprinidae: Barbinae). *Acta Zootaxonomica Sinica* 22(2): 219-223. (In Chinese with English abstract.)

Chen, Y.R., J.X. Yang and Z.M. Chen. 2001a. Species and convergent evolution of cave-dwelling *Sinocyclocheilus* fishes. Memoires de Biospeologie (International Journal of Subterranean Biology), Tome XXVIII : 185-189.

Chen, Y.R., Z.M. Chen and J.X. Yang. 2001b. Cave fishes of china. Memoires de Biospeologie (International Journal of Subterranean Biology), Tome XXVIII : 181-184.

Chen, Y.R., X.L. Chu, Z.Y. Luo and J.Y. Wu. 1988. A new blind cyprinid fish from Yunnan, China with a reference to the evolution of its characters. *Acta Zoologica Sinica* 34(1): 64-70. (In Chinese with English abstract.)

Chen, Y.R., J.X. Yang, B. Sket and G. Aljancic. 1998. A new cave loach of *Paracobitis* with comments of its characters evolution. *Zoological Research* 19(1): 59-63. (In Chinese with English abstract.)

Chen, Z.M., L.X. Li and J.X. Yang. 2010. Description of a new cave-dwelling species of the genus *Triplophysa* (Balitoridae: Nemacheilinae) from Guangxi, China. (Unpublished.)

Chu, X.L. and Y.R. Chen. 1978. The discovery of the first blind fish from China. *Journal of Chinese Nature* 1(6): 343. (In Chinese)

Chu, X.L. and Y.R. Chen. 1979. A new blind cobitid fish (Pisces, Cypriniformes) from subterranean waters in Yunnan, China. *Acta Zoologica Sinica* 25(3): 285-287. (In Chinese with English abstract.)

Chu, X.L. and Y.R. Chen. 1982. A new genus and species of blind cyprinid fish from China with special reference to its relationships. *Acta Zoologica Sinica* 28(4): 383-388. (In Chinese with English abstract.)

Du, L.N., X.Y. Chen and J.X. Yang. 2008. A review of the Nemacheilinae genus *Oreonectes* Günther with descriptions of two new species (Teleostei: Balitoridae). *Zootaxa* 1729: 23-36.

Greenwood, P.H. 1967. Blind cave fishes. *Studies in Speleology* 1(5): 262-275.

Lan, J.H., Y.H. Zhao and C.G. Zhang. 2004. A new species of the genus *Sinocyclocheilus* from China (Cypriniformes, Cyprinidae, Barbinae). *Acta Zootaxonomica Sinica* 29: 377-380. (In Chinese with English abstract.)

Li, W.X. and J.H. Lan. 1992. A new genus and three new species of Cyprinidae from Guangxi, China. Journal of Zhanjiang Fisheries College 12(2): 46-51. (In Chinese with English abstract.)

Li, W.X. and J.N. Tao. 1994. A new species of Cyprinidae from Yunnan— *Sinocyclocheilus rhinocerous. Journal of Zhanjiang Fisheries College* 14(1): 1-3. (In Chinese with English abstract.)

Li, W.X. and Z.G. Zhu. 2000. A new species of *Triplophysa* from cave Yunnan. *Journal of Yunnan University* 22: 396-398. (In Chinese with English abstract.)

Li, W.X. and J.N. Tao. 2002. Local dissection of body of the fish *Sinocyclocheilus rhinocerous. Journal of Yunnan Agricultural University* 17(3): 207-209, 219. (In Chinese with English abstract.)

Li, W.X. and J.N. Tao. 2005. Preliminary research on the cultivation of *Sinocyclocheilus rhinocerous. Yunnan Agricultural Science and Technology* 1: 46. (In Chinese)

Li, W.X., D.F. Wu and A.L. Chen. 1998. Two new species of *Sinocyclocheilus* from Yunnan (Cypriniformes, Cyprinidae). *Journal of Zhanjiang Ocean University* 18(4): 1-5. (In Chinese with English abstract.)

Li, Y.B., J.J. Hou and D.T. Xie. 2002. The recent development of research on karst ecology in Southwest China. *Scientia Geographica Sinica* 22(3): 365-370. (In Chinese with English abstract.)

Li, W.X., J.H. Lan and S.Y. Chen. 2003a. A new species of cave Sinocyclocheilus from Guangxi—*Sinocyclocheilus jiuxuensis. Journal of Guangxi Normal University* 21(4): 83-85. (In Chinese with English abstract.)

Li, W.X., J.C. Ran and H.M. Chen. 2003b. A new species of cave *Sinocyclocheilus* in Guizhou and its adaptation comment. *Journal of Jishou University* (Natural Science Edition) 24(4): 61-63. (In Chinese with English abstract.)

Li, W.X., H. Xiao, R.G. Zan, Z.Y. Luo, C.H. Ban and J.B. Ben. 2003c. A new species of *Sinocyclocheilus* from caves in Guangxi. *Journal of Guangxi Normal University* 21(3): 80-81. (In Chinese with English abstract.)

Li, W.X., H.M. Chen and J.C. Ran. 2004. A new species of Sinocyclocheilus from the cave in Guizhou, China. *Journal of Zhanjiang Ocean University* 24(3): 1-3. (In Chinese with English abstract.)

Li, W.X., D.F. Wu, A.L. Chen and J.N. Tao. 2000. The preliminary investigation of geographical distribution and ecological adaptation for cave environments of *Sinocyclocheilus rhinocerous. Journal of Yunnan Agricultural University* 15(1): 1-4. (In Chinese with English abstract.)

Li, W.X., H.F. Yang, H. Chen, C.P. Tao, S.Q. Qi and F. Han. 2008. A new blind underground species of the genus *Triplophysa* (Balitoridae) from Yunnan, China. *Zoological Research* 29(6): 674-678.

Li, Z.Q., B.C. Guo , J.B. Li, S.P. He and Y.Y. Chen. 2008. Bayesian mixed models and divergence time estimation of Chinese cavefishes (Cyprinidae: *Sinocyclocheilus*). *Chinese Science Bulletin* 53(13): 1560-1569.

Mao, W.N., Z.M. Lu, W.X. Li, H.B. Ma and G. Huang. 2003. A new species of *Sinocyclocheilus* (Cyprinidae) from cave of Yunnan, China. *Journal of Zhanjiang Ocean University* 23(3): 1-3. (In Chinese with English abstract.)

Nelson, J.S. 1994. *Fishes of the World*. John Wiley & Sons, New York.

Proudlove, G.S. 2001. The conservation status of hypogean fishes. *Environmental Biology of Fishes* 62: 201-213.

Ran, J.C. and H.M. Chen. 1998. A survey of speleobiological studies in China. *Carsologica Sinica* 17(2): 151-159.

Romero, A. and K.M. Paulson. 2001. It's a wonderful hypogean life: A guide to the troglomorphic fishes of the world. *Environmental Biology of Fishes* 62: 13-41.

Shan, X.H. and P.Q. Yue. 1994. The study on phylogeny of the *Sinocyclocheilus* fishes. *Zoological Research* 15(Supplement): 36-44. (In Chinese with English abstract.)

Tchang, T.L. 1935. A new genus of loach from Yunnan. *Bulletin of the Fan Memorial Institute of Biology* 7: 17-19.

Trajano, E. 2001. Ecology of subterranean fishes: An overview. *Environmental Biology of Fishes* 62: 133-160.

Wang, D.Z. and J.W. Liao. 1997. A new species of *Sinocyclocheilus* from Guizhou, China (Cypriniformes: Cyprinidae: Barbinae). *Acta Academiae Medicinae Zunyi* 20(2, 3): 1-3. (In Chinese with English abstract.)

Wang, D.Z. and Y.Y. Chen. 2000. The origin and adaptive evolution of the genus *Sinocyclocheilus*. *Acta Hydrobiologica Sinica* 24: 630-634. (In Chinese with English abstract.)

Wang, D.Z. and D.J. Li. 2001. Two new species of the genus *Triplophysa* from Guizhou, China (Cypriniformes: Cobitidae). *Acta Zootaxonomica Sinica* 26: 98-101. (In Chinese with English abstract.)

Wang, S. and Y. Xie. 2004. *China Species Red List*, Vol. I. Higher Education Press, Beijing. (In Chinese)

Wang, D.Z., Y.Y. Chen and X.Y. Li. 1999. An analysis on the phylogeny of the genus *Sinocyclocheilus*. *Acta Academiae Medicinae Zunyi* 22(1): 1-6. (In Chinese with English abstract.)

Xiao, H, S.Y. Chen, Z.M. Liu, R.D. Zhang, W.X. Li, R.G. Zan, Y.P. Zhang. 2005. Molecular phylogeny of *Sinocyclocheilus* (Cypriniformes: Cyprinidae) inferred from mitochondrial DNA sequences. *Molecular Phylogenetics and Evolution* 36: 67-77.

Xie, Y.J. 1541. Travel Notes on Alu Cave (on a stone epigraph located at the entrance of Alugu Dong and dated in Ming Dynasty). (In Chinese)

Yang, J.X. and Y.R. Chen. 1993a. The cavefishes from Doan, Guangxi, China with comments on their adaptation to cave habitats. *Proceedings of XI International Congress of Speleology*: 124-126.

Yang, J.X. and Y.R. Chen. 1993b. The cavefishes from Jiuxiang limestone cave of Yunnan, China with reference to the character evolutions. *Proceedings of XI International Congress of Speleology*: 126-128.

Yang, J.X. and Y.R. Chen. 1994. Characters, exploitation and conservation of the fish resources in karst areas of Yunnan. In: *Proceedings of International Symposium* on the Exploitation and Protection of Karst and Cave Scenic Tourist Resources, L.H. Song (Ed.). Earthquake Press, Beijing, pp. 145-148.

Yue, P.Q. and Y.Y. Chen. 1998. *China Red Data Book of Endangered Animals, Pisces.* Science Press, Beijing.

Zhang, C.G. and D.Y. Dai. 1992. A new species of *Sinocyclocheilus* from Guangxi, China. *Acta Zootaxonomica Sinica* 17(3): 377-379. (In Chinese with English abstract.)

Zhang, C.G. and Y.H. Zhao. 2001. A new fish of *Sinocyclocheilus* from Guangxi, China with a comment on its some biological adaptation (Cypriniformes: Cyprinidae). *Acta Zootaxonomica Sinica* 26: 102-108. (In Chinese with English abstract.)

Zhang, Z.L., Y.H. Zhao and C.G. Zhang. 2006. A new blind loach, *Oreonectes translucens* (Teleostei: Cypriniformes: Nemacheilinae), from Guangxi, China. *Zoological Studies* 45(4): 611-615.

Zhao, Y.H. and C.G. Zhang. 2006a. Cavefishes, concept, diversity and research progress. Biodiversity Sciences 14(5): 451-460. (In Chinese with English abstract.)

Zhao, Y.H., K. Watanabe and C.G. Zhang. 2006b. *Sinocyclocheilus donglanensis,* a new cavefish (Teleostei: Cypriniformes) from Guangxi, China. *Ichthyological Research* 53: 121-128.

Zhao, Y.H., C.G. Zhang and J. Zhou. 2008a. *Sinocyclocheilus guilinensis,* a new species from an endemic cavefish group (Cypriniformes: Cyprinidae) in China. *Environmental Biology of Fishes* DOI 10.1007/s10641-008-9344-8.

Zhao, Y.H., J.H. Lan and C.G. Zhang. 2008b. A new cavefish species, *Sinocyclocheilus brevibarbatus* (Teleostei: Cypriniformes: Cyprinidae), from Guangxi, China. *Environmental Biology of Fishes* DOI 10.1007/s10641-008-9338-6.

Zhou, J. 1985. Fishes of Karst cave in Guangxi. *Carsologica Sinica* 4: 377-385. (In Chinese)

Zhou, J., C.G. Zhang and A.Y. He. 2004. A new species of the genus *Sinocyclocheilus* from Guangxi, China (Cypriniformes, Cyprinidae). *Acta Zootaxonomica Sinica* 29: 591-594. (In Chinese with English abstract.)

Zhou, S.B. and G.L. Li. 1998a. A new species of the genus *Sinocyclocheilus* from cave of Guangxi, China. In: Contributions from Tianjin Natural History Museum. Ocean Press, Beijing, pp. 9-12.

Zhou, S.B. and G.L. Li. 1998b. A new species of *Sinocyclocheilus* from Guangxi (Cypriniformes: Cyprinidae: Barbinae). *Guangxi Sciences* 5: 139-141, 149. (In Chinese with English abstract.)

Zhu, S.Q. and W.X. Cao. 1987. The noemacheiline fishes from Guangdong and Guangxi with descriptions of a new genus and three new species (Cypriniformes: Cobitidae). *Acta Zootaxonomica Sinica* 12: 323-331.

Zhu, Y., Lü, Y.J., Yang, J.X. and S. Zhang. 2008. A new blind underground species of the genus *Protocobitis* (Cobitidae) from Guangxi, China. *Zoological Research* 29(4): 452-454.

CHAPTER 12

Subterranean Fishes of India

Atanu Kumar Pati and Arti Parganiha
School of Life Sciences, Pt. Ravishankar Shukla University
Raipur 492 010, India
E-mail: akpati19@gmail.com

INTRODUCTION

The term "subterranean" is an adjective meaning something/somebody existing or occurring under the earth's surface. It is a larger domain that includes caves, caverns, lava tubes and phreatic spaces. Habitats, such as deep burrows, ant hill and other similar underground biogenic structures, of several burrowing and cryptobiotic animals, could be included in the subterranean category. Manmade underground tunnels, namely concrete pipelines carrying domestic sewage, could also be termed as a type of subterranean niche. In general, subterranean spaces are poorly explored as regards mapping of the biodiversity. The main limitations are: (a) lack of easy access to these underground spaces, (b) prevalence of darkness with high humidity, and (c) the fear of unknown and uncertain things and/or events. In India, so far no concerted efforts have been made to survey the already known subterranean habitats. Further a dedicated research institute/center in the realm of biospeleology (= science of cave/ hypogean organisms) is conspicuously absent.

SEARCH FOR A CAVE (KARST) MAP OF INDIA

We have a hunch that in India more than thousands of caves may be present. Unfortunately for this subcontinent neither a karst nor a cave map is available. Important agencies, such as Geological Survey of India (GSI), National Thematic and Mapping Organization of India (NATMO),

Survey of India and Remote Sensing Organization of India, were contacted through email. Also the websites of these organizations were visited. The result of these efforts was in the negative. The websites of different State Governments of India were also browsed with an emphasis to look into the forest and tourism related data bases, but nothing was found. However, wild life researchers could not be contacted through questionnaire. The later is a daunting task. Nonetheless many caves in India have been explored by amateur cave enthusiasts, biologists, geologists and personnel from the forest departments and NGOs. The only available source of information about the Indian caves is the South Asia Cave Registry compiled by H.D. Gebauer, Germany.

WHY FISHES BELONGING TO CERTAIN FAMILY ARE FOUND IN CAVES?

The organisms belonging to the phylum Arthropod constitute the bulk of the subterranean biodiversity. Among the sub-phylum vertebrata the members of the class Pisces are the most prominent invaders of subterranean niche. The Order Cypriniformes stand out distinctly having the maximum number of reported subterranean species, belonging to the family, namely Cyprinidae, Balitoridae, and Cobitidae (Proudlove 2006). The species belonging to the former two families have been reported abundantly from the subterranean habitats worldwide (Figure 1). The question: What make them to survive in nutrition-deficient and perpetually dark *hypogean* habitats has not yet been answered adequately.

SUBTERRANEAN FISH FAUNA OF INDIA

Although India is the seventh largest country in the world by geographical area, only seven distinctly subterranean fish species have been described so far (Figure 2; Table 1). Of those, three are cave dwellers and the remaining four inhabit shallow phreatic spaces, notably in the network of ground water channels connected to wells in Kerala. Most of the reports on these species deal with their distribution, systematics and phylogenetic relationship. To some extent biology of only two species has been studied. Pati and his colleagues (see Pati 2008) have studied ecophysiological and ethological aspects in *Nemacheilus evezardi* in the last 20 years and have published more than 20 original research papers/reviews. Mercy and her colleagues have studied on various morphological and physiological aspects in *Horaglanis krishnai* and came out with about 9 original research papers (see Proudlove 2006). Biological studies on other five subterranean species have never been carried out. Therefore, this chapter includes

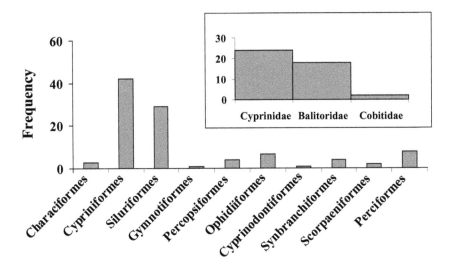

Fig. 1 Subterranean biodiversity among different orders of the class Pisces. Inset represents the maximum number of reported subterranean species, belonging to the family, namely Cyprinidae, Balitoridae, and Cobitidae of the Order Cypriniformes. Based on data given in Proudlove (2006).

mainly a brief review of the work on biology of *N. evezardi* and *H. krishnai*. It also includes brief notes on other five species without going into details regarding their systematics, taxonomy, and conservation status as these aspects have been taken care of firstly by Romero and Paulson (2001) and later by Proudlove (2006) in the recent past.

NEMACHEILUS EVEZARDI (DAY, 1872)

Historical Account

There is complete absence of any scientific report on the exact year of discovery of the loach, *Nemacheilus evezardi* inside the Kotumsar Cave located on the jurisdiction of the Kanger Valley National Park, Bastar, India. There was also no mention about the Kotumsar Cave itself in the Chhattisgarh Feudatory States Gazetteer, 1909 (Tiwari 2000). Possibly it is the geographer Prof. Shankar Tiwari, who spotted the species in 1959 while mapping the cave for the first time between 1958 and 1959 (Tiwari 2000). Subsequently in 1959, Prof. S.M. Agarwal (misspelt as A.M. Agarwal in Proudlove 2006) wrote to G. Thinès about the presence of this species in the Kotumsar Cave (Agarwal – personal communication) and thereafter Thinès (1969) reported about its existence in this cave (Thinès

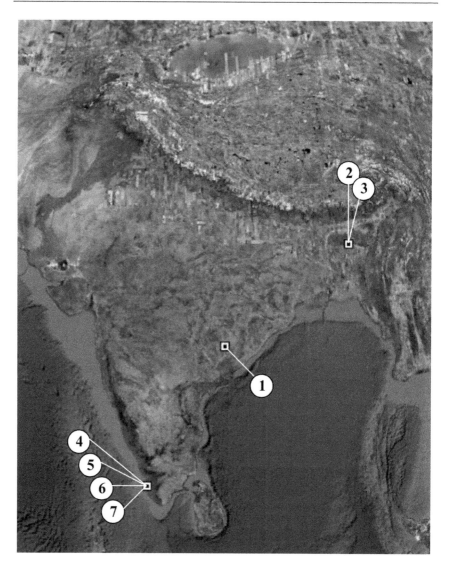

Fig. 2 Distribution of Indian subterranean fishes. Numbers indicate the location of the species found in India. 1: *Nemacheilus evezardi*; 2: *Schistura sijuensis*; 3: *Schistura papulifera*; 4: *Horaglanis krishnai*; 5: *Horaglanis alikunhii*; 6: *Monopterus eapeni*; 7: *Monopterus roseni*.

Color image of this figure appears in the color plate section at the end of the book.

1969). Agarwal also sent specimens of this species to the then Director of the Bombay Natural History Society, E.G. Silas, who identified the species as *Nemacheilus evezardi* Day, 1872 (Agarwal Telephonic Communication in April 2009).

Table 1 List of subterranean fish fauna of India

Name of species	Synonyms	Distribution	Latitude; Longitude	Vulnerability status	Red list
Nemacheilus evezardi	*Indoreonectes evezardi*, *Oreonectes evezardi*	Kotumsar Cave, Bastar district, Chhattisgarh	18°52′09″N; 81°56′05″E	Low vulnerability	Data deficient
Schistura sijuensis	*Noemacheilus sijuensis*, *Mesonoemachilus sijuensis*, *Nemacheilus multifasciatus*	Siju Cave, Garo Hills, Meghalaya	25°21′02″N; 90°41′04″E	Low vulnerability	Vulnerable
Schistura papulifera	None	Krem Synrang Pamiang Cave, Jaintia Hills, Meghalaya	25°12′48″N; 92°21′48″E	Low vulnerability	Not evaluated
Horaglanis krishnai	None	Kottayam district, Kerala	Between 8°04′N, 76°13′E and 10°21′N, 77°38′E	Low vulnerability	Vulnerable
Horaglanis alikunhii	None	Parappukara, Trichur district, Kerala	10°23′N; 76°15′E	Low vulnerability	Not evaluated
Monopterus eapeni	*Monopterus indicus*, *Amphipnous indicus*	Kottayam district, Kerala	9°30′N; 76°33′E	Low vulnerability	Data deficient
Monopterus roseni	None	Periyam village, Kerala	10°38′N; 76°22′E	Low to moderate vulnerability	Not evaluated

Controversy on the Nomenclature and Systematics/ Taxonomic Position of *Nemacheilus evezardi* Day

N. evezardi is a hill stream loach and found in both hypogean (cave) and epigean (river) forms. There have been discrepancies about its nomenclature and systematics/taxonomic position. Kottelat (1990), without giving any supportive reasons, placed it in the genera *Indoreonectes* (Romero and Paulson 2001) and it is still described as *Indoreonectes evezardi* by Proudlove (2001, 2006). Singh and Yazdani (1993) named it as *Oreonectes evezardi*; however, it was valid for a short time only. Pati and colleagues also described it as *Oreonectes* in some of their earlier publications (Biswas *et al.* 1990b, Biswas and Pati 1991) obviously following Singh and Yazdani (1993). Likewise, contradictions were also made for the systematic position of *Nemacheilus evezardi*. Day (1958) included it within the family Cyprinidae. Biswas (1990) placed it in the family Cobitidae, whereas, Eschmeyer (1998) positioned it within the family Balitoridae. Singh (personal communication) kept it within the family Homalopteridae. According to Pati (2008) it belongs to the family Balitoridae. The generic name of the species *N. evezardi* was never spelt as *Noemacheilus* by Day (1958) as has been mentioned by Proudlove (2006). The genus has been subdivided into two groups strictly based on number of barbels and *Nemacheilus evezardi* is the solitary species with eight (8) barbels. Further, surprisingly in Day (1958) there is no mention of the genus *Indoreonectes* as has been reported by Proudlove (2006).

Hypogean versus Epigean Controversy

The both hypogean and epigean forms found abundantly in caves of the Kanger Valley National Park and adjacent river (Kanger Dhara), respectively, continued to be *Nemacheilus evezardi*. Although the exact age of the evolution of the cave population is not known, many distinct differences relating to physiology and behavior do exit when both populations are compared. Prominently, the reduction of eyes, pigmentation and acquisition of air gulping behavior in the hypogean population offer adequate justification for its status as a separate species.

Habitat Ecology

The population of hill stream loach, *N. evezardi* has successfully colonized the Kotumsar Cave on the Kanger Valley National Park. It is found in small water pools and perennial streams of number of other limestone caves located on this Park. It has established itself as a distinct cave population and differs from its epigean population in many ways (Biswas 1990, Pati 2008). The epigean population inhabits rapidly flowing water in hill streams, rivulets and river, "Kanger Dhara," flowing inside the National Park (Biswas 1990,1992). The hypogean *N. evezardi* is a bottom dweller, but prefers to stay in shallow waters.

The Kotumsar Cave – the natural habitat of the cave population, is situated in the left flank of the Kanger River in the Kanger Valley National Park (18° 52′ 09″ N; 81° 56′ 05″ E), at an altitude of 560 m, Jagdalpur, Bastar, India. It is one of the best-studied caves in India and was discovered in 1958. Inside the cave one can notice numerous stalactite and stalagmite formations. The air and water temperatures of the cave remain relatively constant at an annual average of 28.25±1.23°C and 26.33±0.96°C, respectively (Biswas 1990).

Distribution

N. evezardi has an ample distribution in the Godavari and Kolab watersheds, of which the Kanger River is a part. It has also been reported from Pune, India (Day 1958) and may inhabit rivers in the Deccan Plateau.

Morphological Features

N. evezardi has a cylindrical body with dorsally located small-sized eyes. It has eight barbels: one pair is nasal, two rostral and one maxillary. This is the distinguishing character of the species. The lateral line is indistinct. The hypogean form is short (3-4 cm) in size as compared with its epigean one (5-7 cm) (Figure 3). There is also a remarkable difference in the pigment distribution pattern. The cave form is either totally (albinic) or sparsely pigmented. Thus, it lacks the vertically directed small dark blotches of the epigean form. A dark spot at the base of the caudal fin is present in both forms. The hypogean form has reduced eyes as compared with its

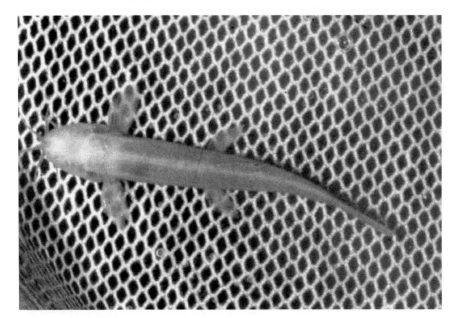

Fig. 3 Albinic subterranean loach, *Nemacheilus evezardi*.
Color image of this figure appears in the color plate section at the end of the book.

epigean counterpart (Biswas 1990). The spines are absent on the head in both forms.

Population Density and Reproduction

N. evezardi does not breed in captive conditions of the laboratory. We failed number of times in our attempt to have effective induced breeding. Therefore, we left with the single alternative option to study its reproductive rhythm in its natural habitat. However, it was extremely difficult to remain inside the cave continuously for longer period of time during and after the monsoon. Further, lack of sexual dimorphism in the species makes the matter more complicated. Biswas *et al.* (1990c) reported occurrence of fingerlings of hypogean fishes in July-August-September. He studied seasonal frequency of adults (pigmented or albino) and juveniles (pigmented or albino) in the water pools located in the first floor of the Kotumsar Cave during four different seasons. He performed the study over two consecutive years to validate the reproducibility of spawning rhythm. He revealed that the total number of hypogean adults increased (approximately doubled) every year soon after the end of monsoon as compared to other seasons of the study. However, juveniles were noticed only during and after the monsoon. Further, it was noticed that at any given time of the year the population of adult albinos does not exceed 10% of the entire hypogean population restricted to the water pools in the first floor of the cave.

The period of July-August-September receives maximum rainfall in the Kanger Valley National Park. Therefore, the spawning in cavefish could be associated with the flooding of the water courses of the caves. The flooding signal might be working as a trigger for the initiation of breeding. Until we observe a free-running rhythm in its reproductive behavior it would be difficult to talk about an underlying circannual clock mechanism. However, in contrast, Biswas (1990) observed an increase in the population of epigean fingerlings in the river in April. It seems that the trigger for the initiation breeding may be different for the epigean population.

Adaptation to Food-deficient Conditions

It is well known that organisms inhabiting caves and caverns develop abilities to withstand prolonged periods of starvation. They exhibit reduced respiratory intensity and decreased activity pattern under food-limited environment. Does hypogean *Nemacheilus evezardi* also exhibit the similar phenomenon? Biswas (1991) performed studies under starvation or restricted feeding schedules in both hypogean and epigean fish to answer this question. Fishes were exposed to these conditions for 30 days and body mass loss in both forms, irrespective of feeding status, was recorded. He demonstrated that reduction in body weight was always greater in epigean fish following starvation or restricted feeding as compared to their hypogean counterparts. On the other hand, hypogean fish gained more weight than epigean fish when food was provided *ad libitum*. He further observed that when food was supplied at the rate of 1 mg feed g^{-1} body weight, the hypogean *N. evezardi* lost less weight as compared to its epigean counterpart. However, the epigean fish perceived the same food regime (1 mg feed g^{-1} body weight) as good as starvation (Figure 4). This clearly suggests that in a food-limited environment the hypogean loach has abilities to optimize energy expenditure.

Energy Metabolism

The hypogean species of several systematic groups exhibit lower metabolic rates than their epigean relatives (Dickson and Franz 1980, Culver 1982, Hüppop 1985, 1986, 1989, Culver *et al.* 1995, Poulson 2001a,b). Perhaps

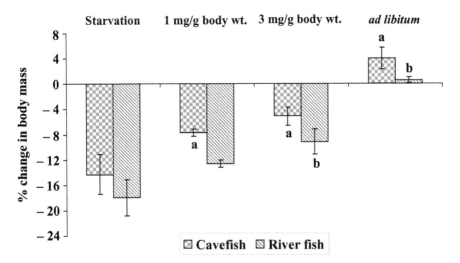

Fig. 4 Percent change in body mass of hypogean (cave) and epigean (river) fish following starvation or restricted feeding. Restricted feeding schedules consisted of provision of 1 mg/g and 3 mg/g body weight food daily during the course of the experiment. The percent change in body mass was computed relative to the initial body weight of an individual fish. a, differs from the mean value of the starved hypogean fish; b, differs from the mean value of the starved epigean fish (based on Biswas, 1991).

lower metabolic rate is one of the important adaptive features among the cavernicoles. This aspect was examined in the hypogean loach *N. evezardi* (Biswas 1991, Biswas and Pati 1991). The metabolic efficiency of both hypogean and epigean population of *N. evezardi* was studied keeping individuals of both populations under either restricted feeding regimens or starved conditions. Effects of T_4 (thyroxine) and T_3 (3,5,3' triiodothyronine) on the rate of muscle tissue (that constitutes 70% of total body mass) respiration were also investigated in both populations. Results indicate that the rate of oxygen consumption by muscle tissue of the hypogean *N. evezardi* is remarkably lower (about 2.7 times) than the epigean fish (Biswas 1991, Biswas and Pati 1991). Further, period of one week starvation was found to have significant effect on muscle tissue respiration in epigean population as compared to the respiratory rate in fully-fed condition. Moreover, T_4 or T_3 *in vivo* did not induce any significant change in the rate of oxygen consumption by muscle tissue of hypogean population. In contrary, in epigean population, both T_4 and T_3 significantly elevated the rate of oxygen consumption by the muscle tissue (Biswas and Pati 1991). Could it be that the muscle tissue of the hypogean fish lost its sensitivity to thyroid hormones during adaptation to the subterranean environment? The second possible explanation is that there may be some other hormone(s) and/or substance(s) that may oppose the action of thyroid hormones. This speculation looks more appropriate as it is supported by the results obtained from *in vitro* study in that the rate of oxygen consumption by the muscle tissue of both hypogean and epigean fish increased significantly, irrespective of dose level employed. The hypogean population of *N. evezardi* might have evolved some specialized endocrine mechanism(s) to ensure that the rate of oxidative metabolism does not elevate remarkably even if levels of thyroid hormones shoot up in the circulation abruptly.

Cryptobiotic Behavior

Burying behavior in epigean forms is a common phenomenon that protects them from predators, strong light, and extreme temperatures (Rita Kumari *et al.* 1979). Biswas *et al.* (1990b) examined this behavior in both epigean (under natural day length, NDL) and hypogean (under constant darkness, DD) populations of *N. evezardi* as function of time of the day, type of substratum, water current and body size. In general they concluded that: (i) frequency of burying varies as function of time of the day in both forms, (ii) the frequency of burying is considerably less in the hypogean fish as compared to its epigean counterpart, (iii) fish of both populations prefer to bury themselves under the stones; however, hypogean forms also bury themselves in the mud (Figure 5), (iv) the pattern of burying behavior

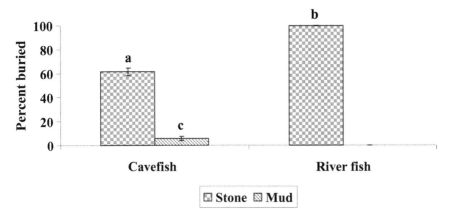

Fig. 5 Burying behavior in relation to substrata choice. Histograms represent mean percent buried in each substrata ± 1 SE. Histograms bearing dissimilar superscripts differ statistically significantly at $p < 0.05$ level (based on Biswas *et al.* 1990b).

with reference to water current is different in hypogean loach than that of epigean ones, and (v) burying behavior has direct relationship with body size in the epigean loach. Exhibition of burying behavior in hypogean *N. evezardi* may also reflect its resting phase. Furthermore, behavior of burying in mud for the hypogean loach may be a reflection of food-searching behavior. It is known to feed upon soil nematodes present in the water pools inside the cave.

In addition, a statistically significant circadian variation in the rate of burying behavior in both populations of *N. evezardi* has been demonstrated (Biswas *et al.* 1990b). The peak burying time was found to be early hours of the day in both (Figure 6). Could it be that the predator pressure is more intense at this time of the day?

Surfacing Activity/Behavior

The phenomenon of surfacing activity has been reported in hypogean fish *N. evezardi* (Biswas *et al.* 1990a). This activity in fish is one of the examples of adaptation to its environment for survival. The term 'surfacing activity' is synonymous with other terms, namely air-breathing activity, air-breathing behavior, air-gulping behavior, surfacing behavior that are frequently used in the literature (Maheshwari 1998). It is characterized sequentially by a fast upward movement, a quick air gulping at the water-air interface and an equally swift descent.

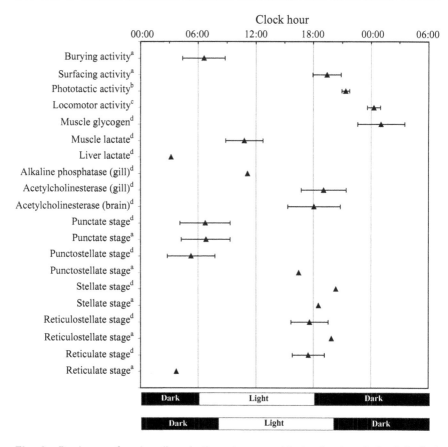

Fig. 6 Peak map for circadian rhythms in several behavioral and physiological variables in the cave loach, *N. evezardi* acclimated to LD 12:12 or DD. Solid triangle represents estimated acrophase (peak) of a given variable. The horizontal line on either side of the solid triangle defines the 95% confidence limits of acrophase. Absence of horizontal line indicates non-detection of statistically validated circadian rhythm by Cosinor. [a]DD; [b]choice chamber consisting of both photic (LL) and aphotic (DD) zones; [c]LD 12:12 (Lights-on at 06:00); [d]LD 12:12 (Lights-on at 08:00) (based on Pradhan 1984, Pradhan et al. 1989, Biswas 1990, Biswas et al. 1990a, b, Pradhan and Biswas 1994b, Pati, 2001b).

Biswas (1990) reported that, the epigean *N. evezardi,* unlike its hypogean counterpart, never exhibits air-gulping behavior either in its natural habitat or under laboratory conditions. It seems that the hypogean loaches have undergone behavioral divergence. Probably this behavior is of immense adaptive significance for the species. This unique behavior probably enables the hypogean loach to survive in stagnant water pools of the cave that becomes sporadically hypoxic over an annual time scale. The habit of gulping atmospheric air has also been reported in many other epigean tropical fish species those inhabit shallow ponds and streams that dry up periodically or in stagnant water pools that become oxygen deficient by decaying vegetation during summer. Thus, air-gulping behavior is indispensable for survival of the species and helps it to overcome the problem of poor availability of oxygen in the stagnant water pools.

For the first time Biswas *et al.* (1990a) documented the multifrequency rhythms in air-gulping behavior in hypogean *N. evezardi* both under laboratory and natural cave conditions (Figure 6). They recorded circadian variation in air-gulping activity once in every month for 14 consecutive months under carefully simulated cave conditions in the laboratory in respect of darkness, temperature and humidity. A statistically significant circadian rhythm in air-gulping activity was validated during nine out of 14 months of observations. The peak activity was recorded between 14:00 and 18:00 (Biswas *et al.* 1988). Furthermore, a 6-monthly rhythm in circadian Mesor was detected, whereas circannual rhythms were witnessed for circadian amplitude and acrophase. Two prominent peaks for circadian Mesor were observed, one in September and the other in March. They further documented that probably air-gulping behavior exhibited free-running rhythm. However, the method employed to record air-gulping activity was subjective and the duration of study was short, i.e., two consecutive cycles (48-h span) every month. Therefore, at this juncture it is difficult to conclude that air-gulping behavior in hypogean *N. evezardi* is endogenous in nature.

Biswas (1990) made an attempt to study air-gulping behavior in natural conditions of the cave only over a single 24-h transverse scale in the month of April. In this condition fish exhibited a statistically significant circadian rhythm in air-gulping activity. The circadian Mesor was remarkably more than that obtained under the laboratory conditions in April. In another study effect of photoperiods (DD, LL, LD 12:12 and DL 12:12) on the air-gulping behavior was also investigated (Biswas *et al.* 1990a, Pradhan and Biswas 1994a). They demonstrated that this behavior was significantly influenced by different photoperiods.

Phototactic Behavior

Phototaxis refers to the movement of an organism toward (positive phototaxis) or away (negative phototaxis) from the source of light. The study of phototactic behavior in cavernicolous fish is of great interest, especially because they undergo depigmentation and have either regressed or vestigial eyes. This behavior has been studied extensively in the hypogean

N. evezardi (Pradhan *et al.* 1989, Biswas 1993, Pradhan and Biswas 1994a). Biswas (1993) documented that the propensity of photophilic nature is strong in troglophilic (sparsely pigmented) *N. evezardi* than its troglobitic (depigmented) counterparts.

A statistically significant *24-h rhythm* in phototactic behavior has been documented in *N. evezardi* (Figure 6) (Pradhan *et al.* 1989, Biswas 1993, Parganiha *et al.* 2006). Pradhan *et al.* (1989) documented a high amplitude circadian rhythm in phototactic behavior in hypogean population of *N. evezardi*. It is suggested that locomotor activity may reflect phototactic behavior of organisms as phototaxis is associated with some degree of locomotion. However, a difference between peaks of phototactic behavior (Pradhan *et al.* 1989) and locomotor activity rhythm (Biswas 1990) was documented. Therefore, phototactic behavior is established as an independent behavior. Furthermore, the population of hypogean loach fed strictly between 07:00 and 11:00 exhibited a peak in phototactic rhythm at 17:00, whereas peak occurred at 02:42 in those fed between 19:00 and 23:00. This clearly suggests that periodic restricted feeding schedule may modulate the rhythm in phototactic behavior. Pati and Agrawal (2002) also witnessed a significant circadian rhythm in phototactic behavior. They examined phototactic rhythm in the population of a cave loach for eight consecutive days and documented that significant circadian rhythm in phototactic behavior persisted for about five days only. They demonstrated a free-running pattern in the occurrence of peaks. The disappearance of rhythm in phototaxis after 5th day could be explained as a consequence of starvation and it could be a strategy to conserve energy when food is not available. In another experiment they reported photophobia in the same population of cave loach under starved condition (Agrawal and Pati 2002). The availability of food seems to be an important modifier of phototactic behavior in hypogean *N. evezardi* (Pati and Agrawal 2002).

Parganiha *et al.* (2006) reproduced comparable results on phototactic rhythm in an experiment conducted under different light intensities (1- or 10- or 100-lux). The animals were not provided with the food during the entire period of the study. A statistically significant circadian rhythm in the phototacic activity of the population under 1- and 10-lux conditions, but not under 100-lux condition, was demonstrated. The mean level of activity gradually declined from 1-lux condition to 100-lux condition. In general, amplitude declined and peak advanced as function of increase in light intensity. In another study choice made by the population was recorded every five minutes over a 90-minute session either at midday or at midnight under all three 1-, 10- and 100-lux conditions from a remotely placed monitor connected to a video camera. The Phototactic Index (PI) was computed (Camassa 2001). Results obtained for phototactic index supported the findings on the characters of rhythmic activity of the population of loach in the light zone. The loaches were more negatively phototactic in the midnight (versus midday) and at 100-lux condition (versus 1- or 10-lux conditions). These results corroborate earlier findings reported from our laboratory in that *Nemacheilus evezardi* exhibits a distinct circadian rhythm in phototactic rhythm. Parganiha *et al.* (2006) concluded that the background light intensity in the photic zone could alter circadian rhythm parameters of phototactic rhythm considerably.

Rhythmic Behavior in Locomotor Activity

Number of studies has been performed by Pati and colleagues on locomotor activity in *N. evezardi*. Pati (2001) reported a bimodal pattern in locomotor activity rhythm in the hypogean population of cave loach kept under LD 12:12 cycles. The onset and cessation of activity coincided with the lights-off (18:00) and lights-on (06:00) timings, respectively. However, it was more precise at the onset of darkness. The phase angle (ψ) was computed with reference to both time points; it was found to be almost zero to lights-off time. This indicates harmonization between beginning of activity and onset of darkness, whereas such phenomenon could not be validated with reference to lights-on timing. Most of the test fishes exhibited a large amount of activity before and after lights-on time (Pati 2001). The bimodal pattern was further confirmed from the results of Cosinor rhythmometry. Although a statistically significant 24-h rhythm in locomotor activity was discerned in this hypogean form (Figure 6), a 12-h rhythm was more pronounced. It can be explained in terms of peak spread and amplitude. Peak spread was large (about 11.7 h) when computed at 24-h time window, whereas it reduced to 1.1 h when 12-h time window was considered for the analysis. Likewise, difference was also observed for amplitude. Amplitude was higher in 12-h rhythm than that of 24-h rhythm. Further, results of spectral analysis also complemented the above findings. In some specimens the locomotor activity rhythm exhibited 12-h period as the prominent component of its waveform. Interestingly, this bimodality in locomotor activity pattern was absent in epigean loach though they exhibited comparable activity rhythm. The onset and cessation of activity were equally precise in epigean loach. In contrast to hypogean loach, the population of epigean loach exhibited shorter peak spread in 24-h fitted period as compared to that with 12-h period. Further, amplitude of 24-h rhythm was also observed to be higher than that of 12-h rhythm. Circadian period (τ = 24 h) was the prominent feature in epigean loach (Pati 2001). This noticeable difference between hypogean and epigean forms concerning bimodality and characteristics of rhythm parameters of locomotor activity is difficult to explain at this moment.

Pati and colleagues conducted a number of experiments exposing the hypogean and epigean populations of *N. evezardi* to LD 12:12 for about six to eight days and thereafter to constant darkness *(DD)* or continuous light *(LL)* for varying length of time. In one of the experiments the majority of the hypogean loach exhibited free-running rhythms in locomotor activity following their transfer from LD 12:12 to DD (Pati 2001). The population of epigean loach also exhibited free-running rhythms in locomotor activity. However, in both cases, it was difficult to locate the accurate timings of activity onset, activity cessation or activity midpoints in the actogram.

A diminished 24-h mean activity level was observed in hypogean loach following their transfer from LD 12:12 to LL (Pati 2008). A weak and indistinct free-running rhythm with 24-h or nearly 24-h period in locomotor activity was noticed in this loach under LL. These findings clearly suggest that the hypogean loach, *N. evezardi* has an endogenous circadian rhythm that is characterized by persistence of locomotor activity rhythm when exposed to either DD or LL. Further, it strongly suggests that the hypogean fish still possesses a functional circadian oscillator underlying its overt circadian rhythm in locomotor activity under LD 12:12. It could be concluded that the circadian machinery is still conserved in the population of the hypogean loach.

In another experiment Pati (2008) documented higher level of activity in epigean population following their exposure to DD for 18 days. The 24-h mean locomotor activity, expressed as % change from 24-h mean activity in LD, increased many folds on individual basis as well as at group level after onset of darkness. However, the significant changes in mean activity level were observed only after 10th day. Surprisingly, hypogean loaches did not exhibit any noticeable changes in the average level of locomotor activity under DD. Another interesting significant difference was noticed between the two populations regarding occurrence of peaks of locomotor activity rhythm under DD. The hypogean fish always exhibited phase advance, whereas the epigean fish exhibited phase delay under constant darkness (Figure 7). Such differences between the hypogean and the epigean loaches are good examples of divergence in the behavioral trait following colonization of subterranean habitat.

Color Change Behavior

Chromatophore dependent color change behavior is a physiological function that is present in many organisms, such as fish, frog, lizard, etc. It has been reported that the epigean population of *N. evezardi* is able to darken its body color very quickly with little excitement. Pradhan and Biswas (1994b) demonstrated that this chromatophore dependent color

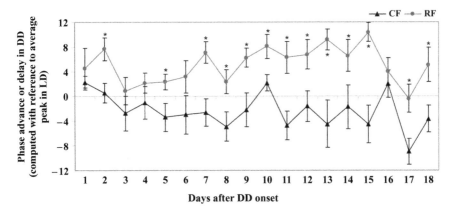

Fig. 7 Day-to-day change in peak of the locomotor activity of the hypogean loach (solid triangle) and the epigean loach (solid circle) in DD, computed from their respective overall 24-h mean peak in LD. An asterisk indicates statistically significant difference in the response to DD between the populations of the hypogean and the epigean loaches, based on *t*-tests ($p<0.05$). Vertical line on either side of the mean value (either solid triangle or solid circle) represents 1 standard error. The hypogean fish always exhibited phase advance, whereas the epigean fish exhibited phase delay in locomotor activity rhythm under constant darkness (DD).

change phenomenon is also functional in the hypogean population kept under LD 12:12 photoperiod. Further, they observed that different stages of chromatophore, such as punctate (stage-I), punctostellate (stage-II), reticulostellate (stage-IV) and reticulate (stage-V) vary as a function of time, except stellate (stage-III) (Figure 6). Pigments dispersed maximally during later half of the light phase, whereas concentration was the maximum during dark span of the LD 12:12 cycles (Pradhan 1984, Pradhan and Biswas 1994b). Conversely, a statistically significant rhythm was validated only in the occurrence of stage-I chromatophore (punctate type) with the peak located at 06:48 under DD. Non-detection of significant rhythm in other types of chromatophore in DD suggests absence of any particular advantage of color change in subterranean mode of life. Notwithstanding subterranean life inside the cave, chromatophore of the hypogean loach still retained their ability to respond to light and darkness.

Bhargava *et al*. (1984) reported that the cavefish *N. evezardi* is extremely sensitive to its background color/illumination. Under black background it becomes dark, whereas on a white background it becomes pale. In addition, they also demonstrated that the rate of color change is rapid initially; later on it becomes slow and gradual, irrespective of its responses to background (black or white).

Biochemical Constituent

A statistically significant circadian variation in a number of biochemical constituents of different tissues has been documented in the hypogean *N. evezardi* (Figure 6) (Pradhan 1984). Particularly content of muscle glycogen, muscle lactate and acetylcholinesterase enzyme activities of gill and brain tissues of the hypogean form exhibited a significant 24-h rhythm when exposed to LD 12:12 cycles. However, this phenomenon could not be validated for content of liver lactate or alkaline phosphatase activity of gill tissue. Nonetheless, it suggests that biochemical clocks of this fish could be synchronized by the artificially maintained photoperiods despite the fact that it has been adapted to lifestyle inside the perpetually dark cave. However, these biochemical variables in the hypogean population of loach have not yet been studied under constant conditions (DD or LL).

SCHISTURA SIJUENSIS (MENON, 1987)

This fish belonging to the family Balitoridae is known only from its type locality, the Siju Cave, Garo Hills, Meghalaya, India (25°21′N; 90°41′E). There are number of papers that describe its systematic and phylogenetic relationships (Kemp and Chopra 1924, Hora 1924, 1935, Pillai and Yazdani 1977, Menon 1987, Kottelat, 1990, Talwar and Jhingran 1992, Brooks and Smart 1995). Interestingly, these authors describe this species under various synonyms (refer Proudlove 2006). Further, none of these papers deals with the study on its system biology or physiology or ethology.

SCHISTURA PAPULIFERA (MENON, 1987)

This species has been described (new species) recently from the Krem Synrang Pamiang system in the Jaintia Hills, eastern Meghalaya (between 25°11′14″N, 92°21′03″E and 25°12′48″N, 92°21′48″E) (Kottelat *et al.* 2007). The authors have claimed that this is the first strictly hypogean loach from India and that it is not known if it has an epigean ancestor. It displays few troglomorphic characters, such as regression in the pigment system and vestigial lateral eyes (Kottelat *et al.* 2007). It differs from other known species of Schistura characterized by the lower half of the head covered by small skin projections and presence of five pores in the supratemporal canal of the cephalic lateral-line system (Kottelat *et al.* 2007). This species has been reported to be light insensitive (Kottelat *et al.* 2007), although so far no empirical studies have been conducted to support its photoneutral behavior.

HORAGLANIS KRISHNAI (MENON 1950)

Horaglanis krishnai is a unique freshwater fish belonging to the Family Clariidae. It inhabits subterranean channels that anastomose with an underground network of wells in the Kottayam district of Kerala, India (between 8°04′N, 76°13′E and 10°21′N, 77°38′E). It has been reported to migrate from one well to another through this network of subterranean channels. It has also been collected from a locality at Ettumanur, which is about 12 km from Kottayam town (Subhash Babu and Nayar 2004). It could be considered endemic to Kottayam district. The systematics and phylogenetic relationships of this species has been outlined elsewhere (Romero and Paulson 2001, Proudlove 2006). There are two genera of the family Clariidae in which the eyes are absent; they are *Uegitglanis* Gianferrari and *Horaglanis* Menon. The former has been reported from the wells in the village Uegit, Southern Somalia. *Horaglanis* is different from *Uegitglanis* in having relatively shorter dorsal and anal fins, vestigial pectoral fins without spines and gill membranes united with isthmus (Subhash Babu and Nayar 2004).

H. krishnai (Maximum size: 4.2 cm) is blind and it is presumed that the absence of the eyes could be attributed to the degeneration of the optic lobes. There are no scales on the skin (Figure 8). It possesses bulbous stomach that facilitates storage of food. It has specialized ileo-sphincter that retains the digested food to maximize the rate of absorption (Mercy and Pillai 1984). It has been speculated that it feeds on insects and crustaceans available in the phreatic spaces. The alimentary canal has been suitably modified to suit to its carnivorous habit. The skull of this fish is strong and bones are firmly articulated. Perhaps this is essential for its migratory

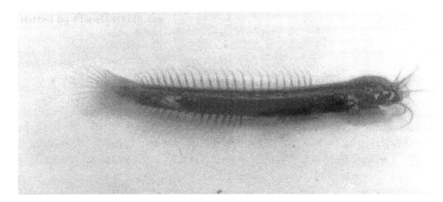

Fig. 8 Subterranean Clariidae, *Horaglanis krishnai* (Source: www.planetcatfish. com).
Color image of this figure appears in the color plate section at the end of the book.

habit inside narrow subterranean passages. Mercy *et al.* (1983) studied the oxidative metabolism in this species and reported a linear relationship between respiratory metabolism and body weight.

HORAGLANIS ALIKUNHII (SUBHASH BABU AND NAYAR 2004)

H. alikunhii has an elongated eel-like body (3.2 cm) belonging to the family Clariidae. It was discovered in a well in the void laterite soil of Parappukara (10°23′N, 76°15′E), Trichur district, Kerala. It has four pairs of barbels. Eyes are completely absent. Gill membranes united at the anterior end near the lower jaw. The *H. alikunhii* differs from *H. krishnai* remarkably with reference to shape of the head (elongated vs. globular), pectoral fins (highly vestigial), and number of caudal fin rays (30 vs. 24) (Subhash Babu and Nayar 2004). The red color of the live fish has been attributed to the abundance of erythrocytes in the superficial blood capillaries and is explained as an adaptation to the oxygen-deficient waters in its habitat (Subhash Babu and Nayar 2004). However, so far no physiological or behavioral studies have been conducted on this species.

MONOPTERUS EAPENI (TALWAR IN TALWAR AND JHINGRAN 1992)

It is commonly known as Malabar swampeel and belongs to the family Synbranchidae. The major synonym of this species is *Monopterus indicus* (Eapen 1963). It is distributed in wells of Kottayam district (9°30′N, 76°33′E). The troglomorphic fish, *H. krishnai* and *M. eapeni* coexists in the same wells. The fish is scaleless and has sunken eyes (Romero and Paulson 2001). The fish lives in the water table up to 10 m below the surface (Proudlove 2006).

MONOPTERUS ROSENI (BAILEY AND GANS 1998)

It belongs to the family Synbranchidae. It was found in a well at Periyam village (10°38′N, 76°22′E), Kerala, India, about 100 km away from the type locality of the *M. eapeni*. No details of its habitat and biology are available except that it differs significantly from the other subterranean *Monopterus* species, *M. eapeni* (Bailey and Gans 1998).

CONCLUSIONS AND FUTURE DIRECTIONS

Many other fish species have been reported from different caves of India. However, they are presumably nonstygobitic fish fauna (Table 2; Figure 9) without any distinct troglomorphic features. These fish species might have been accidentally discovered in the caves. There are many rivers that flow through the caves. The fish fauna found in the part of the river inside the cave are not necessarily truly subterranean fauna. In general, biological studies on subterranean fish fauna of India are restricted to *Nemacheilus evezardi* and *Horaglanis krishnai* only. Further, the subterranean *Monopterus* species, *M. eapeni* and *M. roseni* are known only from single individuals. The biological data available on subterranean species from India are meager. Although there may be thousands of caves located on the Indian subcontinent, adequate efforts have never been made to carry out biospeleological surveys. There is an urgent need to lay emphasis on research on caves and cave fauna of India. The funding agencies of the country should keep biospeleology in the list of the thrust areas of research. Fully dedicated research centers should be created in the already identified karstic areas of the country to pursue biospeleological research.

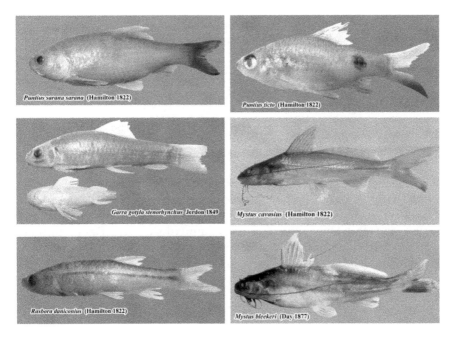

Fig. 9 Nonstygobitic fish species collected from the Nelabhilum Cave, Andhra Pradesh (courtesy: Ranga Reddy).
Color image of this figure appears in the color plate section at the end of the book.

Table 2 Suspected cave fauna from Siju cave and other caves of India[*]

Scientific name	Species authority	Cave	State
Barilius barna	Hamilton, 1822	Siju cave	Meghalaya[1]
Barilius bendelisis	Hamilton, 1807	Siju cave	Meghalaya[1]
Danio aequipinnatus	McClelland, 1839	Siju cave	Meghalaya[1]
Acrossocheilus hexagonolepis	McClelland, 1839	Siju cave	Meghalaya[1]
Tor tor	Hamilton, 1839	Siju cave	Meghalaya[1]
Noemacheilus multifasciatus	Day, 1878	Siju cave	Meghalaya[1]
Channa orientalis	Bloch & Schneider, 1801	Siju cave	Meghalaya[1]
Ambassis nama	Hamilton, 1822	Siju cave	Meghalaya[1]
Schizothoracine fish	Not known	A cave near Udaipur	Rajasthan[2]
Tiny pigmented fish	Not known	Gupta Godavari Gufa (lower cave)	Madhya Pradesh[3]
Small pigmented fish	Not known	Gupta Godavari Gufa (upper cave)	Madhya Pradesh[3]
Puntius sarana sarana	Hamilton, 1822	Nelabhilum	Andhra Pradesh[3]
Puntius sarana subnasutus	Valenciennes, 1842	Nelabhilum	Andhra Pradesh[3]
Channa orientalis	Bloch & Schneider, 1801	Nelabhilum	Andhra Pradesh[3]

[*]Based on: [1]Pillai and Yazdani (1977), [2]Tehsin *et al.* (1988), [3]Reddy (2009 – Personal communication)

Acknowledgements

We thank Dr. Rohit Kumar Pradhan and Dr. Jayant Biswas, for permitting us to use data from their Ph.D. theses submitted to Pt. Ravishankar Shukla University, Raipur, India. We thank Dr. Ranga Reddy for providing us with the pictures of nonstygobitic fish fauna of Nelabhilum Cave, Andhra Pradesh. Some of the findings of the researches reported in this chapter have been supported by the Department of Science & Technology, New Delhi, and the University Grants Commission, New Delhi, through its DRS-SAP sanctioned to the School of Life Sciences, Pt. Ravishankar Shukla University, Raipur, India, in the thrust of area of research in Chronobiology.

References

Agrawal, A. and A.K. Pati. 2002. Feeding schedule modulates phototactic responses in cave populations of *Nemacheilus evezardi*. *XVI* International Symposium of Biospeleology, Museo Civico di Storia Naturale di Verona, September 8-15: 27.

Bailey, R.M. and C. Gans. 1998. Two new synbranchid fishes, *Monopterus roseni* from penisular India and *M. desilvai* from Sri Lanka. *Occasional Papers of the Museum of Zoology of the University of Michigan* 726: 1-18.

Bhargava, H.N., A.K. Jain and D. Singh. 1984. On background related chromatic response in the cave fish, *Nemacheilus evezardi* (Day). *Journal of Animal Morphology and Physiology* 31: 203-209.

Biswas, J. 1990. Biospeleology: The behavioral and physiologic adaptations in a cavernicole. Ph.D. Thesis, Pandit Ravishankar Shukla University, Raipur, India, 136 pp.

Biswas, J. 1991. Metabolic efficiency and regulation of body weight: A comparison between life in hypogean and epigean ecosystems. *International Journal of Speleology* 20: 15-22.

Biswas, J. 1992. Influence of epigean environmental stress on a subterranean cave ecosystem: Kotumsar. *Biome* 5: 39-43.

Biswas, J. 1993. Constructive evolution: Phylogenetic age related visual sensibility in the hypogean fish of Kotumsar Cave. *Proceedings of the National Academy of Sciences India* B63: 181-187.

Biswas, J. and A.K. Pati. 1991. Influence of thyroid hormones on muscle tissue respiration in hypogean and epigean populations of loach *Oreonectes evezardi*. *Indian Journal of Experimental Biology* 29: 933-936.

Biswas, J., A.K. Pati and R.K. Pradhan. 1990a. Circadian and circannual rhythms in air-gulping behavior of cave fish. *Journal of Interdisciplinary Cycle Research* 21: 257-268.

Biswas, J., R.K. Pradhan and A.K. Pati. 1990b. Studies on burying behavior in epigean and hypogean fish, *Oreonectes evezardi*: An example of behavioral divergence. *Mémoires de Biospéléologie* 17: 33-41.

Biswas, J., A.K. Pati, R.K. Pradhan and R.S. Kanoje. 1990c. Comparative aspects of reproductive phase dependent adjustments in behavioral circadian rhythms of epigean and hypogean fish. *Comparative Physiology and Ecology* 15: 134-139.

Biswas, J., A.K. Pati, R.K. Pradhan and R.S. Kanoje. 1988. Reproductive phase dependent adjustments in behavioral circadian rhythm characteristics: An example of constructive evolution in the hypogean fish. Symposium on Recent Trends in Comparative Endocrinology, Centre of Advanced Study in Zoology, Banaras Hindu University, Varanasi, March 11-13: 18.

Brooks, S.J. and C.M. Smart. 1995. *Caving in the Abode of the Clouds. The Caves and Karst of Meghalaya, North East India.* Bristol Exploration Club and Orpheus Caving Club, Somerset and Derbyshire.

Camassa, M.M. 2001. Responses to light in epigean and hypogean populations of *Gambusia affinis* (Cyprinodontiformes: Poeciliidae). *Environmental Biology of Fishes* 62: 115-118.

Culver, D.C. 1982. *Cave Life: Evolution and Ecology*. Harvard University Press, Cambridge.

Culver, D.C., T.C. Kane and D.W. Fong. 1995. *Adaptation and Natural Selection in Caves: The Evolution of* Gammarus minus. Harvard University Press, Cambridge.

Day, F. 1872. Monograph of Indian Cyprinidae, Part IV. *Journal of the Asiatic Society* 41: 171-198.

Day, F. 1958. *The Fishes of India; Being a Natural History of the Fishes Known to Inhabit the Seas and Fresh Waters of India, Burma and Ceylon*. William Dawson & Sons Ltd., London, Vol. 1, 778 pp.

Dickson, G.W. and R. Franz. 1980. Respiration rates, ATP turnover and adenylate energy charge in excised gills of surface and cave crayfishes. *Comparative Biochemistry and Physiology* A65: 375-379.

Eapen, K.C. 1963. A new species of *Monopterus* from South India. *Bulletin of the Department of Marine Biology and Oceanography, University of Kerala* 1: 129-132.

Eschmeyer, W.N. 1998. *Catalog of Fishes*. California Academy of Sciences, San Francisco.

Hora, S.L. 1924. Fish of the Siju cave, Garo Hills, Assam. *Records of the Indian Museum* 26: 27-31.

Hora, S.L. 1935. Notes on fishes in the Indian Museum. XXIV. Loaches of the genus *Nemacheilus* from eastern Himalayas, with the description of a new species from Burma and Siam. *Records of the Indian Museum* 37: 49-68.

Hüppop, K. 1985. The role of metabolism in the evolution of cave animals. *Bulletin of the National Speleological Society* 47: 136-146.

Hüppop, K. 1986. Oxygen consumption of *Astynax fasciatus* (Characidae, Pisces): A comparison of epigean and hypogean populations. *Environmental Biology of Fishes* 17: 299-308.

Hüppop, K. 1989. Genetic analysis of oxygen consumption rate in cave and surface fish of *Astyanax fasciatus* (Characidae, Pisces): Further support for the neutral mutation theory. *Mémoires de Biospéleologie* 16: 163-168.

Kemp, S. and B. Chopra. 1924. The Siju Cave, Garo Hills, Assam Part 1. Introduction. *Records of the Indian Museum* 26: 3-22.

Kottelat, M. 1990. New species and populations of cave nemacheilines in south and south-east Asia (Osteichthyes, Balitoridae). *Mémoires de Biospéleologie* 17: 49-55.

Kottelat, M., D.R. Harries and G.S. Proudlove. 2007. *Schistura papulifera*, a new species of cave loach from Meghalaya, India (Teleostei: Balitoridae). *Zootaxa* 1393: 35-44.

Maheshwari, R. 1998. An analysis of the air-gulping behavior of the catfish, *Heteropneustes fossilis*, with reference to hormonal regulation. Ph.D. Thesis, Pandit Ravishankar Shukla University, Raipur, India, 112 pp.

Menon, A.G.K. 1950. On a remarkable blind siluroid fish of the family Clariidae from Kerala (India). *Records of the Indian Museum* 48: 59-66.

Menon, A.G.K. 1987. *Noemacheilus sijuensis* sp. nov. The fauna of India and adjacent countries. Pisces, Zoological Survey of India, Calcutta. IV, Teleostei–Cobitidae, Part 1, Homalopteridea, 175.

Mercy, T.V.A. 1981. The unique disposition of the *valvula cerebelli* in *Horaglanis krishnai* Menon (Teleostei). *Geobios* 8: 95-96.

Mercy, T.V.A. 1982. On the presence of a forked barbel in the blind catfish *Horaglanis krishnai* Menon. *Geobios New Reports* 1: 60-61.

Mercy, T.V.A. 1983. Studies on the air bladder of the blind clariid *Horaglanis krishnai* Menon. *Geobios New Reports* 2: 111-115.

Mercy, T.V.A. 2000. Biology of an endemic blind catfish *Horaglanis krishnai* Menon. In: *Endemic Fish Diversity of Western Ghats*, A.G. Ponnaih and A. Gopalakrishnan (eds.). NBFGR/NATP Publishers, Lucknow, India, pp. 278-279.

Mercy, T.V.A. and N.K. Pillai. 1984. The anatomy and histology of the alimentary tract of the blind catfish *Horaglanis krishnai*. *International Journal of Speleology* 14: 69-85.

Mercy, T.V.A. and N.K. Pillai. 2001. Studies on the cranial osteology of the blind catfish *Horaglanis krishnai* Menon (Pisces, Clariidae). *International Journal of Speleology* 30: 1-14.

Mercy, T.V.A., K.G. Padmanabhan and N.K. Pillai. 1982. Morphological studies on the oocytes of the blind catfish *Horaglanis krishnai* Menon. *Zoologischer Anzeiger* 209: 211-223.

Mercy, T.V.A., N.K. Pillai and N.K. Balasubramonian. 1983. Studies on the oxygen consumption of the blind catfish *Horaglanis krishnai* Menon in light and dark conditions. *Matsya* 9 and 10.

Mercy, T.V.A., N.K. Pillai and N.K. Balasubramonian. 2001. Studies on certain aspects of behavior in the blind catfish *Horaglanis krishnai* Menon. *International Journal of Speleology* 30: 57-68.

Parganiha, A., A. Ramteke, A. Kar and A.K. Pati. 2006. Background light intensity modulates circadian rhythm in phototactic behavior in the cave loach, *Nemacheilus evezardi*. 18th International Symposium of Biospeleology, Cluj-Napoca, Romania, July 10-15: 51-52.

Pati, A.K. 2001. Temporal organization in locomotor activity of the hypogean loach, *Nemacheilus evezardi*, and its epigean ancestor. *Environmental Biology of Fishes* 62: 119-129.

Pati, A.K. 2008. Circadian rhythms in hypogean fish: With special reference to the cave loach, *Nemacheilus evezardi*. In: *Fish Life in Special Environments* P. Sébert, D.W. Onyango and B.G. Kapoor (eds.). Science Publishers, New Hampshire, pp. 83-130.

Pati, A.K. and A. Agrawal. 2002. Circadian rhythm in phototactic behavior in cave loach *Nemacheilus evezardi*. XVI International Symposium of Biospeleology, Museo Civico di Storia Naturale di Verona, September 8-15: 58.

Pillai, R.S. and G.M. Yazdani. 1977. Ichthyo-fauna of Garo Hills, Meghalaya (India). *Records of the Zoological Survey of India* 72: 1-22.

Poulson, T.L. 2001a. Morphological and physiological correlates of evolutionary reduction of metabolic rate among amblyopsid cave fishes. *Environmental Biology of Fishes* 62: 239-249.

Poulson, T.L. 2001b. Adaptations of cave fishes with some comparisons to deep-sea fishes. *Environmental Biology of Fishes* 62: 345-364.

Pradhan, R.K. 1984. Biochemical studies of some tissues of *Nemacheilus evezardi* Day from Kotumsar Cave. Ph.D. Thesis, Pandit Ravishankar Shukla University, Raipur, India, 195 pp.

Pradhan, R.K. and J. Biswas. 1994a. The influence of photoperiods on the air-gulping behavior of the cave fish *Nemacheilus evezardi* (Day). *Proceedings of the National Academy of Sciences India* B 64: 373-380.

Pradhan, R.K. and J. Biswas. 1994b. Towards regressive evolution: The periodic color change behavior of a troglophilic fish *Nemacheilus evezardi* (Day). *International Journal of Speleology* 23: 191-201.

Pradhan, R.K., A.K. Pati and S.M. Agarwal. 1989. Meal scheduling modulation of circadian rhythm of phototactic behavior in cave dwelling fish. *Chronobiology International* 6: 245-249.

Proudlove, G.S. 2001. The conservation status of hypogean fishes. *Environmental Biology of Fishes* 62: 201-213.

Proudlove, G.S. 2006. *Subterranean Fishes of the World. An Account of the Subterranean (Hypogean) Fishes Described up to 2003 with a Bibliography 1541–2004.* International Society for Subterranean Biology, Moulis.

Rita Kumari, S.D., N.B. Nair and N.K. Balasubramanium. 1979. On the burying behavior of the tropical loach, *Lepidocephalus thermalis* (Cuv. & Val.). *Proceedings of the Indian National Science Academy* B 45: 129-141.

Romero, A. and K.M. Paulson. 2001. It's a wonderful hypogean life: A guide to the troglomorphic fishes of the world. *Environmental Biology of Fishes* 62: 13-41.

Singh, D.F. and G.M. Yazdani. 1993. Studies on the ichthyofauna of Nasik District, Maharashtra, India. *Records of the Zoological Survey of India* 90: 195-201.

Subhash Babu, K.K. and C.K.G. Nayar. 2004. A new species of the blind fish *Horaglanis* Menon (Siluroidea: Clariidae) from Parappukara (Trichur District) and a new report of *Horaglanis krishnai* Menon from Ettumanur (Kottayam district) Kerala. *Journal of the Bombay Natural History Society* 101: 296-298.

Talwar, P.K. and A.G. Jhingran. 1992. *Monopterus* (*Monopterus*) *eapeni* Talwar, nova nom. In: *Inland Fishes of India and Adjacent Countries*, P.K. Talwar and A.G. Jhingran (eds.). A.A. Balkema, Rotterdam.

Tehsin, R., V.S. Durve and M. Kulshreshtha. 1988. Occurrence of a Schizothoracine fish (snow trout) in a subterranean cave near Udaipur, Rajasthan. *Journal of the Bombay Natural History Society* 85: 211-212.

Thinès, G. 1969. *L'Evolution regressive des poissons cavernicoles et abyssaux.* Mason et Cie, Paris.

Tiwari, P. 2000. *Bastar ki Kotumsar Guphayein.* in Hindi. English translation, *The Caves of Bastar.* Sharda Publishing House, New Delhi, 22 pp.

Subject Index

380, 382-386, 397, 398, 400, 402, 404, 406, 407, 415-422, 424, 426, 428, 430-436

Subterranean communities 65

Subterranean ecosystems 1

Subterranean fish 416

Subterranean ichthyofauna 331, 335, 336, 353

Swampfish 170, 172, 174, 175, 181, 182, 193, 200, 202, 214, 228, 232-234, 242, 246

Swimming activity 34, 96, 100, 237, 249, 363

Sympatry 23, 385

Systematics 174, 178, 361, 365, 370, 374, 377, 381, 383, 386, 416, 417, 419, 433

T

Tabasco 284, 285, 379

Tactile stimuli 107

Taste 6, 7, 12, 87, 119, 142, 143, 206, 293, 362

Taste buds 6, 7, 12, 87, 119, 142, 147, 206, 293, 362

Taxonomy 141, 171, 174, 175, 180, 187, 332, 339, 417

Temperature tolerance 237

Tennessee 15, 129, 169, 172, 175-177, 183-185, 229, 247, 256, 260

Territoriality 3, 241

Territory 21, 105, 369

Tetra characin 336

Thailand 49, 55-58, 85, 86, 90, 94, 109, 120, 121, 123

Thigmotaxis 212, 213, 240, 254

Threats 66, 67, 76, 171, 173, 243, 257-259, 261, 263, 264, 408

Thyroid activity 236

Thyroxine 424

Tiggy winkle hedgehog gene 154

Time-control mechanism 108

Topographic orientation 109, 363

Tradeoffs 2, 5, 17, 220, 221, 255, 257

Transcription factors 128, 130, 153, 154, 156

Triiodothyronine 424

Trinidad 49, 58, 306

Troglobitic 3, 9-12, 14, 16-23, 26, 32, 65-70, 72-75, 77, 82-92, 94-96, 102, 103, 106-108, 120, 169, 172, 174, 175, 177, 181, 183-186, 188, 191, 193, 198, 199, 205, 206, 211, 213, 218, 220, 222, 224, 228, 230, 232-235, 237-239, 241-246, 256-259, 266, 294, 331-333, 335-340, 342, 344, 347, 348, 352, 398, 406, 432, 434, 435

Troglobitic fish 74, 332, 374

Troglomorphic 8, 12, 15, 16, 18, 19, 24, 28, 29, 36, 42, 53, 66, 67, 85, 87, 115, 116, 119, 120, 122, 124, 127, 171-173, 178, 180, 190, 205, 234, 257, 331, 332, 335-339, 343, 344, 347, 398, 406, 432, 434, 435

Troglomorphy 5, 8, 11-13, 15-19, 24, 25, 28, 30-33, 42, 53, 90, 117, 123, 135, 170, 172, 173, 191, 198, 199, 204, 206, 251, 254

Troglophilic 17, 29, 32, 65, 68, 84, 88, 106, 120, 171, 191, 198, 199, 211, 218, 244, 332, 337-339, 347, 428

Trogloxenic 78, 332, 337

Turkmenia 57

Type locality 45-47, 53, 69, 73, 176, 336, 339, 364, 365, 369, 370, 372, 373, 376, 380, 383, 385, 432, 434

Tyrosinase 159-161

U

UAE 53, 54

Units of evolution 3

USA 19, 42, 49, 57, 60-62, 78, 115, 141, 169, 281, 336

Use of water 66

Index of Scientific Names

Author Index

About the Editors

Eleonora Trajano, zoologist and speleobiologist, is Professor at the Instituto de Biociências da Universidade de São Paulo, Brazil, dedicating to the study of subterranean fishes since her Doctorate in the 1980´s. She is a leading researcher on Subterranean Biology in Brazil, coauthoring more than 120 papers published in specialized refereed journals, books, annals of scientific meetings and in the popular media, focusing mainly on the biology, ecology and evolution of bats, fishes and subterranean invertebrates, also including faunistic surveys on caves from several karst areas in Brazil and abroad. E. Trajano is an active member of the International Society for Subterranean Biology and other international scientific associations, and a member of several editorial boards. The conservation of subterranean habitats is currently one of her major areas.

Maria Elina Bichuette, zoologist and speleobiologist, is Professor at the Departamento de Ecologia e Biologia Evolutiva da Universidade Federal de São Carlos, Brazil, since 2006. Her research area is the study of subterranean fishes, mainly population ecology and behaviour. She is author of many Brazilian cave fish descriptions, papers published in refereed journals, book chapters focusing on threatened cave fauna, annals of scientific meetings focusing mainly on the systematics, population ecology and evolution of fishes, besides subterranean invertebrates. M. E. Bichuette is starting work in some Brazilian regions which have never been studied until now; she is also council member of the International Society for Subterranean Biology and other national scientific and speleological associations.

B.G. Kapoor was formerly Professor and Head of Zoology in Jodhpur University (India). Dr. Kapoor has co-edited 20 books published by Science Publishers, Enfield, NH, USA. The most recent ones are Fish Defenses (Volume 1), 2009 (with Giacomo Zaccone, José Meseguer Peñalver and Alfonsa García-Ayala); The Biology of Blennies, 2009 (with Robert A. Patzner, Emanuel J. Gonçalves and Philip A. Hastings); Development of Non-Teleost Fishes, 2009 (with Yvette W. Kunz and Carl A. Luer); Fish Defenses (Volume 2), 2009 (with Giacomo Zaccone, C. Perrière and A. Mathis); Fish Locomotion: An Eco-ethological Perspective, 2010 (with Paolo Domenici). Dr. Kapoor has also co-edited The Senses of Fish: Adaptations for the Reception of Natural Stimuli, 2004 (with G. von der Emde and J. Mogdans); and co-authored Ichthyology Handbook, 2004 (with B. Khanna), both from Springer, Heidelberg. He has also been a contributor in books from Academic Press, London (1969, 1975 and 2001). His E-mail is: <u>bhagatgopal. kapoor@rediffmail.com</u>

Color Plate Section

Foreword

Photograph of Tom Poulson reproduced by Courtesy of Dante Fenolio

Chapter 2

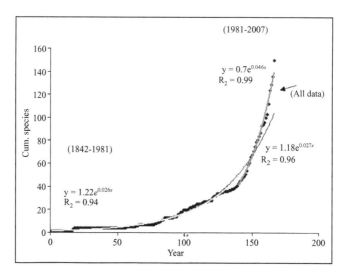

Fig. 1   An arithmetic plot of the cumulative number of subterranean fish species versus the date they were described. Year 0 = 1842 when the first subterranean fish was described.

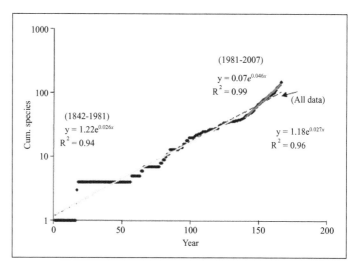

Fig. 2   A logarithmic plot of the cumulative number of subterranean fish species versus the date they were described. Year 0 = 1842 when the first subterranean fish was described.

Chapter 5

Fig. 1 Individuals of *Astyanax mexicanus* from a surface population and from the Molino and Pachón cave populations. The cave populations independently converged on eye reduction, albinism, and decreased numbers of melanophores.

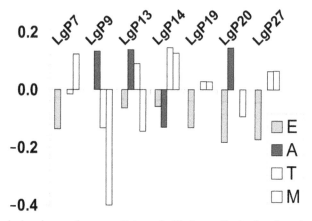

Fig. 5 Scaled values of cave allele substitution effects for four traits in QTL clusters on seven linkage groups in *Astyanax mexicanus*: Amino acid sensitivity (A), number of maxillary teeth (T), relative eye size (E), and melanophore count (M). Substitution effects were scaled against phenotypic range in the mapping F_2 (Pachon × Surface cross). Positive values are phenotypic increases, negative values are decreases. (From Protas *et al.* 2008.)

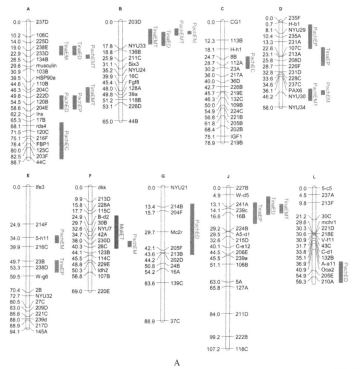

A

Fig. 6 Contd..

Fig. 6 Contd..

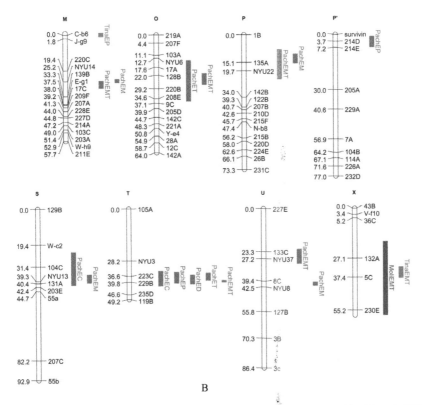

B

Fig. 6 Genetic map of *Astyanax mexicanus* showing alignment of multi-trait QTL detected in the mapping progenies of crosses between surface fish and cave fish from Pachón, Tinaja, or Molino. QTL were detected as previously described (Protas *et al.* 2008) and hand placed on an integrated map, based on positions determined on cross specific maps. The bars cover ± one LOD intervals. Multitrait QTL were based on combinations of two or three traits: E = eye size; M = melanophore count; T = number of maxillary teeth; D = depth of the caudal peduncle; P = placement of the dorsal fin. QTL are coded, PACH, MOLI and TINA to identify the cross in which they were detected. Traits are fully defined and methods are detailed in (Protas *et al.* 2008).

Chapter 7

Fig. 1 Members of the Amblyopsidae. The family includes (A) the surface-dwelling swampfish (*Chologaster cornuta*), (B) the troglophile spring cavefish (*Forbesichthys agassizii*), and four troglobites: (C) the southern cavefish (*Typhlichthys subterraneus*), (D) the northern cavefish (*Amblyopsis spelaea*), (E) the Ozark cavefish (*A. rosae*), and (F) the Alabama cavefish (*Speoplatyrhinus poulsoni*). Photos courtesy of Uland Thomas (A), Dante Fenolio (E), and Richard Mayden (F).

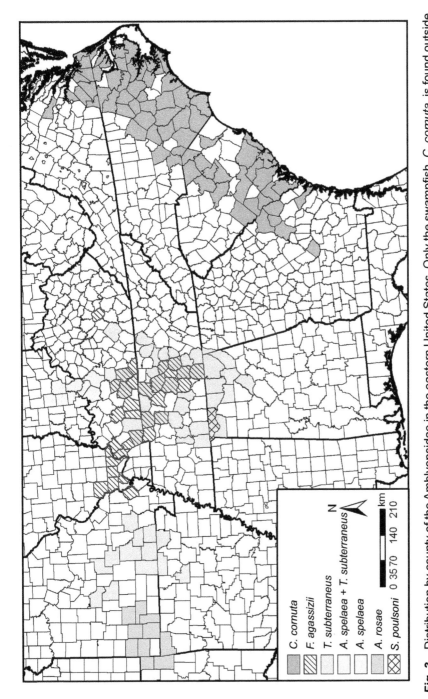

Fig. 3 Distribution by county of the Amblyopsidae in the eastern United States. Only the swampfish, *C. cornuta*, is found outside the Interior Low Plateau or Ozark Plateau.

Chapter 8

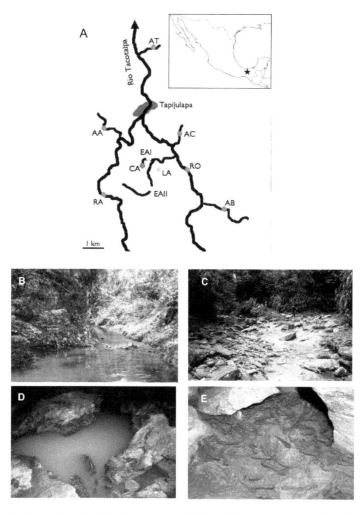

Fig. 1 (A) Location of field sites near Tapijulapa, Tabasco, Mexico. Blue indicates non-sulfidic surface sites (AT Arroyo Tres, AA Arroyo Tacubaya, RA Río Amatan, AC Arroyo Cristal, RO Río Oxolotan, AB Arroyo Bonita), yellow indicates sulfidic surface sites (EA I, II El Azufre, sites I and II, red: the sulfidic Cueva del Azufre, brown: the non-sulfidic Cueva Luna Azufre. The inserted figure shows the location of the Cueva del Azufre system in Mexico. (B-E) Habitat types inhabited by *P. mexicana* in the Cueva del Azufre system. (B) non-sulfidic surface habitat (Arroyo Bonita), (C) sulfur sources in the springhead area of El Azufre, (D) Cueva del Azufre. (E) *P. mexicana* occur at high abundance inside the Cueva del Azufre.

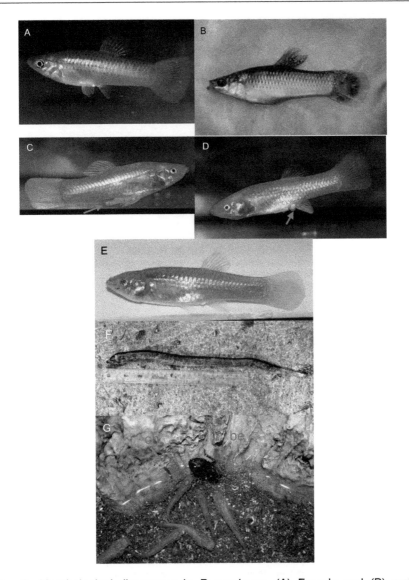

Fig. 7 Morphological divergence in *P. mexicana*. (A) Female and (B) male from a non-sulfidic surface site (Arroyo Cristal). Note the bright coloration of large,dominant males in (B). (C) Male from chamber XI of the Cueva del Azufre, "golden morph". Arrow shows the gonopodium, the male copulatory organ. (D) female from chamber XI with a pronounced genital pad (arrow) around the genital pore. (E) Cave molly female from chamber V, "sunken eye type". (F) The role of *Ophisternon aenigmaticum* as a molly predator in the Cueva del Azufre remains to be examined. This dead individual was found in El Azufre and had probably been flushed out of the cave. The ruler is 30 cm long. (G) Inside the Cueva del Azufre, cave mollies face predation by giant water bugs (*Belostoma* sp., be). ch: chironomid larvae, one of the food sources of cave mollies.

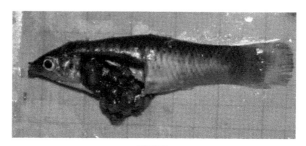

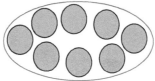

Non-sulfidic surface sites

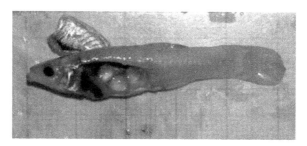

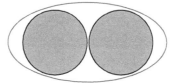

Cueva del Azufre

Fig. 13 Life history trait evolution in *P. mexicana*. Assuming that the volume of the ovary constrains the number of oocytes a female can produce, *P. mexicana* females have two options: they can either produce numerous small oocytes (above, female from Arroyo Cristal) or few, but large oocytes containing more yolk (below, cave molly female from cave chamber X of the Cueva del Azufre (compare Trexler and DeAngelis, 2003).

Chapter 9

Fig. 1 Contd..

Fig. 1 Contd..

Fig. 1 Brazilian troglobitic fishes, showing inter-specific differences in degree of reduction of eyes and melanic pigmentation: a. *Pimelodella spelaea* (Photo: Maria Elina Bichuette); b. *Eigenmannia vicentespelaea* (Photo: José Sabino); c. *Rhamdiopsis* sp. 1 (Photo: Adriano Gambarini); d. *Rhamdiopsis* sp. 2 (Photo: José Sabino); e. *Ancistrus formoso* (Photo: José Sabino); f. *Stygichthys typhlops* (Photo: Cristiano Moreira); g. *Glaphyropoma spinosum* (Photo: Adriano Gambarini).

Fig. 2 Contd..

Fig. 2 Contd..

Fig. 2 Intra-specific variability in the degree of reduction of eyes and pigmentation in *Trichomycterus itacarambiensis*; a-b: pigmented fish with variably sized eyes; c-d: light pigmented *versus* albino specimens. (Photos: José Sabino).

(a)

(b)

Fig. 3 Contd..

Fig. 3 Contd..

Fig. 3 Habitat and population study methods for Brazilian cavefishes (Photos: Adriano Gambarini): a-b. habitat of *Glaphyropoma spinosum* from Chapada Diamantina, northeastern Brazil; c. collecting procedure for mark-recapture studies and; d. measuring standard length for mark-recapture studies.

Chapter 10

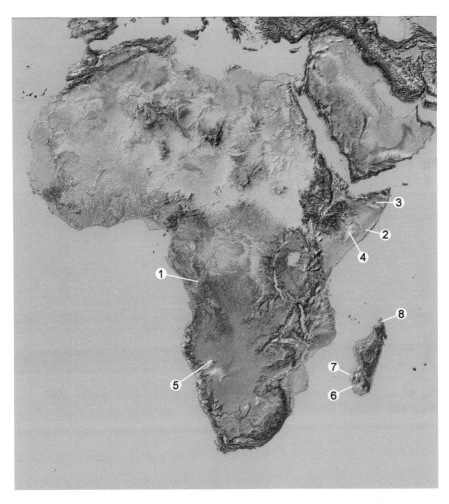

Fig. 1 Distribution of African subterranean fishes. The type localities of the eight species are indicated. 1: *Caecobarbus geertsii*; 2: *Phreatichthys andruzzii*; 3: *Barbopsis devecchii*; 4: *Uegitglanis zammaranoi*; 5: *Clarias cavernicola*; 6: *Typhleotris madagascariensis*; 7: *Typhleotris pauliani*; 8: *Glossogobius ankaranensis*.

Fig. 3 *Phreatichthys andruzzii* (photo by R. Innocenti).

Fig. 5 *Uegitglanis zammaranoi* (photo by R. Berti; from Ercolini *et al.* 1982, modified).

Fig. 9 *Glossogobius ankaranensis* (photo by Jane Wilson).

Chapter 12

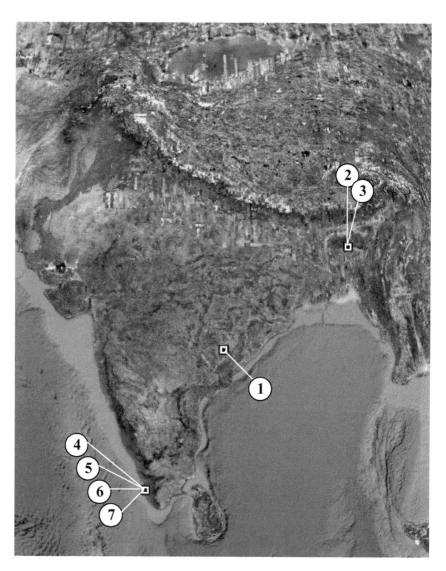

Fig. 2 Distribution of Indian subterranean fishes. Numbers indicate the location of the species found in India. 1: *Nemacheilus evezardi*; 2: *Schistura sijuensis*; 3: *Schistura papulifera*; 4: *Horaglanis krishnai*; 5: *Horaglanis alikunhii*; 6: *Monopterus eapeni*; 7: *Monopterus roseni*.

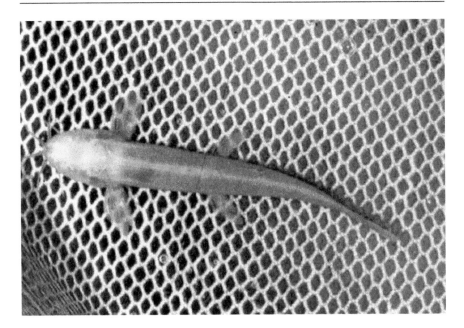

Fig. 3 Albinic subterranean loach, *Nemacheilus evezardi*.

Fig. 8 Subterranean Clariidae, *Horaglanis krishnai* (Source: www.planetcatfish.com).

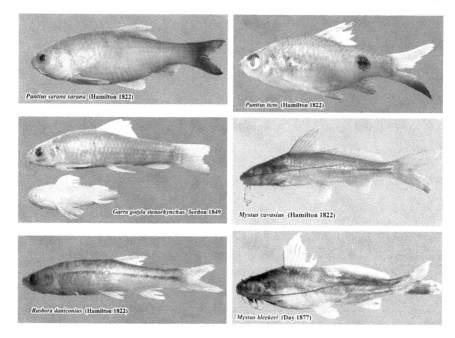

Fig. 9 Nonstygobitic fish species collected from the Nelabhilum Cave, Andhra Pradesh (courtesy: Ranga Reddy).